The fifth international seminar, "Application of Science in Examination of Works of Art" was dedicated to emeritus head of the Research Laboratory William J. Young, who was in attendance.

The Research Laboratory was started by Mr. Young in 1928 and it was through his dedication and exemplary service over a period of forty-nine years as head of the department that the Research Laboratory achieved international recognition and was accorded world-class status. Although Bill Young reached emeritus status in 1976, he has remained in constant contact with the Research Laboratory, and gives generously of his bountiful knowledge and wisdom. He is admired as a museum scientist and revered as a warm hearted friend. His years of retirement have been blessed by his marriage in 1981 to his long-time associate Florence Whitmore.

Application of Science in Examination of Works of Art

Editors:

Pamela A. England and Lambertus van Zelst

Proceedings of the Seminar:

September 7-9, 1983

The Research Laboratory

Museum of Fine Arts

Boston, Massachusetts

Foreword

The 1983 international seminar "Application of Science in Examination of Works of Art" is the fifth in this series of seminars, which was started in 1958 by the head of the Research Laboratory, William J. Young. The success of this seminar, which has by now become a Boston tradition, is evident from the fact that 183 scientists from thirteen countries traveled to Boston to participate in the events, which included twenty-six papers and eighteen poster representations.

While much of the value of these meetings cannot be quantified because it resides in the discussions, the exchanges of information, and the establishment of personal contacts with colleagues having similar interests, the publication of the proceedings will be of lasting value. Without a permanent record of the scientific results, the significance of such international manifestations would be greatly reduced.

It is a great pleasure, therefore, to record here our indebtedness to AT&T for its generosity in printing the proceedings on our behalf. To Lambertus van Zelst, director of research, Museum of Fine Arts, and to Pamela A. England, research scientist, we express our gratitude for the organization of these remarkable events.

JAN FONTEIN
Director

Introduction

The fifth international seminar "Application of Science in Examination of Works of Art," held in September 1983, was dedicated to William J. Young, emeritus director of the Research Laboratory of the Museum of Fine Arts, who started this seminar series in 1958. The dedication was made in recognition of Bill Young's pioneering work in the field of museum science, his large contributions to this field, and the international standing his work brought to the laboratory of which he was the founder and longtime director.

The decision to revive the series was met with such enthusiasm that it was evident that the fifth installment was long overdue. It was decided to return to the earlier, more general title used for the first three seminars, while at the same time allowing for review presentations of principally educational value as well as research communications. A number of papers were invited, and the call for research papers was enthusiastically answered. The submitted abstracts were reviewed by a program committee comprised of Pamela A. England, Edward V. Sayre, and myself.

Practical arrangements for the organization of sessions, social programs, housing, transportation, etc., were carried out by a number of enthusiastic staff members among whom Pamela England, especially, deserves mention. In addition, members of the Ladies Committee of the Museum were generous in many helpful ways. Gratefully acknowledged financial support from the Polaroid Foundation, the Fisher Scientific Company, Mr. and Mrs. Frederick Brech, and Mr. and Mrs. Paul Bernat kept the expenses for seminar participants appreciably below actual cost.

The AT&T Corporation has generously supported the publication of these proceedings; the lion's share of their editing has been done by Pamela England. We hope this volume will be as successful a publication as were its predecessors in the series.

LAMBERTUS VAN ZELST
Director of Research

Contents

Poster Presentations

G. HARBOTTLE, E.V. SAYRE, and R.W. STOENNER

The Use of Miniature Carbon-14 Counters in Dating and Authentication in the Museum*

In the course of our work with carbon-14 in the museum setting, we have found that curators and others whose specialties are in the fine arts and art history often do not understand the basis for the ^{14}C method of age determination and, consequently, its limitations. The purpose of this paper, is to fill in that background for them by describing the underlying principles, and how these must influence the interpretation of the radiocarbon measurement of a museum specimen in terms of "age." Several techniques for that measurement will be described briefly, focusing on the miniature proportional counters that we developed for the Smithsonian Institution. We shall describe three actual museum problems that we have attacked, and some interesting further research on museum objects that has been proposed. Finally we shall conclude with some projections on the future applications of the different methods, in the context of museum work.

Carbon-14 is produced in the upper atmosphere by the steady bombardment of nitrogen atoms by cosmic-ray produced neutrons. The ^{14}C atoms so generated are oxidized quickly to carbon dioxide, which then completely mixes over a period of a few years with the carbon dioxide already present in the atmosphere. Thus the atmospheric carbon dioxide—about one part in 3000 of air—is uniformly "labeled" with radioactive carbon, which is almost indistinguishable chemically and biologically from ordinary carbon. The ^{14}C concentration amounts to about one atom of ^{14}C to every million million of normal carbon. That concentration is what we measure when we wish to determine the "radiocarbon age" of something. Plants, trees, animals, etc. have about that concentration of ^{14}C when they are alive; when they die, however, the radiocarbon, being a radioisotope with a half life of 5720 years, begins to die away, according to the law

$$C = C_o \exp(-\lambda t) \tag{1}$$

where C is the measured activity of the sample of carbon, C_o the activity when the carbon was living matter, λ a decay constant that is the reciprocal mean life of a ^{14}C nucleus, and t the "age." Rearranging the equation to give the age:

$$t = (1/\lambda)\ln(C_o/C) \tag{2}$$

This idealized equation would let us calculate the age t of something if only C could be measured accurately, and if C_o, the assumed value of C when the carbon was alive, could be known as well. In the early years of ^{14}C dating, it was assumed that *modern* carbon, i.e., living plants, carbon from carbon dioxide in the atmosphere, etc., could be measured for ^{14}C and this would give you not merely the C_o for today but for all previous epochs as well. Then it was discovered that the extensive burning of coal during the Industrial Revolution and later, diluted out the "living" natural ^{14}C with "dead" carbon—coal has no ^{14}C in it. Hence C_o could change from year to year and century to century. Finally, after a great deal of study, C_o has been found to vary over the last seven or eight millennia in two ways—a smooth, long-term variation and a very erratic short-term fluctuation around the smooth long-term trend. What is done now is that an artificial C_o based on an agreed standard is taken, C *is measured and* C_o/C calculated. An arbitrary value of λ is then fed in to equation (2) and the resultant t, called the "conventional radiocarbon age," is taken as a rough indication of the age, subject to corrections that can be read off curves or found in tables.[1,2]

These corrections apply equally to the counting or to the newer accelerator method of ^{14}C age determination. In the latter, which is discussed in another paper, (see Long et al., this volume), the ratio C_o/C is determined not by radioactivity measurements but rather by direct counting of ^{14}C ions in the standard and unknown samples. In both cases C actually measures the *ratio of* ^{14}C atoms to normal carbon (^{12}C) in the sample. Figure 1, which is taken from a paper by Stuiver[1] illustrates the variation in C_o. In trees, the rings tell you the age of the wood, and measurement of ^{14}C in tree-ring dated wood samples gives you an absolute age calibration that is reliable. In figure 1, after Stuiver[1] the tree-ring (dendrochronological) age of the wood is plotted horizontally against the conventional radiocarbon age as determined by equation (2) under the assumption of constant standard C_o. If C_o were forever constant, forever equal to the modern value, this wiggly line would be straight and the conventional radiocarbon age would everywhere almost equal the dendrochronological age—almost, because the value of λ used in calculating the conventional age by equation (2) is not the right one, but about 3% off. So you can see that C_o fluctuates rapidly and significantly. What does this mean to the curator?

Note especially the portion of this calibration curve (fig. 1) from 1500 to 1950. By 1960 the radiocarbon age again rises to a peak of about 200 radiocarbon years. Consider a hypothetical museum laboratory problem, the authentication of a supposed Stradivarius violin. Stradivari made violins over roughly the period 1680-1730. Therefore you would expect the wood in a Stradivarius to have a real age of typically 250 years: note however that its radiocarbon age would be only ca. 100-110 years, and that this "age" is repeated at several times from about 1815 up to 1880 and again at about 1915. And this is with zero error in the radiocarbon determination. With a little error in this measurement (± 80 years, for example, which is the 1% level of precision) you can see that radiocarbon would not help with this problem and indeed is useless for dating anything after about 1660. Even before that date there are ambiguities in the calibration curve that often prevent the determination of an unambiguous calendar date.

Since it is unlikely that one would need to date a Stradivarius violin, consider another example, where ^{14}C authentication was actually proposed. The authenticity of the Georges de la Tour painting *The Fortune Teller* at the Metropolitan Museum, New York, was the subject of a recent vitriolic controversy, the museum regarding it as an authentic production of ca. 1630 and the critics alleging it to be a forgery of 1940 or thereabouts. If the radiocarbon could have been very precisely measured on a sample of the canvas or oil, the Stuiver curve predicts a radiocarbon age of 300 years, and if that were what one observed, then the painting, or at least the canvas, would be authentic. But suppose the radiocarbon age were measured as 250 $\pm$ 100 years. This age would be consistent with a date of 1630, but also with 1730-1810 and 1930-1940. And you would be as much in doubt as before, except that now both sides would have hard evidence to support their position.

In spite of these difficulties there *are* many museum studies in dating and authentication upon which radiocarbon could shed considerable light. However, until about 1975, conventional ^{14}C measure-

*This research was carried out under contract with the U.S. Department of Energy.

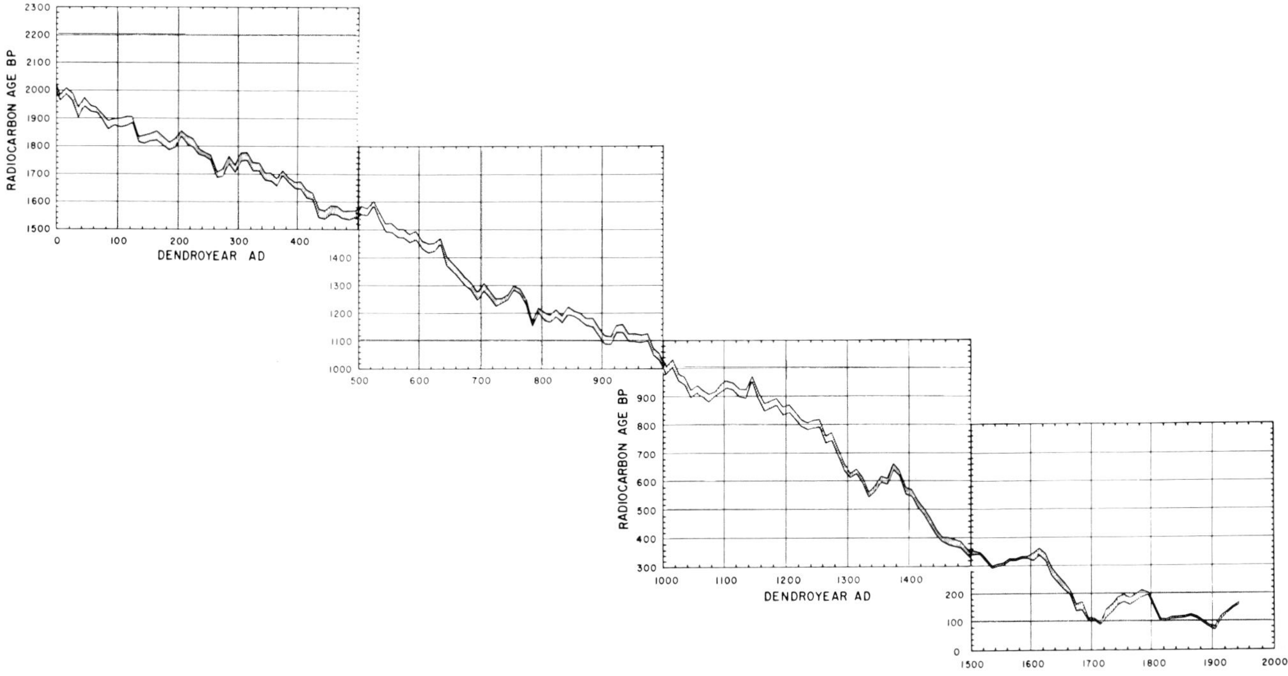

Fig. 1. The Stuiver[1] calibration curve for conventional radiocarbon dates.

ments required rather large samples — 1 to 10g of carbon, which translates into roughly 3-30g of wood, cloth or paper, for example. Certainly museum work *was* performed, but the relatively large sample size must have been inhibiting to curators, and all but impossibly large for many classes of fine art objects. At the Brookhaven National Laboratory (BNL), realizing the need for drastic reduction in sample size, and having a great deal of experience through our neutrino detection project with miniature counters and low-background counting, we turned to a simple miniaturization of the proportional ^{14}C counter, aiming at a sample of 10mg of carbon. In a number of other laboratories, as was mentioned earlier, an entirely different approach, ^{14}C ion counting in an accelerated carbon beam, was developed.

Our work actually began in 1975: the Conservation-Analytical Laboratory of the Smithsonian Institution, knowing the advantages of milligram-scale dating in general in the museum, but also having several specific problems in mind that demanded small-sample measurements, contracted with us to develop a suitable miniature counter. Although the concept was simple and the electronics relatively straightforward, special techniques of counter construction were required. We succeeded[3-5] and now several laboratories, in addition to BNL, are either using our counters or miniature counters of similar design. The design improvements included constructing the counters essentially totally out of highly purified quartz and metals. Glass is present in them only in the last centimeter of the graded seal at the extremities of legs, extending some distance from the counting chamber, which are necessary to seal in the external electrical leads. Construction was such as to permit a thorough high-temperature baking out of the detectors under vacuum. The detectors are filled to a pressure of four atmospheres with the carefully purified carbon dioxide prepared by combustion of the sample being measured. The purity of this gas sample is checked by analyzing the response of the counter to exposure to test radiation, and the stability of a counter is checked throughout each run by periodic examination of its output. The best reduction of background counts, arising from the effect of cosmic ray bombardment, was achieved by inserting the counters into a well inside large single crystal sodium iodide detectors which serve as anticoincidence shields. With such a shield, ratios of the counting rate of the detector filled with "contemporary" carbon dioxide to the background counting rate of the detectors filled with "dead" carbon dioxide as high as 8.9 were obtained. The highest such ratios for similar small counters reported previous to our work had been 1.7.

Advanced proportional counter technology is all very well, but the payoff is whether worthwhile work can be done in a museum. We would now like to present briefly three such cases.

An Eskimo Mask

An Eskimo mask was brought into the American Center for Conservation of Art and Antiquities in New York (Director, Dr. K. J. Linsner). Although the mask belonged stylistically to the Recent Prehistoric Culture (1600s to early 1700s A.D.) it had been found at the level of the much earlier Okvik Eskimo Culture (300 B.C. to A.D. 500) raising the question of whether this style had actually begun earlier than previously thought. A sample of about 100mg was removed, and the radiocarbon age determined. Dendrocorrected, it gave a date of A.D. 1610 ± 105 years, which agreed with the stylistic assignment.[6] Here, a clearcut choice was presented, and a clearcut answer given: the possible earlier date could be rejected.

The Frobisher Bloom

The second study pertained to the "Frobisher Bloom," a roughly 9kg half-sphere of raw iron, found by the American C.F. Hall in the course of his explorations around the coasts of Baffin Island in the Canadian Arctic in around 1860. On a tiny Island called "Kodlunarn" by the Inuit, Hall observed many remains of previous non-Eskimo visitors— ceramics, coal, flints, glass, and wooden timbers. On questioning the natives he was told that they had been left there by white men who had come on three successive summers many years ago, first in two, then three, then a great many ships. In fact "Kodlunarn" meant "white man's" island in the Inuit language. Hall realized that they were describing the three voyages of Sir Martin Frobisher who came seeking gold and a Northwest Passage in 1576, '77, and '78.[7]

For many years the Smithsonian had exhibited and loaned this object (Accession No. 49459), labelling it "Iron smelted by the Frobisher Expedition A.D. 1578," but after the discovery of unmistakable Viking remains at L'Anse aux Meadows in Newfoundland by the Ingstads in 1960, Dr. Melvin Jackson suggested that "the bloom might not be a Frobisher relic—since there was no mention of a smelter in any of the extensive records written by Frobisher." Hence the desire of the Smithsonian to obtain a date for the bloom.

Fortunately, N. van der Merwe had perfected a technique for carbon-dating iron in the 1960s.[8] The procedure involves extraction and radioactivity measurement of carbon from the iron, and the assumption that the carbon was "living," i.e. in equilibrium with the biosphere at the time the iron was smelted. Such would be the case if the iron was smelted from ore by means of charcoal obtained from recently dead trees, but would not be if, for example, coal or coke had been used. Unfortunately, bloomery iron is characterized by a very low carbon content—typically 0.1% or less: thus, the entire Frobisher bloom would have yielded barely enough carbon for a conventional ^{14}C determination. For this reason, the development of a miniature counter method was sponsored by the Conservation-Analytical Laboratory of the Smithsonian Institution.

We have published two dates based on minicarbon samples from the bloom which, when corrected by means of the Stuiver curve, are 1293 ± 133 and 1262 ± 107, average A.D. 1278 ± 121[4,5] Stimulated by these findings, the Smithsonian Institution sponsored an expedition to Kodlunarn Island in the summer of 1981, led by William Fitzhugh. Three more "Frobisher" blooms were found, which are now in a Canadian museum in the process of conservation. Adhering to the rough outer surface of one of the new blooms were bits of charcoal that appeared to have remained there from the smelting. These minute bits have been dated using our small counters. The Stuiver correction curve yields three possible dates: 1392, 1322, and 1337, all ± 150 years, quite consistent with the dates from the other bloom. All seem too early for the Frobisher expedition unless old wood, for example driftwood, was employed for smelting. Research on the blooms is continuing, and obviously the final conclusions are

yet to be drawn, but even at this point we can say that we have opened up the whole question of Viking origin with our minicarbon dates.

The Stavelot Triptych

The third example involves measurements made on the Stavelot Triptych in the Pierpont Morgan Library, New York, with the collaboration of Dr. William N. Voelkle, curator, and Dr. A. J. Yow, conservator.[9] The Stavelot Triptych (fig. 2), a masterpiece of Romanesque art, is decorated with champlevé enamels that are among the finest and best preserved of those made during the twelfth century in the Mosan region of Belgium. It encloses at its center two Byzantine triptychs, dating to the late eleventh or early twelfth century, and carrying splendid cloisonné enamels. The larger of the two Byzantine triptychs centers on a relic of the True Cross which provides, in fact, one of the themes for the entire Stavelot Triptych.

Art-historical research has linked the Stavelot Triptych to that monastery's great abbot Wibald (d. 1158), although documentation is lacking.[9] Wibald traveled to Constantinople twice during the 1150s as ambassador for the Emperor Frederick Barbarossa, and it is presumed that the Byzantine triptychs were brought back by him at this time. Only at the time of the French Revolution does there begin to be a written record: carrying the triptych with him, the last Abbot of Stavelot fled before Napoleon's troops to Hanau where he died in 1796. Nearly a century later it was discovered in the house of the Waltz family, who possessed it until 1909, when it was offered for sale by the Durlacher Brothers of London. In the summer of 1910, on the advice of Sir Charles Read of the British Museum, it was purchased by J. P. Morgan. It remained on loan at the British Museum until 1913, when, after Morgan's death, it was brought to New York and its present home.

We were approached by Dr. Voelkle of the Pierpont Morgan Library in the hope that minicarbon dating might shed a little light on the rather obscure history of this object. But there was something more than the history of the triptych itself, interesting as that might be. It had always been known that one of the two themes of the Stavelot Triptych was the discovery and excavation of the True Cross by the Empress Helena, mother of Constantine, during her pilgrimage to the Holy Land, and the reference would seem to have been to the small relic of the Holy Cross displayed at the center of the larger Byzantine triptych. When, however, in 1973 the two Byzantine triptychs were removed for examination by the Library's conservation department, a surprise was in store. The removal of the Crucifixion enamel of the small triptych revealed an unexpected cavity hollowed out in the underlying wooden support. In it was a small pouch, 1½ inches long, tightly wrapped in red silk thread; the pouch consists of a piece of yellow and brown Byzantine silk, on which was woven the head of a griffin. In addition there was a small pin, perhaps regarded as a relic of a Holy Nail.

When the pouch was opened there was a piece of parchment inside, stating in Latin that the contents of the pouch came from the Lord's wood (*De ligno domini*), from the Lord's sepulcher (*De sepulchro domini*) and from the garment of the Blessed Virgin (*De vestimento sancte Marie virginis*). The contents referred to on this parchment ticket were a tiny fragment of wood, a pinch of some uni-

Fig. 2. The Stavelot Triptych (photo courtesy Pierpont Morgan Library)

dentified debris, and a small piece of fine white silk. Unfortunately, there was no mention of the Abbot Wibald nor any date, nor any clue as to the source of the holy relics so cunningly concealed within the body of the triptych.

Using the milligram-scale ^{14}C counters we have dated the red silk thread that wrapped the pouch, the "Robe of the Virgin Mary," and the fragment of the True Cross. To gain information on the history of the triptych itself we also dated the oak supports of the right half of the center panel and of the right wing. The results are as follows, after fully correcting the counting data for isotopic and dendrochron-ological effects: the wood of the right wing is essentially modern; the wood of the center panel is A.D. 1405 $\pm$ 115; the red silk thread had the same date, A.D. 1401 $\pm$ 120; the wood of the True Cross is A.D. 595 $\pm$ 115; the Robe of the Virgin Mary is 1260 $\pm$ 150.

These dates vary in some cases from those given by E. V. Sayre at the Milwaukee meeting of the American Institute of Conservation in 1982: the changes arise from applying the whole range of dendro-corrections and in one case a revised counter background.

What do these dates tell us about the art history of this extraordi-narily important museum object? Although some of the dates lead to fairly clear conclusions, others must give rise to speculation.

First, the modern wood of the right wing strongly suggests that that portion was restored by the Durlacher Brothers at about the begin-ning of the twentieth century. The center panel of A.D. 1405 $\pm$ 115 seems too late to be part of the original construction, if that took place in the era of Wibald, about 1160, although there is a small chance that this might still be so. If the red silk thread wrapping the relics has the same date, it suggests the possibility that when the center panel was rebuilt, the relics were removed, "refreshed," as was often done, and restored to their rightful place underneath the Byzantine triptych. The possibility that they were there from the origi-nal construction of the Stavelot Triptych is attested by the Byzantine pouch and the mid twelfth-century Latin script on the vellum ticket.[9]

The date of the "Robe of the Virgin Mary" is in agreement with this, A.D. 1260 $\pm$ 150 being consistent within experimental error with Wibald, his trips to Constantinople, and the presumed date of construction of the triptych. But the date of the robe is clearly not consistent with that of the famous robe of the Virgin Mary which had been revered in Constantinople at the Church of the Virgin Mary at Blachernae since at least the year A.D. 626[10] Several medieval trav-elers and the histories of Byzantium place it there up to the sack of Constantinople in 1204, at which time it disappeared.

What then are we to make of the date of the True Cross? From the Crusades onward, Europe was awash with holy nails, thorns from the crown of Christ, and pieces of the True Cross but our date is too early for *ad hoc* relics of that time. It seems too late however to be "authentic," whatever that may mean. We are five standard devia-tions from the birth of Christ and His crucifixion: if the piece of wood we dated were actually from the early part of the first century, truly, it would be a miracle.

In the final days of Byzantium in 1204, Robert de Clari, a French-man who accompanied the Fourth Crusade, wrote of the collection of relics that it included "two pieces of the True Cross as large as the leg of a man"[11] and that it was kept in the Pharos Chapel of the royal palace. It is tempting to identify these pieces of the True Cross with the relic looted from Jerusalem by the Persian Chosroes II in A.D.

615, which according to the "Golden Legend"[12] had been left there by Helena after her discovery, as commemorated in the right wing of the Stavelot Triptych. The Byzantine Emperor Heraclius (575-642) liberated these relics and is thought to have returned with them to Constantinople around A.D. 629.[12,13] Although Gibbon[14] tells us that the seals on the box containing the True Cross were not tampered with during the time it spent as Chosroes's prize in Ctesiphon, one still wonders. Our date of A.D. 595 $\pm$ 115 is more consistent with it being produced for the benefit of Heraclius than as the cross of the Savior's martyrdom. In any case, it is entirely consistent with the Stavelot wood fragment being a gift to Wibald on his visit to Constantinople.

It is however, also possible that the holy wood simply came to Stavelot as a result of the looting of Constantinople in 1204 on the occasion of the Fourth Crusade. Runciman, in describing the after-effects of the debacle, writes that "throughout the West there were paeans of praise and the enthusiasm mounted when precious relics began to arrive for the churches of France and Belgium."[15]

Further Work

With milligram scale ^{14}C dating and authentication now available as a routine tool for the curator and conservator, one may ask, what kinds of problems can be attacked? We would like to give a few examples, where ^{14}C has been, or could be used.

The Grolier Codex, whose place of writing and details of discovery are both unknown, is a document of extraordinary importance to Mayan study, in that it is only the fourth known pre-Conquest codex,[16] and the only Mayan codex resident in the country of its pre-sumed origin, Mexico. Its iconography and astronomical content (Venus calendar) have been the subject of scholarly examination[16,17] and a recent study by Carlson summarizes the arguments for its au-thenticity—an authenticity that had been rejected by the Mayanist Eric Thompson, largely on stylistic grounds.[18] The codex was found with three unpainted sheets of bark paper, and these had a conven-tional ^{14}C date of A.D. 1230 $\pm$ 130[19] which works out to an unambig-uous A.D. 1280 if dendrocorrections are applied. A ^{14}C date of the codex itself, based on milligram sampling, should help to remove any lingering doubt as to its authenticity.

A second study, feasible only with small-sample ^{14}C analysis, is the famous Vinland Map, which belongs to Yale and concerning which the university brought out a book *The Vinland Map and the Tartar Relation* in 1965.[20] The map, thought to be of ca. 1440, pur-ports to show Iceland, Greenland, and "Vinland" resembling Baffin Island. It has been attacked as a forgery, defended, and exhaustively discussed[21,22] but the strongest piece of physical evidence was Wal-ter C. McCrone's discovery that the pigment of the ink contained tita-nium dioxide in the anatase form, which was not generally used until the 1920s.[23] However, others have suggested that anatase might have occurred in genuine ancient iron-gall inks if the mineral ilmen-ite, an iron titanium oxide, had been used for their preparation.

It should be noted that in both these cases what one really authen-ticates with ^{14}C is the *paper* or *parchment* and not the date of the original written production. Thus, a ^{14}C measurement may prove that something is a forgery: it cannot *prove* authenticity. In the case of the Grolier Codex, it was suggested that forgers found old bark paper sheets in a cave, and forged the written codex on these.

A third example of work on what may be called a "museum object" is, of course, the Shroud of Turin. This long strip of linen cloth, which bears the imprint of a man's face and his scourged body, is held by many to be the burial shroud of Christ, or at least, associated in some way with that event. We shall not go into its description or history—that has often been done, best by Wilson[10] who has complete scholarly references. Recently a team of physical scientists under the acronym STURP, for Shroud of Turin Project, have published a book on the very extensive examination made in 1978 of the Shroud itself.[24]

Wilson believes that the Shroud might be identical with the Holy Image of Edessa—the so-called Veronica of Christ. This image-bearing cloth was known in Edessa from around A.D. 525 until 944, when it was taken to Constantinople and where it remained until 1204. In the sack of the city by the Crusaders, the Veronica disappeared. The present Shroud of Turin has an uninterrupted history from 1357 onward, when it was first exhibited and also denounced as a fake.

Suppose, then, that one could carry out a ^{14}C measurement of the Shroud, at a precision level of $\pm$ 100 years. What could it tell you? An examination of the Stuiver curve (fig. 1) shows that, if made at the time of Christ's crucifixion, the linen should have a radiocarbon age close to its true age, with little correction or ambiguity. So there is every reason to believe that the ^{14}C test could reliably *disprove* the authenticity, if such proved to be the case. In particular, if the corrected date came out fourteenth century, it would give great weight to the pronouncement of Pierre d'Arcis, Bishop of Troyes (France) in 1389, who said in a letter to Pope Clement VII, that his predecessor, Henry of Poitiers "after diligent inquiry and examination...discovered the fraud and how the said cloth had been cunningly painted, the truth being attested by the artist who had painted it, to wit, that it was a work of human skill and not miraculously wrought or bestowed."[25] Or, if the date came out sixth century, it would lend considerable support to Ian Wilson's theory of the identity of the Shroud of Turin with the Veronica Image of Edessa.

As of today, (September 1983), the status is that the ^{14}C committee of STURP has sent in a proposal for minicarbon measurement, and is also conducting rather delicate negotiations with the Church concerning this proposal. At this point, there is not the slightest indication that the experiment will take place in the near future, but if it does, it is likely that the measurements will involve a consortium of ^{14}C laboratories.

By these examples we hope to have shown that ^{14}C measurement on the milligram scale is a technique that is rapidly becoming a routine practice and, as such, is available to the curator. The most obvious use at the moment would seem to be authentication: besides wood, materials like paper, leather, cloth of various kinds, and bone can be dated. The use of ^{14}C measurement in art history is a little more sophisticated and may require series of dates to be taken to correlate the development of a style, for example.

Carbon-14 measurements on the milligram scale are available commercially. An outstanding example of service to the art world was the United Kingdom Atomic Energy Research Establishment's Harwell Laboratory's recent authentication of a gold treasure, thought to be part of the Vrap hoard, attributed to the Avar period. Harwell's client was Sotheby's, who placed a value of about a million dollars on the 122 pieces totalling 7.3 pounds of gold and 11.7 pounds of silver. Harwell charged their usual fee, a few hundred dollars, to measure ^{14}C in the fragments of flax adhering to the meshes of a massive gold belt. Using Brookhaven counters, the date was determined to be A.D. 700 $\pm$ 100, just right for the Avars and in agreement with the markings of the Byzantine Emperor Constans II (641-651 and 659-663) on two plates found with the gold.

Conclusion

We are often asked whether the small counter can hope to compete with the much faster accelerator method for ^{14}C measurement on the milligram scale. Several points need to be made.

1) The accelerators are so expensive that very few ^{14}C laboratories will be able to afford them. The counters are inexpensive enough to be installed in existing laboratories and require no new technology.

2) The total length of time required before a date can be obtained includes the sample processing time, which appears to be comparable for the two methods, and queuing time, which is utterly unpredictable.

3) Although the counters take longer, the operator is not doing anything during the extra time: the counters are simply accumulating counts by themselves. Therefore, if many counters are available and are set up to run in parallel, many samples can be running at once, and the limiting factor again tends to become the sample preparation. Harwell will soon have thirty counters running simultaneously.

4) In many cases, if authentication, or a rough date, is all that is desired, the counting time can be drastically shortened.

5) If the sample available is larger, a larger counter can be used, and the time also shortened. Harwell has found that, for archaeological work, a 30ml counter holding 60mg of carbon is especially useful. For museum work, smaller samples will probably be the rule. Curators should remember that as much as half the sample may vanish during pretreatment, and only about one-third of wood is carbon hence 10mg of carbon may require 60mg of wood or cloth as a sample to be taken from the object for a single determination.

6) For the smallest samples, the accelerator will be the unique method. There is some indication that samples as small as 50-100µg of carbon may be accessible to dating, using accelerators. This would be achieved with truly minute samples, which will please painting conservators, among others.

7) In any case, and whatever method is used, curators would be well advised, before spending museum money on ^{14}C tests, first to see, with reference to the Stuiver curve, whether or not the specified measurement at the expected level of precision can possibly be made to yield the kind of answer required.

To summarize, we are still actively gaining experience in a very new field, and there would appear to be plenty of work available for all of us—curators, art historians, conservators, and physical scientists.

References

1. M. Stuiver, "A High Precision Calibration of the A.D. Radiocarbon Time Scale," *Radiocarbon, 24* (1982): 1.

2. J. Klein, J. C. Lerman, P. E. Damon and E. K. Ralph, "Calibration of Radiocarbon Dates," *Radiocarbon, 24* (1982): 103.

3. G. Harbottle, E. V. Sayre and R. W. Stoenner, "Carbon-14 Dating of Small Samples by Proportional Counting," *Science,* 206 (1979): 683.

4. E. V. Sayre, G. Harbottle, R. W. Stoenner, W. Washburn, J. S. Olin and W. Fitzhugh, "The Carbon-14 Dating of an Iron Bloom Associated with the Voyages of Sir Martin Frobisher, in L. A. Currie, ed., *Nuclear and Chemical Dating Techniques: Interpreting the Environmental Record*, A.C.S. Symp. Series 176 (1983) p. 441.

5. E. V. Sayre, G. Harbottle, R. W. Stoenner, R. L. Otlet and G. V. Evans, "The Use of Small Gas Proportional Counters for the Carbon-14 Measurement of Very Small Samples," *IAEA Proceedings on Methods of Low Level Counting and Spectrometry*, Vienna (1981) p. 393.

6. K. J. Linsner, "Less Permits More ... Expanding an Analytical Dating Method," *Technology and Conservation,* 6 (1981): 5.

7. C. F. Hall, *Arctic Researches and Life Among the Esquimaux* (New York: Harper and Brothers, 1865).

8. N. J. Van der Merwe, *The Carbon-14 Dating of Iron* (Chicago: University of Chicago Press, 1969).

9. W. N. Voelkle, *The Stavelot Triptych* (New York: The Pierpont Morgan Library, 1980).

10. I. Wilson, *The Shroud of Turin: The Burial Cloth of Jesus Christ?* (Garden City: Doubleday, 1978) p. 129.

11. Robert de Clari, *The Conquest of Constantinople*, trans. E. H. McNeal (New York: Columbia University Press, 1936).

12. Jacobus de Voragine, *The Golden Legend* adapted by G. Ryan and H. Ripperger (New York: Arno Press, 1969) excerpted in Voelkle, ref. 9, pp. 39-44.

13. *Encyclopedia Britannica* "Heraclius" (Chicago: University of Chicago Press, 1947).

14. E. Gibbon, *The History of the Decline and Fall of the Roman Empire,* Chap. XLVI footnote 86 (New York: Heritage Press, 1946) p. 1551.

15. S. Runciman, *A History of the Crusades*, Vol. III (Cambridge: Cambridge University Press, 1955) p. 128.

16. M. Coe, *The Maya Scribe and His World* (New York: The Grolier Club, 1973) p. 150.

17. J. B. Carlson, "The Grolier Codex: A Preliminary Report on the Content and Authenticity of a Thirteenth Century Maya Venus Almanac." Proc. 44th International Congress of Americanists (1982) in press.

18. J. E. S. Thompson, "The Grolier Codex," John A. Graham ed., *Studies in Ancient Mesoamerica II* (Berkeley: Univ. California Press, 1975).

19. J. Buckley, "Isotopes' Radiocarbon Measurements X," *Radiocarbon,* 15 (1973): 293.

20. R. A. Skelton, T. E. Marston and G. D. Painter, *The Vinland Map and the Tartar Relation* (New Haven: Yale University Press, 1965).

21. S. E. Morison, *The European Discovery of America. The Northern Voyages A.D. 500-1600* (New York: Oxford University Press, 1971) pp. 69-72.

22. H. Ingstad, *Westward to Vinland* (New York: St. Martin's Press, 1969) pp. 87-89.

23. W. C. McCrone Associates, Inc. Report to Yale University Library, Chemical Analytical Study of the Vinland Map. January 22, 1974 (unpublished).

24. J. H. Heller, *Report on the Shroud of Turin* (New York: Houghton Mifflin, 1983).

25. I. Wilson, ibid., text of a letter of Pierre d'Arcis, Bishop of Troyes, to Pope Clement VII, pp. 230-235.

A. LONG, D. J. DONAHUE, A. J. T. JULL, and T. ZABEL

Tandem Accelerator Mass Spectrometry Applied to Archaeology and Art History

Direct ion-counting using a tandem accelerator mass spectrometer (TAMS), enables $\pm 2\%$ precision dating of one milligram of carbon. In about one hour we can assign a carbon-14 age to a snippet of canvas from a painting or to a fragment of cloth from a garment, to a splinter from a carving or to a bone fragment from a skeleton. Although all studies dependent on radiocarbon dating will benefit from direct ion-counting, this paper emphasizes its possibilities in archaeology, geology, paleoecology, and art history. With current technology, art forgeries less than thirty years old are easily detectable by ^{14}C analysis of hardened linseed oil. As technology continues to develop, dating objects created before the atmospheric testing of atomic bombs (the bomb pulse) will become more and more precise.

With direct dating of milligram-sized samples, scientists can now determine the age of materials more directly related to prehistorical and anthropological questions, for example, early cultigens (corn, wheat, squash, bean seeds) and organic matter indigenous to bones. Illustrations show how ages of these materials inferred from associated charcoal dates have been in error. Preliminary experiments indicate that pinhead-sized fragments of cast iron artifacts can be dated directly with a precision adequate for many investigations.

Carbon-14 dating has proved to be the most important and widely used of the radiometric age-determination methods. This is due in part to its application to a wide spectrum of disciplines and to its usable time range during a large portion of human cultural activity. The Nobel Prize Committee acknowledged the impact of ^{14}C- dating on science in 1962 when it recognized W. F. Libby for his work in developing the method.[1]

Living organisms, because they are part of the carbon cycle, contain ^{14}C produced continuously by cosmic-ray processes in the atmosphere. All land plants take carbon directly from the atmosphere and animals receive this carbon through the food chain in an essentially constant ratio of ^{14}C-to-total carbon $(^{14}C/CT)$. This $^{14}C/CT$ has remained surprisingly constant for at least the past 8000 years[2] (see Harbottle, this volume, for more details). When the organism no longer takes in carbon-cycle carbon, the $^{14}C/CT$ ratio decreases at a statistically predictable rate through the radioactive decay of ^{14}C. The time since the organism's death can then be calculated from its $^{14}C/CT$ ratio. The useful time range for ^{14}C dating is limited by the sensitivity of the ^{14}C measurement technique and by the possibility that ^{14}C was more recently added to the sample. The maximum useful dating range is from 40,000 to 60,000 years BP. Isotope enrichment can extend this to 75,000 years.[3] Precision equivalent to 8 years ($\pm 0.1\%$) is possible for certain large younger samples counted for long periods in stable, low-background systems.

Basic Equations

For the purposes of this discussion, we have conveniently combined all methods of conventional ^{14}C dating into "β-counting techniques" as distinguished from "ion-counting." Equation (1) is the basic radioactive decay equation; it illustrates not only the relationship between the rate of decay and the ^{14}C count, but also the relationship between ion-counting and β-counting:

$$\frac{d[^{14}C/CT]}{dt} = \frac{-[^{14}C/CT]}{8033} \tag{1}$$

The time increment, dt, is very short compared to the decay half-life. The term on the left is the ^{14}C decay rate for a certain amount of carbon (CT). It is the measured parameter in β-counting. The constant term on the right side of (1) derives from the half-life* and linearly relates the count rate to the ^{14}C CT. Ion-counting measures the $(^{14}C/CT)$ term directly. After integration with appropriate limits, equation (1) becomes:

$$^{14}C \text{ Age} = -8033 \ln \frac{[^{14}C \text{ } CT]}{[^{14}C \text{ } CT]} \tag{2}$$

where $(^{14}C/CT)_i$ is the initial ratio, or "modern" standard. Both ion-counting and β-counting techniques use equation (2) for age computation. Because $(^{14}C/CT)_i$ is an assumed value and corresponds to the actual initial ratio only to a first approximation, the ^{14}C age and the true age often differ significantly. (See Harbottle this volume for a detailed account of these differences and the consequent ambiguities in inferring true age from ^{14}C measurements.) These equations illustrate that ion-counting is subject to the same fundamental accuracy limitations as β-counting, but as elaborated in the next section, the accelerator measures $^{14}C/^{13}C$ (a value directly related to $^{14}C/CT$) directly, rather than

$$\frac{d[^{14}C/CT]}{dt}$$

as measured in β-counting.

The University of Arizona-NSF National Accelerator Facility

Direct ion-counting of ^{14}C is a formidable task, and only within the past few years have accelerator physics and electronic technologies developed to the necessary level. Consider that the maximum natural (pre-bomb) value of ^{14}C/CT is about 10^{-12}. For older samples, this value approached 10^{-15}. The needle-in-the-haystack analogy is no exaggeration here if you can imagine one ^{14}C atom per 10^{15} carbon atoms.

Figure 1 shows the physical layout of the Arizona tandem accelerator mass spectrometer (TAMS), viewed from above. In the present source design the carbon must start out as a solid; we use either iron carbide or graphite. This sample, now called the target, is placed in the source at the beginning of the long ion path through the evacuated tube. During analysis, a beam of cesium ions bombards, coats, and heats up the target. Carbon ions, some negative, are created at the target surface, and a strong (30kV) electric field draws the negative ions toward the magnet at the 90-degree bend in the flight tube. The right combination of electric and magnetic field strengths allows selection of mass 12, 13, or 14 (the three isotopes of carbon) to pass into the accelerator section.

Sulfur hexafluoride gas in the T-shaped tank at 8 atmospheres pressure encloses and insulates the positive 2-million-volt terminal, accelerating tube, and high-voltage generating system. The positive voltage attracts carbon negative ions, either mass 13 or 14 as selected at the 90-degree injector magnet. Waiting for the C⁻ ions at the point of highest voltage is a veil of argon atoms, which the negative carbon ions impact with sufficient energy to strip electrons from the

---

*This illustration uses the "conventional" half-life. A more recently determined vaue increases this constant by about 3%.

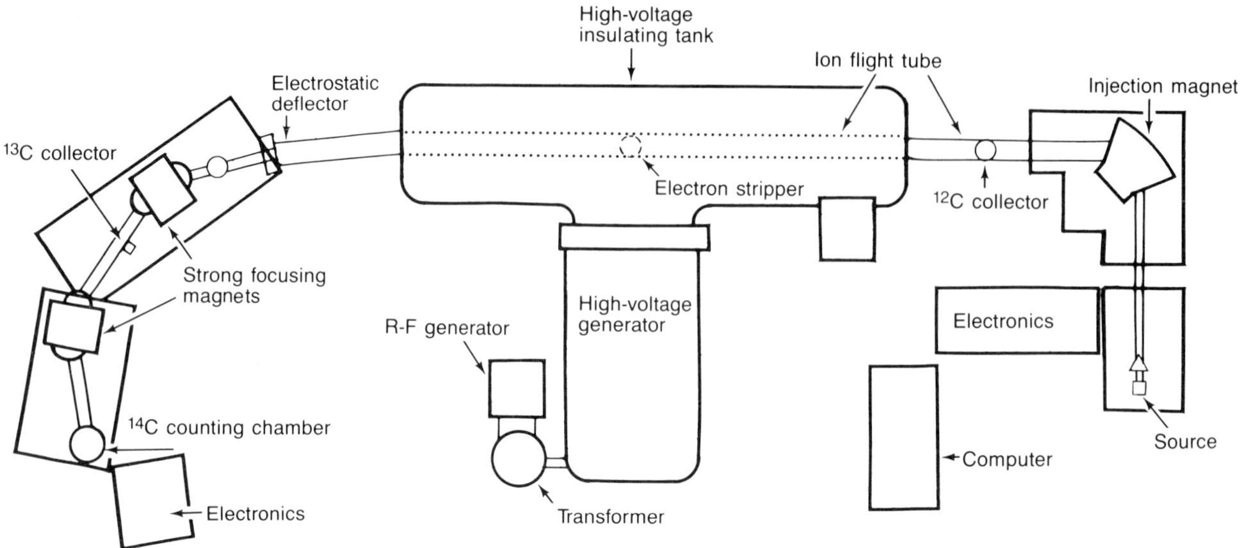

Fig. 1. Simplified layout of tandem accelerator mass spectrometer.

carbon ions. The resulting positive carbon ions are then repelled by the positively charged terminal and accelerate to even greater energies back to ground potential.

An electric analyzer selects the C^{3+} ions and these, now 8MeV particles, proceed through a 45-degree magnetic analyzer and to a Faraday cup ion collector for ^{13}C, or, for ^{14}C, to another 45-degree magnet and finally into an energy-discriminating ion counter.

Fortunately, ^{14}N seems not to form negative ions. This potential problem is virtually eliminated in the first acceleration stage. ^{13}C-H^{-1} and ^{12}C-H_2^{-1} ions (also mass 14 ions) cannot survive the electron stripping to 3 + charge state. The energy-discriminating ion counter removes the last remnants of ^{13}C that manage to pass through the other analyzers.

During an actual run, the Faraday cup and ion counter register ^{13}C and ^{14}C for a particular sample alternately sequenced by varying the potential on the first leg of the pathway. This normalizes for differences in ion-production efficiency from one target to another. The target holder contains six targets that can be rotated into position in sequence. Typically, four are unknown, one is a blank or background, and one is a ^{14}C standard such as the National Bureau of Standards issues. It takes five hours to complete our routine four cycles around the target wheel. This procedure provides the opportunity to evaluate both the long-term and short-term stability of the instrument and target system. Each sample thus is "counted" for ^{14}C for at least 2000 seconds. See Donahue et al. (1983)[4] and Zabel et al. (1983)[5] for detailed description of the Arizona TAMS, its specifications, and operational characteristics.

Donahue et al. (1984)[6] have discussed the target preparation procedure, so we present here only a brief review. The ^{14}C sample consists of an iron carbide bead referred to as "the target." The bead is roughly spherical or oblate with a diameter of about a millimeter. This bead is produced simply and quickly by heating a 12:1 (by weight) mix of iron and elemental carbon for a few seconds at melting point in an inert atmosphere. Production of the elemental carbon may be accomplished by pyrolysis for samples of reasonably pure carbon or carbohydrate. Others require combustion to CO_2, then reduction to elemental carbon using a process similar to the original Libby (1955)[7] technique.

Comparison of Ion-Counting With β-Counting

The critical determinants of the true precision of measurement by either technique are the number of ^{14}C events counted and the stability of the measurement system. By convention, precision of ^{14}C dating analyses considers only the number of ^{14}C events. In this comparison we follow the convention and calculate the statistical uncertainty from the number of ^{14}C events counted. Figure 2 shows the ^{14}C-decay curve (this time using the assumed best half-life of ^{14}C) with age plotted against ^{14}C on the vertical logarithmic scale in various units. With the following accommodations, the two scales on the right side of figure 2 enable comparison of the number of ^{14}C events counted using ion-counting and β-counting. The ion-counting calculation must reflect the percent consumption of sample during analysis and total efficiency of ion transfer through the accelerator flight tube; the β-counting computation must account for event-detection efficiency (usually close to 100%), weight of carbon actually in the counter or scintillation vial, and time spent counting. Thus in β-counting, to convert the far right-hand column to a measure of efficiency, one multiplies the number corresponding to the age by sample weight in grams and by counting time in minutes. The longer the counting time, the greater the concerns about system stability. For larger samples, counting time is typically 2000 or 3000 minutes, but for very small samples counting time may be weeks (see Harbottle, this volume).

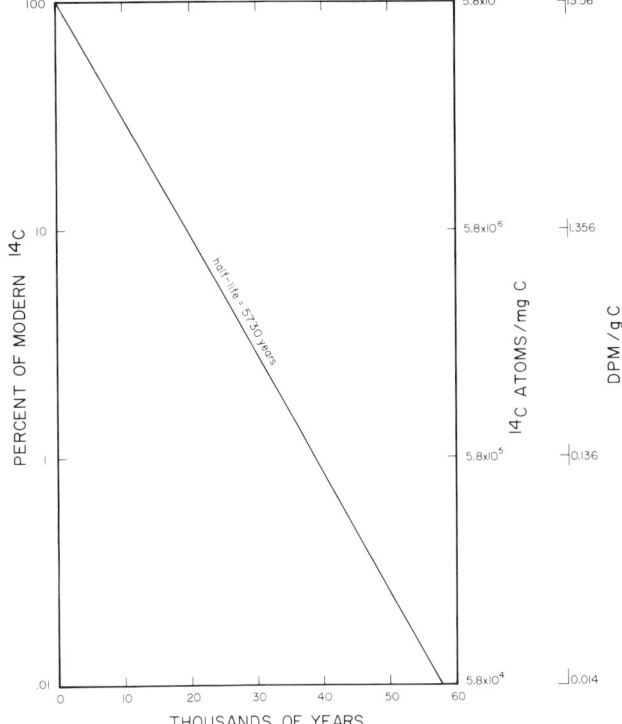

Fig. 2. Comparison of conventional and accelerator dating systems. If systems are stable, detection limits and precision depend on the total number of ^{14}C ions, or β disintegrations detected. To compare two systems, select age on horizontal coordinate. Where age intersects the diagonal line indicates the ^{14}C content or activity on vertical scale. For an accelerator, multiply the ^{14}C content by the weight of carbon (in mg) in the target and by the total analysis efficiency. For β-counting systems, multiply the DPM/g by the weight of carbon (in grams), by the counting time in minutes, and by the counting efficiency. Systems with similar resulting figures will have similar detection limits and precisions.

Table 1
Zea mays from Archaeological Sites in the Southwestern United States

Site	Conventional ^{14}C Date (years BP)	95% Probability Calibrated Date*
Tumamoc Hill, Ariz.	2580 ± 270	1370 to 155 B.C.
Fresnal Shelter, N. Mex.	2540 ± 200	1095 to 205 B.C.
Jemez Cave, N. Mex.	2410 ± 360	1205 B.C. to A.D. 45
Bat Cave, N. Mex.	2340 ± 420	1035 B.C. to A.D. 210
Chaco Canyon, N. Mex.	2250 ± 280	865 B.C. to A.D. 235
Chaco Canyon, N. Mex.	2150 ± 170	750 B.C. to A.D. 220
Matty Canyon, Ariz.	2050 ± 140	400 B.C. to A.D. 230
Fresnal Shelter, N. Mex.	1990 ± 320	750 B.C. to A.D. 575
Tumamoc Hill, Ariz.	1660 ± 210	A.D. 1 to 635
Tumamoc Hill, Ariz.	1490 ± 270	A.D. 10 to 1030

*From Klein et al. (1982). Note that tables give no values beyond ± 300 years. Thus if this figure is greater than 300, 95% probability is a minimum.

Table 2
Squash (*Cucurbita*) from Archaeological Sites

Site and Location	^{14}C Date (years BP)	95% Probability Calibrated Date*
Koster, Ill.	6820 ± 240	6300 to 5235 B.C.
Carlston Annis, Ken.	5730 ± 640	5180 to 3885 B.C.
Cloud-Splitter, Ken.	4700 ± 250	3925 to 2880 B.C.
Phillips, Ill.		
Sample 1	4800 ± 890	4325 to 3055 B.C.
Sample 2	4970 ± 1260	4400 to 3175 B.C.
Bowles, Ken.	4060 ± 220	3140 to 2160 B.C.
Chaco Canyon, N. Mex.		
Sample 1	2820 ± 220	1665 to 405 B.C.
Sample 2	2130 ± 280	795 B.C. to A.D. 410

*From Klein et al. (1982).

In ion-counting, although efficiencies are much lower, only a small portion of the ^{14}C atoms in a sample actually decay (produce measurable β-particles) during typical counting periods. Only about one part per million decays in a few days. Consequently accelerators require much less carbon than β-counters do. In our TAMS, the modern standard may yield 2000 to 4000 ^{14}C ion "counts" during the usual 2000-second counting time, yet consume only a few hundred micrograms of carbon. A sample 40,000 years old would be down two orders of magnitude in count rate (fig. 2). This is about the count rate of our blank; thus 40,000 years is a conservative estimate of the maximum age-detection limit using the TAMS. Improved target efficiency will increase this limit, and development work on this continues in our laboratory.

The ion-counting technique can theoretically produce results with ± 0.5% precision on near-modern samples if only $^{1}/_{1000}$ of the ^{14}C atoms in 1 mg of carbon are counted and all systems are stable. To accomplish this within a reasonable time frame, a higher ion yield is necessary. The TAMS system can produce precision of ± 1.5% on younger samples. For various reasons, this figure is often higher. Results reported here have plus-or-minus figures that include both the number of ^{14}C ions counted and the stability of the system.

Archaeology-Ethnobotany Examples

Among the first unknowns dated by Libby were archaeological samples. This new dating technique was revolutionary.[7,8] Archaeologists who have been major beneficiaries of this dating technique are now renewing their enthusiasm since the TAMS permits the dating of samples that were too small for the classical ^{14}C method.

Cultigens are often too scarce or too valuable as study specimens to sacrifice for ^{14}C analysis. Determining when maize (corn) was first cultivated in the southwestern United States is a problem particularly suitable for the TAMS in that the oldest dates (ca. 6000 years BP) for maize cultivation are based not on the cultigen itself, but on the charcoal associated with the maize. We have begun a systematic search for the oldest cultivated maize by studying samples from sites in Arizona and New Mexico, where with archaelogists' cooperation we are reevaluating cultigen dates directly (table 1). So far we have been unable to confirm the inferred 6000 BP date for maize cultivation in the southwestern United States.[9]

On the other hand, we were able to confirm the inferred 7000 BP date* for squash in the midwestern United States (table 2). Interestingly, maize kernels associated with the 7000-year-old squash are not nearly that old. These results are consistent with the current belief (R. I. Ford, pers. comm.) that squash cultivation began in the Midwest long before maize cultivation migrated from Mexico into the Southwest.

It is apparent from these dates (table 2) that small kernels of corn, squash seeds, and rind fragments can migrate within a stratigraphic section, perhaps by bioturbation, and leave virtually no evidence of translocation. Therefore dates of small objects based only on stratigraphic position are not as reliable as direct dates on the subject.

Bone Dating

Bones are a classic ^{14}C dating problem ideally suited to solution by combining recent developments in analytical chemistry and accelerator physics. The problem with bone ^{14}C dating is that the longer the bone has been in the ground, the greater the degree of degradation of original carbon, and also the greater the chance of adsorption of soil organic matter.

The solution is high-performance liquid-chromatographic (HPLC) separation of bone collagen, collagen fragments, and amino acids derived from collagen for TAMS ^{14}C dating. Stafford et al. (1984)[10] separated numerous organic fractions from a mammoth bone whose age is well known from cultural, stratigraphic, and extinction arguments. TAMS ^{14}C dates on these fractions (table 3) show that previous conventional ^{14}C dates on bone material from the same animal were "too young" probably due to contamination with younger humic material. Only with TAMS analysis of column chromatographic separates is it possible to obtain ^{14}C dates on original bone carbon without sacrificing large amounts of valuable paleontologic specimens.

Prehistoric Cast Iron

Cast iron contains enough carbon to serve as target without further processing. If the carbon used in the smelting process is contemporary charcoal, a small (1mm diameter) fragment of cast iron should be sufficient. We anticipate testing this direct process on samples of both known and unknown provenances.

Pot Sherds

A 100mg fragment of organic-matter–tempered prehistoric ceramic yielded a 1250 ± 160 BP ^{14}C age. This ^{14}C age calibrates to A.D. 610-990 (Stuiver, 1982).[11] This result, analyzed as a "blind" test, is well within the expected range (P. Dean, pers. comm.). Other ceramic samples are now being analyzed.

Art History and Forgery Detection

One advantage of the TAMS is the ability to anlayze very small portions of carbon-containing material. Although a centimeter or two of thread from a canvas or a splinter from a panel provides sufficient carbon for a TAMS analysis, this may not date the painting itself. Linseed oil and egg, which must be fresh to be useful pigment binders, are ideal ^{14}C samples for dating paintings. One must of course make

*Determined on associated charcoal in the Koster site by β-counting at the Illinois State Geological Survey.

Table 3
Various Chemical Fractions from the Domebo Mammoth Bone

Description of Sample	^{14}C Date (years BP)
Wood associated with bone	11,450 ± 320
Alpha-hydroxyacids	11,350 ± 340
XAD-purified hydrosylate	11,450 ± 350
Raw bone after acetic acid treatment	10,950 ± 450
XAD-purified gelatin	10,850 ± 400
0.6N HCl-insoluble organics	10,800 ± 300
Amino acids	10,500 ± 350
Gelatin	10,450 ± 400
0.6N HCl-soluble organics	9,500 ± 400
Inorganic CO_2 from acetic acid leaching	9,350 ± 320
CO_2 from combustion of raw bone	8,000 ± 450
Bone carbonate	7,250 ± 320
Fulvic acids	5,100 ± 200

sure that the pigment is inorganic, or at least if organic that it is composed of carbon contemporary with the painting and not "fossil" carbon. Restoration may complicate the sampling procedure, but it is often of interest to know what portions of a painting were restored at what dates in the past. Additions, alterations, or forgeries within the past thirty years are especially easy to detect because of the dramatic ^{14}C increases due to nuclear bomb tests.[12]

To illustrate the application of TAMS for forgery detection, we sampled a fleck of paint from a recent painting the date of which was known. The result was 136 ± 10% modern (that is 36% greater than the A.D. 1950 ^{14}C activity), clearly post-modern, most probably either post-1972 or 1961-62. A few splinters from a presumed Mississippian Indian mask were sufficient for the TAMS to demonstrate that the mask was constructed from wood grown after bomb testing, and could not have been made by Mississippian Indians several hundred years ago.

The Waukegan Horn

A TAMS ^{14}C analysis clearly distinguished between two proposed dates for a supposed artifact, the "Waukegan Horn." This nicely carved and inscribed bovine horn was discovered in 1951 near Waukegan, Illinois, in a vacant lot that had been recently graded and rained on. A physicist, Dr. Landsverk, who also was head of the Landsverk Foundation for Research on Norsemen in America, with the help of another Norwegian-American and retired U.S. Army cryptographer, decoded the inscriptions and concluded that the horn was carved on December 15, 1317. Landsverk maintained that this horn, like the Vinland Map, was authentic and provided evidence of medieval Norse contact with North America.

His comparison with the Vinland Map was unwittingly prophetic, as our ^{14}C analysis gave a radiocarbon age of 20 ± 240 years BP, a result not clearly post-bomb, but certainly not medieval. Subsequent research by M. L. LeVan disclosed a paper by Haugen[13] which convincingly argued that the horn was carved ca. 1920. The TAMS analysis is entirely consistent with that supposition. Rather than some obscure code, the inscriptions are evidently modern Icelandic in runic characters.

Future Outlook

During this decade we expect to see significantly more precise ^{14}C analyses. Ion-production rate from the source currently limits the precision. Higher ion output sources now under development may eventually bring TAMS precision in line with that of the better conventional laboratories.

Other isotopes detectable by the TAMS technique, such as ^{10}Be and ^{36}Cl, are primarily applied to geological and hydrological problems, although ^{10}Be could also help to "fingerprint" ceramics for source identification. TAMS's ability to measure natural levels of ^{14}C in very small samples of carbon will continue to be its primary contribution to museum science.

References

1. W. F. Libby, "Radiocarbon Dating," *Science*, 133 (1961): 621-629.

2. J. Klein, J. C. Lerman, P. E. Damon, and E. K. Ralph, "Calibration of Radiocarbon Dates: Tables Based on the Consensus Data of the Workshop on Calibration of the Radiocarbon Time Scale," *Radiocarbon*, 24 (l982): 103-150.

3. M. Stuiver, C. J. Heusser, and I. C. Yang, "North American Glacial History Extended to 75,000 Years Ago," *Science*, 200 (1978): 60-21.

4. D. J. Donahue, et al., "The Use of Accelerators for Archaeological Dating," *Nucl. Instrum. Methods*, 218 (1983): 425-428.

5. T. H. Zabel, et al., "Quantitative Radioisotope Measurements with the NSF-Arizona Regional Accelerator Facility," *IEEE Transactions on Nuclear Science*, NS-30 (2) (1983): 1371-1373.

6. D. J. Donahue, et al., "Results of Radioisotope Measurements at the NSF-University of Arizona Tandem Accelerator Mass Spectrometer Facility," *Proc. Accel. Mass Spec. Conf.*, Zurich, April 1984 (in press).

7. W. F. Libby, *Radiocarbon Dating*, 2nd ed., (Chicago: University of Chicago Press, 1955).

8. W. F. Libby, *Radiocarbon Dating*, (Chicago: University of Chicago Press, 1952).

9. H. W. Dick, "Bat Cave," *School of American Research Monograph 27*, 1965.

10. T. W. Stafford, Jr., et al., "Radiocarbon Dating of Bone Collagen and Organic Fractions by TAMS," *Proc. Accel. Mass Spec. Conf.*, Zurich, April 1984, poster session.

11. M. Stuiver, "A High-Precision Calibration of the A.D. Radiocarbon Time Scale," *Radiocarbon*, 24 (1982): 1-26.

12. R. Nydal, and K. Loveth, "Tracing Bomb ^{14}C in the Atmosphere 1962-1980," *Journal of Geophysical Research*, 88 (1983): 3621-3642.

13. Haugen, E., "The Youngest Runes; From Oppdal to Waukegan," *Michigan Germanic Studies*, (1981) 7:1.

PETER KLEIN

Dendrochronological Studies on Oak Panels of Rogier van der Weyden and Robert Campin

Dendrochronology is a recent discipline of the biological sciences that enables the age of wooden objects to be determined. Primarily employed for dating archaeological and architectural objects, the procedures have also been applied to art historical problems. In 1965 the Institute of Wood Biology, University of Hamburg, began to apply dendrochronology to the dating of oak panels and more than 800 Dutch, English, Flemish, French and German panels and wood carvings of oak, beech, lime, and poplarwood have been investigated.[2,8,12]

In this context, the objective is to achieve, at least, a "terminus post quem" for a painting by determining the felling date of the tree from which the panel was derived. The greatest amount of information can be obtained by studying the whole oeuvre of one artist or of a group of loosely related painters rather than by investigating single paintings. Such conclusions were corroborated for example by the investigations of panels by Rubens and Rembrandt.[2,3]

There is a special interest in panels of the fifteenth century because, generally, these are neither signed nor dated. In this context, the analysis of the works of Rogier van der Weyden (1399/1400-1464) and of Robert Campin (1375-1444) was very important, because not a single painting was unequivocably attributed.[4,9] The tree ring analysis of the panels resulted in new criteria concerning their attribution.[10]

Besides the determination of the felling date, further information can be derived, not only of the storage period of the wood before its use for a painting, but also for the provenance of the trees themselves. For oakwood this origin can be traced to different regions of the Netherlands, Belgium, or Germany.

Firstly a new master chronology for the region around Brussels had to be established which was used for the determination of the felling date of the trees used as panels by van der Weyden and by Campin.[10] Since the dendrochronology technique has been described in detail elsewhere,[1,5,7,11] only the specific aspects pertaining to these oak panels will be explained and supplemented by actual datings of works by Rogier van der Weyden and Robert Campin.

Special Features for the Dating Procedure

The oak panels were usually cut from the tree trunk with an approximately radial orientation and therefore contain a large number of growth rings. The key information to be determined for the dating of panels is the felling year of the tree. Oak has light-colored, perishable sapwood of 20 ± 5 growth rings. These "missing" years must be accounted for in the dating of a panel. Panels containing a few sapwood rings enable a determination of the felling date with an accuracy of ± 5 years. In the case of missing sapwood, the felling date can only be approximated by adding a minimum of fifteen sapwood rings to the last measured growth ring. Therefore, only an approximation of the possible felling date can be derived that is illustrated by the term 20 ± 5,x. In this expression the $20\text{-}5$ accounts for the sapwood rings possibly involved, while the x stands for an unknown number of removed heartwood rings.

Examples of the Dendrochronological Determination of Panel Paintings by van der Weyden and Campin

The first dating examples are of five altarpieces from Berlin, Frankfurt, and Munich, originally attributed to Rogier van der Weyden, il-lustrated in figure 1.

Based on the assumptions for the sapwood, the *Columba Altarpiece* of Munich can be dated, to the year 1439 ± 5, x and the *St. John Altarpiece* of Frankfurt to the year 1498 ± 5, x. Obviously, the last altarpiece must be a copy made about fifty years after the death of van der Weyden.

In addition to the determination of the felling date other interesting observations can be made. The analysis of the *Columba Altarpiece* (fig. 2) showed that the right and the left wings were made of four boards and the center of six boards. Only two boards of the right wing, those marked by identical symbols, came from the same tree.

In contrast, analysis of the *Bladelin Altarpiece* (fig. 3) shows that all the boards came from only three trees, and furthermore that two boards marked with symbols (xx) came from adjacent positions in the same tree. One was used in its entire width for manufacturing part of the left wing. The other board was sawn down the center; one half made part of the left wing, the other made part of the right. A saw cut of 2-3mm, based on the width of two missing growth rings, has been calculated (fig. 4).

Beside the five altarpieces, other paintings attributed to Rogier van der Weyden and his circle were examined in Antwerp, Berlin, The Hague, Munich, Paris and Vienna (fig. 5). The dendrochronological analysis 1425 ± 5, x of the painting *Woman with a White Headdress* (Gemäldegalerie Berlin-Dahlem) confirms its art historical attribution to the year 1435.[4] Analysis of the two small panels *St. Catherine* and *The Virgin* showed that the panels came from the same tree. The panels contain only ninety growth rings and a felling date can be derived for the year 1392 ± 5, x. In this case the panels probably came from the center of the tree.

Furthermore it is obvious that the panel painting of *St. Luke Painting the Virgin* (fig. 6) (Alte Pinakothek, Munich) is an early copy with a felling date of 1475 ± 5, x. The panel painting *St. Luke Painting the Virgin* in the Museum of Fine Arts, Boston, was recently analyzed and a felling date of 1434 ± 5, x was derived. This date supports its attribution as an original. Clearly a dendrochronological analysis of the *St. Luke Painting the Virgin* painting in Leningrad would be of great interest.

The dendrochronological analysis of the two copies of *Philip the Good* in Antwerp and Paris confirms the art historical attribution for both panels. Accounting for the sapwood and the storage time assumptions, the portraits were painted at the beginning of the sixteenth century.

Five paintings by Robert Campin (1375-1444) were investigated (fig. 7). None of the panels contains sapwood rings, and accordingly only the earliest possible felling date can be determined. The examinations indicate that the felling date for the painted panel *The Madonna of Humility* can be dated to the year 1383 ± 5, x and those for the three panels *St. Veronica, The Trinity,* and *The Virgin with the Children* can be dated to the years 1411 ± 5, x, 1410 ± 5, x and 1404 ± 5, x. These dates agree with the art historical attributions. Additional information from the dendrochronological dating can be derived for the panels *St. Veronica* and *The Trinity* (fig. 8). The assumption that the painting of *The Trinity* had previously been on the back of the panel *Veronica* can no longer be upheld. The one panel consists of two boards and the other panel of three boards. Furthermore, neither is it possible that *The Trinity* was on the back of

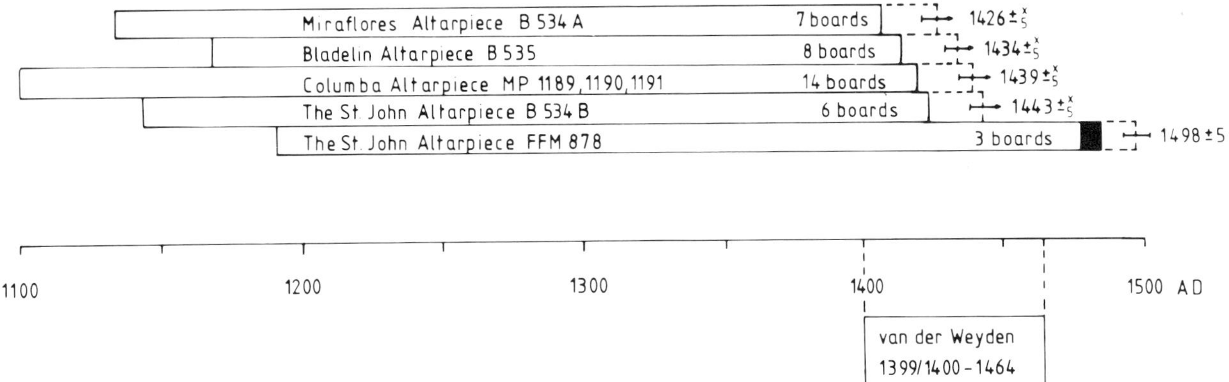

Fig. 1. Dendrochronological dating of oak panels from altarpieces originally attributed to Rogier van der Weyden. □-heartwood, ▮-sapwood, ⊢→-felling date years; ⊬-felling date ± 5, x years: B-Gemäldegalerie Berlin, West Germany, FFM-Städelsches Kunstinstitut Frankfurt, West Germany, MP-Alte Pinakothek Munich, West Germany.

Fig. 2. Rogier van der Weyden, *Columba Altarpiece*, Alte Pinakothek Munich. Construction of the altarpiece from different boards; o-boards from the same tree; i-toward the pith; o-toward the bark.

Fig. 3. Rogier van der Weyden, *Bladelin Altarpiece*, Gemäldegalerie Berlin-Dahlem. Construction of the altarpiece from different boards; x, xx, xxx-boards from the same tree; i-toward the pith; a-toward the bark.

the panel *The Virgin with the Children* because the growth ring curves of the boards are too different from each other. Dendrochronological studies have, therefore, indicated that panels comprising more than one board include boards from the same tree. This observation does not extend beyond individual paintings.

The determination of the felling date of the tree also gives information as to the storage time of the wood prior to its use for paintings. For oak panels of the seventeenth century, in most cases the interval between tree felling and painting was between two and eight years.[3] The investigations carried out with signed and dated Flemish panels do not yet permit such a close estimate.[11] However, present studies indicate a storage time of ten to fifteen years which agrees with analyses performed on panels of the School of Cologne from the fifteenth century.[5] With assumptions of a storage time of ten years and of twenty sapwood rings—most of the trees of the investigated panels contain 200 and more growth rings—we have postulated dates for some panels and altarpieces of Rogier van der Weyden in table 1 (see also fig. 5).

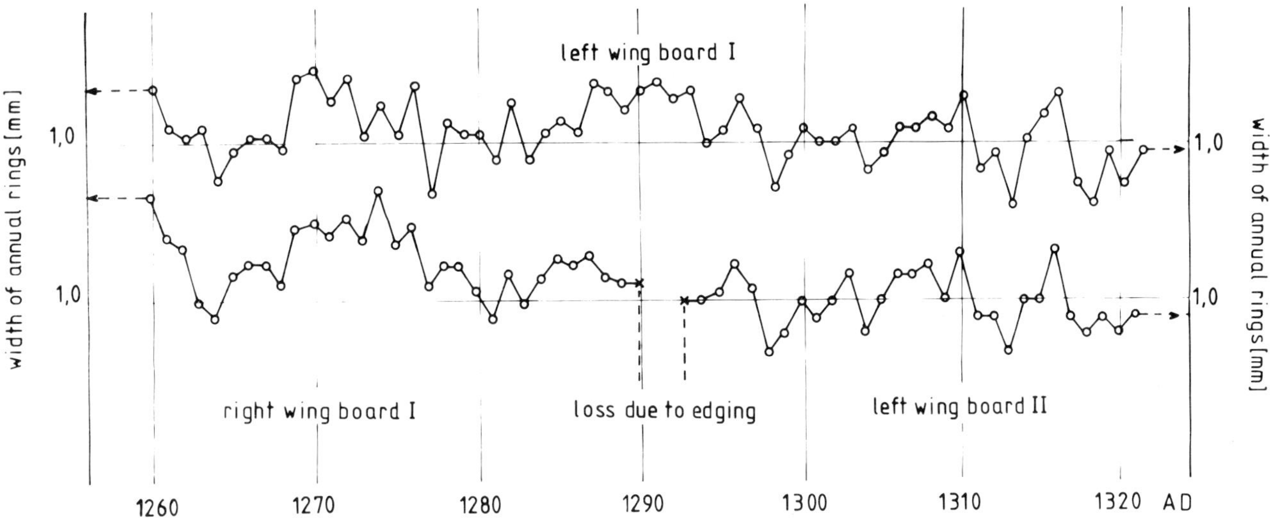

Fig. 4. *Bladelin Altarpiece*: comparison of growth ring curves derived from board I, II (left wing) and board I (right wing).

It will be seen from the Table that the *Miraflores Altarpiece* in Berlin is an early van der Weyden painting, which confirms the recent art historical attribution derived by Grosshans in the Berliner Jahrbuch.[9] A dendrochronological analysis of *The Altarpiece of the Virgin*, the so called "Granada Altarpiece" (left wing and central panel in the Capilla Real of the Cathedral of Granada, the right wing in the Metropolitan Museum in New York) has not yet been executed, but such an investigation could be helpful in distinguishing between the original and a copy.

Conclusions

Dendrochronological studies of panels can give a "terminus post quem" for the component boards of panels. Furthermore a comparison of the growth ring series of the single boards often enables the determination as to whether or not they originate from the same tree and consequently, permits an attribution to a particular panel workshop. By the statistical analysis of a number of dated and signed paintings, the period between the felling date of the tree and the creation of the painting can be established, and finally the different regional master chronologies in Europe often permit the determination of the provenance of the trees and, consequently, the origin of the panels.

It would be of value for dendrochronological analyses to be conducted on other paintings of Rogier van der Weyden and Robert Campin to form a study of their complete works. From this information art historians may find corroborative evidence for temporal attributions.

Table 1

Postulated Attribution Date of Rogier van der Weyden Paintings and Altarpieces

Paintings	Felling date of the trees (Sapwood: 20 years)	Presumed date of creation of the paintings (Storage time: 10 years)
Annunciation (PL)	1421	1431
St. Luke Painting the Virgin	1424	1434
Woman with Headdress	1425	1435
Miraflores Altarpiece (B)	1426	1436
Pieta (DH)	1427	1437
Bladelin Altarpiece (B)	1434	1444
Columba Altarpiece (MP)	1439	1449
The St. John Altarpiece (B)	1443	1453
St. Luke Painting the Virgin (MP)	1473	1483
St. John Altarpiece (FFM)	1498	1508

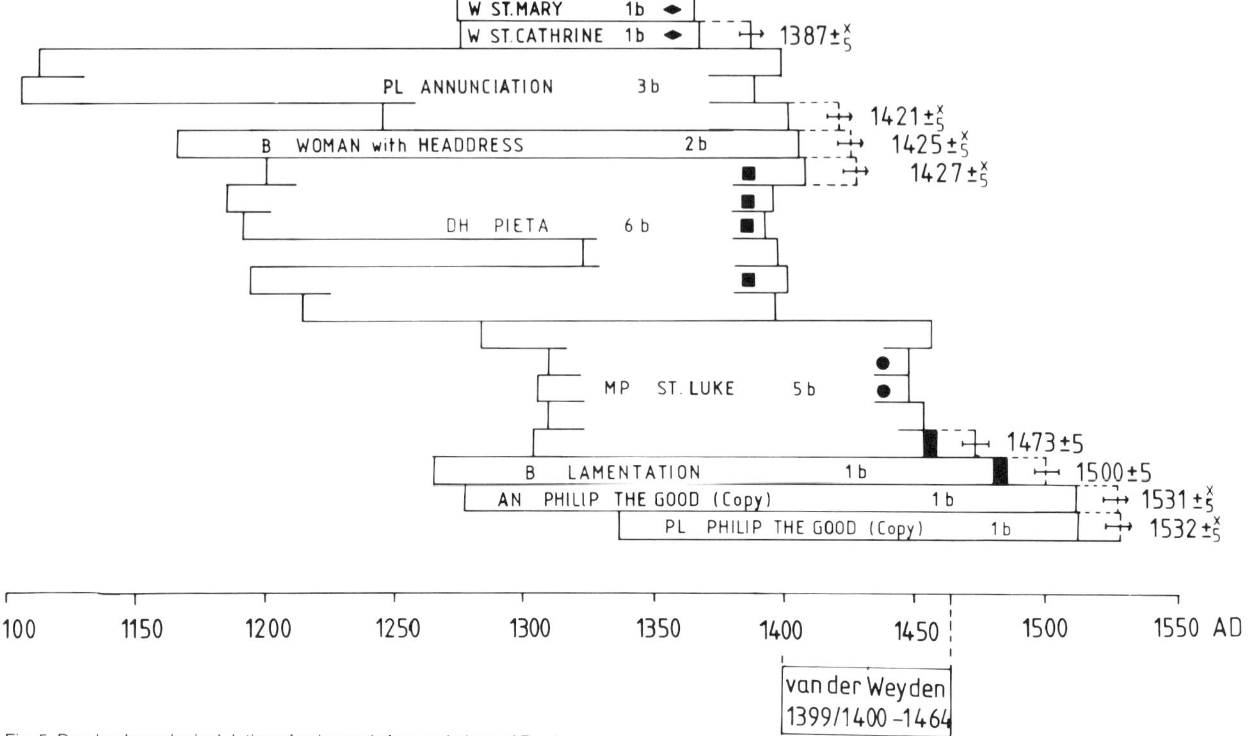

Fig. 5. Dendrochronological dating of oak panels from paintings of Rogier van der Weyden and his circle. ☐-heartwood; ▮-sapwood; ⊢↦ felling date ±5, x years; ⊬⊢felling date ±5, x years; ▪, ●, ◆, boards from the same tree; AN-Koninklijk Museum, Antwerp, Belgium; b-number of boards,; B-Gemäldegalerie Berlin-Dahlem, West Germany; DH-Mauritshuis, The Hague, Netherlands; MP-Alte Pinakothek Munich, West Germany, PL-Musée du Louvre, Paris, France, W-Kunsthistorisches Museum, Vienna, Austria.

Fig. 6. Copy after Rogier van der Weyden, *St. Luke Painting the Virgin*, Gemäldegalerie Berlin-Dahlem, West Germany. Construction of the altar from different boards; ☐-boards from the same tree; i-toward the pith; o-toward the bark.

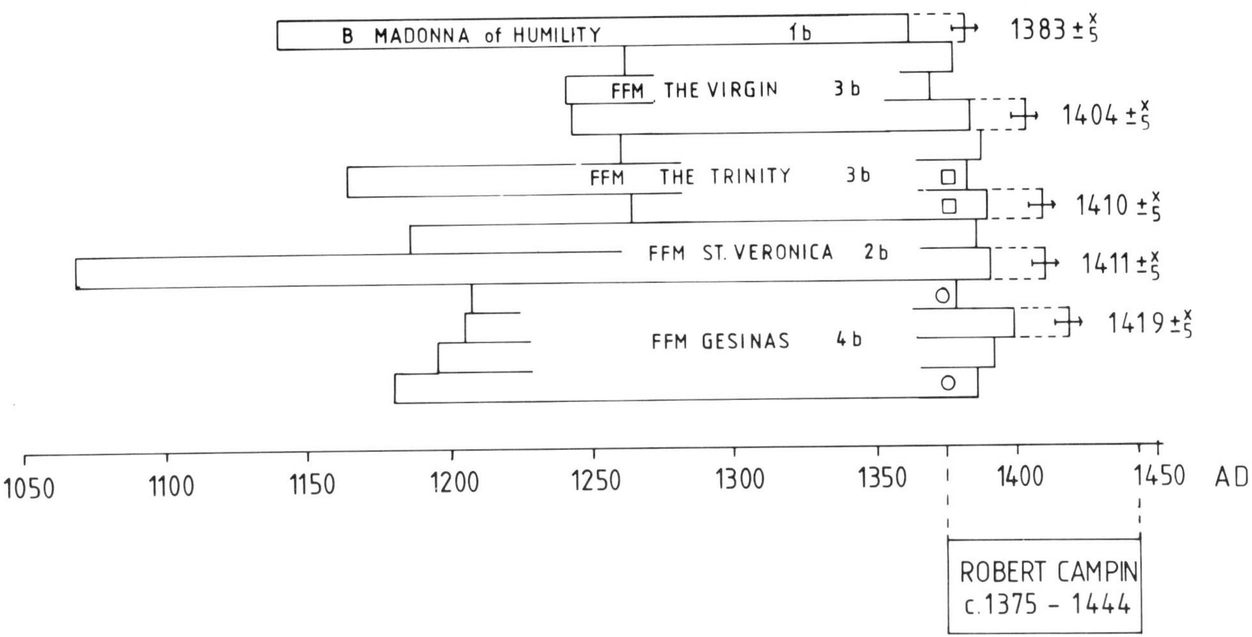

Fig. 7. Dendrochronological dating of oak panels from paintings of Robert Campin. heartwood; felling date ± 5, x years; □, ○ -boards from the same tree; b-number of boards; B-Gemäldegalerie Berlin-Dahlem, West Germany; FFM-Städelsches Kunstinstitut, Frankfurt, West Germany.

Fig. 8. Robert Campin, *St. Veronica* and *The Trinity*. Städelsches Kunstinstitut Frankfurt, West Germany. Construction of the two paintings from different boards; o, ◇-boards from the same tree; i-toward the pith; o-toward the bark.

References

1. M. G. L. Baillie, *Tree-Ring Dating and Archaeology* (Chicago: The University of Chicago Press, 1982).

2. J. Bauch and D. Eckstein, "Woodbiological Investigations on Panels of Rembrandt Paintings," *Wood Sci. Technol.* 15 (1981): 251.

3. J. Bauch, D. Eckstein, and G. Brauner, "Dendrochronologische Untersuchungen an Eichenholztafeln von Rubensgemälden, *Jahrbuch Berliner Museen* 20 (1978): 209.

4. M. Davies, *Rogier van der Weyden* (Munich: C. H. Beck, 1972).

5. D. Eckstein and J. Bauch, "Dendrochronologische Altersbestimmung von Bildtafeln," *Die Kölner Maler von 1300 – 1430* (Cologne: Wallraf-Richarts-Museum, 1974).

6. D. Eckstein, S. Wrobel, and R. W. Aniol, ed., *Dendrochronology and Ecology in Europe* (Hamburg: Mitt. Bundesforschungsanstalt f. Forst- und Holzwirtschaft, 1983).

7. J. Fletcher, ed., *Dendrochronology in Europe* (Oxford: BAR Intern. Series No. 51, 1978).

8. J. M. Fletcher, *Tree-Ring Dating of Tudor Portraits*, Proc. Royal Inst. of Great Britain 52 (1980):81.

9. R. Grosshans, "Rogier van der Weyden, Der Marienaltar aus der Kartause Miraflores," *Jahrbuch der Berliner Museen*, 23 (1981): 113.

10. P. Klein, "Dendrochronologische Untersuchungen an Eichenholztafeln von Rogier van der Weyden," *Jahrbuch der Berliner Museen*, 23 (1981): 49.

11. P. Klein, "Grundlagen der Dendrochronologie und ihre Anwendung für kunstgeschichtliche Fragestellungen," *Berliner Beiträge zur Archäometrie* 7 (1982): 253.

12. P. Klein and J. Bauch, "Aufbau einer Jahrringchronologie für Buchenholz und ihre Anwendung für die Datierung von Gemälden, *Holzforschung* 37 (1983): 35.

JOHN S. MILLS and RAYMOND WHITE

Organic Analysis in the Examination of Museum Objects

The identification of the minor organic components of objects of art and archaeology has, in the past, usually taken second place to the study of the inorganic materials. They often play a less conspicuous role in the structure of such objects. Nonetheless interest in them has always existed[1] and attempts have been made at analysis since the early nineteenth century but the means have been lacking to make certain identifications. Only very small samples may be available, too small for the older spectrometric methods to be applicable even if such methods were potentially capable of giving useful results. Often they would not be, for the composition of the natural materials concerned may be complex to start with and change with time in ways that, until recently, have been little studied and understood.

In recent years there have been great advances in analytical methods and instrumentation, and this has led to an increased understanding of oxidation processes and products. Thus it is now feasible to undertake the examination of materials such as paint media, varnishes, adhesives, impregnants, and the adventitious materials found in or on vessels or sherds, with the reasonable expectation that much can be determined regarding their chemical composition, and with the good hope that their original make-up, in terms of natural products, can be inferred from this. In the Scientific Department of The National Gallery, London, it has long been our conviction that the only practicable approach to such analyses, given the probable complexity and the smallness of the samples, must be on the basis of separation and identification of the individual components. In practice this means the adoption of some form of chromatography, at least as a first step. Chromatographic methods are well known and need no description here. Gas chromatography (GLC) has the great advantages over the cheaper methods of paper and thin-layer chromatography of sensitivity, reproducibility, and the ability to quantify the individual components. These and other advantages more than compensate for the added expense and for the need for manipulation of the sample to convert it to derivatives sufficiently volatile to be subjected to GLC.

When very complex chromatograms are obtained, and there are peaks on them that cannot be identified, the next step is to resort to such methods as combined gas-chromatography-mass-spectrometry (GLC-MS). This yields the mass spectrum of the compound responsible for the peaks and, in many cases, allows their identification. Rather recently the linked method of gas chromatography with Fourier transform infrared spectrometry has been put forward as an alternative approach but it does not, *a priori*, appear to be as informative as GLC-MS. For example, members of a homologous series such as the fatty acid esters, or the hydrocarbons in waxes, have closely similar infrared spectra and would be hard to distinguish one from another whereas MS would at once identify them by their molecular weights.

GLC-MS is the mainstay of our own approach to analysis,[2] and most of the examples that follow are based on it or on GLC alone. A few words should, however, be said regarding its difficulties and limitations, as well as alternative approaches. A general problem when examining very small samples is their proclivity to contamination from dirty glassware, impurities in solvents, and cross-contamination from sample to sample (for example, in the syringe used to inject the sample into the GLC); great care has to be taken

to avoid this (or, when one gets an unusual result, the worry that it may have occurred). It must be realized also that maintaining a GLC-MS system in good working order is no small or carefree undertaking: it has to be kept running continuously, not just switched on and off as needed, and much time will undoubtedly be taken up in adjustments and troubleshooting. Its limitations are essentially those of GLC itself, namely the limitations imposed by the need for the sample's components to be sufficiently volatile to pass through a gas chromatograph column at a reasonable temperature. This eliminates all polymers from examination, as well as compounds with molecular weights much in excess of 500, or those that are involatile for other reasons, for example, extensive functionalization as occurs in most organic dyestuffs of natural origin. For these, other analytical methods have had to be applied. High-performance liquid chromatography has been found useful and we are currently engaged in optimizing experimental conditions for its application to the analysis of the dyestuffs present in the lake pigments used in oil paintings.

It is not possible in the space of this paper to survey the state of the analyst's art as applied to the whole range of materials that may be encountered. We will content ourselves with the briefest of remarks on proteins and waxes, say rather more on the subject of oils and fats (particularly as regards their use as paint media), and then illustrate the application of mass spectrometry to the detection of resins in a variety of environments.

Proteins

Despite the above-mentioned liability of organic materials to change with time, some materials are relatively stable and resist oxidation. Proteins are such a material, at least as regards their basic amino acid composition. Unfortunately the various proteins—collagen, albumin, casein, etc.—do not differ very much in amino acid make-up and their identification requires a rather laborious and careful gas-chromatographic determination of the relative proportions of these. This is generally precluded in the study of paint media by the smallness of available samples but it can be rewarding in the case of other samples where size is no problem. Gas chromatography alone using well-established derivatization techniques is the method adopted. A paper by one of us outlining the methods and a number of results is in preparation.[3]

Waxes

Waxes, largely made up of saturated compounds, are generally highly resistant to oxidative change. This stability allows identification by infrared spectrometry.[4,5] Even so gas chromatography still has great advantages in that it is applicable to far smaller samples.[6] Moreover it permits the detection of mixtures of waxes, particularly in the case of the presence of small quantities of one wax in large quantities of another, which would be quite unobservable in the infrared spectrum. Wax analysis can be quite rewarding, where several different mixtures have been used for making casts and models. These various mixtures can sometimes help to establish date and provenance; and with greater knowledge, this approach will become more useful.

Oils and Fats

Oils and fats enter into the composition of a great range of materials and few samples are encountered that are entirely free of them, as fatty material is a ubiquitous contaminant. This can be troublesome in the examination of very small samples and care has to be taken to ensure clean solvents and glassware. We will say nothing here of their incorporation into wax mixtures, quite a common occurrence, but concentrate on the analysis of paint media by fatty-acid determination, which has been a particular study at The National Gallery for a number of years.

Drying oils are, by their nature, very reactive chemical compounds and are indeed required to change chemically and physically if they are to fulfill their function of forming a solid binding medium. Their final composition is therefore very different from their original one; nonetheless the pattern of fatty acids that results is fairly simple and stable.[7] Its determination by GLC is a straightforward procedure well suited as a routine method for building up a body of data on the media used by different painters and schools. Because fatty materials are so common, fatty-acid analysis often gives a lead with samples of all types and is routinely applied as a first step to most unknown samples.

A few words should perhaps first be said regarding sampling. Generally, samples are only removed from paintings undergoing conservation treatment, for it is only with these that one can be confident of knowing the limits of original paint and be sure of the absence of retouching or varnish that could affect the results. Samples are usually taken from the edges of the painting or losses, preferably the latter. They can be extremely small, but must be visible to the naked eye. Sampling is probably the most tiresome part of the whole analytical procedure for it is difficult to be sure that one is taking material from a single paint layer (as is desirable) and the choice of the right spots can be critical. The work-up of the samples has been adequately described before,[7,8] and need not be discussed here but we must re-emphasize the need to use freshly distilled ether for the extraction and methylation operations to avoid contamination. Methylation using diazomethane is, we believe, the only practicable method with such small samples. Diazomethane is toxic and this step *must* be conducted in a hood.

As has been explained in previous articles,[1,7,9] fatty acid analysis permits the distinction between egg tempera and oil media by the amount of azelaic acid detected in the sample: tempera films show very little (relative to palmitic and stearic acids), oil films show a large amount. Different drying oils, or at least linseed, walnut, and poppyseed oils, may be distinguished with a reasonable degree of assurance from a measurement of their palmitate-stearate ratios. In the case of linseed and walnut oils these overlap to some extent, in the region of around 2.1 to 2.3, but the histogram in figure 1 shows that there is a genuinely bimodal distribution of the ratios corresponding to the two different oils. This is really quite remarkable given the number of factors that could affect the results. Results so far suggest that walnut oil was the preferred medium of the first oil painters in Italy, whereas linseed oil, used from the start in northern Europe, was, in general, adopted later by Italian artists. Our analysis of early English paintings from the thirteenth and fourteenth centuries, and that by others of Norwegian art,[10] has shown that oil painting was

practiced in these countries long before it was adopted in Italy. Poppyseed oil has regularly been detected only in French paintings, especially those of the later nineteenth century when it must have been the usual medium incorporated by artists' colormen into their tube colors.

The results of such an analysis are not necessarily confined to the above points as others may emerge. Thus, in some samples (from overpaint or varnish) a peak corresponding to dimethyl phthalate has been observed, from the use of *alkyd resins*, an indication that the sample dates from the twentieth century. Other types of compound also sometimes appear such as the components of diterpenoid resins, though these are often present in such small amounts that they may not be positively detectable except with the aid of mass spectrometry.

Fig. 1. Histogram Showing the Distribution of the Palmitate-Stearate Ratios for Paint Samples from Italian Oil Paintings Between 1450 and 1800

Palmitate/ Stearate	No. of Results	
1	0	
1.1	1	*
1.2	4	****
1.3	4	****
1.4	10	**********
1.5	12	************
1.6	10	**********
1.7	18	******************
1.8	15	***************
1.9	12	************
2	12	************
2.1	8	********
2.2	5	*****
2.3	4	****
2.4	9	*********
2.5	11	***********
2.6	13	*************
2.7	9	*********
2.8	10	**********
2.9	8	********
3	12	************
3.1	6	******
3.2	3	***
3.3	2	**
3.4	2	**
3.5	2	**
3.6	1	*
3.7	0	
3.8	0	
3.9	0	
4	0	

Natural Resins

The use of mass spectrometry comes into its own in the detection of natural resins by identification of their characteristic components. Some natural materials contain homologous series of chemical compounds (the fatty acid esters of oils, or the hydrocarbons of waxes) that emerge as regularly spaced peaks on a chromatogram, making them relatively easy to recognize. With the natural resins, however, the components differ by skeletal structure or functionalization and emerge on a chromatogram as a jumble of irregularly spaced peaks, the pattern of which will only be occasionally recognizable to give a clue as to individual identities. Even this pattern may be concealed when in admixture with other natural materials, or it may have been altered by aging.

Resins fall into two groups,[11] those containing diterpenoids (the conifer resins and the leguminous African and South American copals) and those containing triterpenoids (including the familiar varnish resins dammar and mastic and many others). Resins have been used in paint media, varnishes, and adhesives but they are also encountered as the bulk material of carvings and other objects, as many of them (notably the semi-fossil copals and the fossil ambers) are found naturally as large pieces that can be worked. Their identification can thus be of help in the placing of ethnographic and other objects. This is still far from being achievable with exactitude in many cases for there are very large numbers of different resin types. With the dammars, for example, which come from different areas of South East Asia and from many different species of the Dipterocarpaceae, it would be difficult to pin down an exact origin. For certain other resins (kauri from New Zealand, for example) identification is easy and indicates a very specific origin for the object.

The identification of resins in objects or materials of European production might be expected to involve the whole range of imported materials, so many of which are mentioned in early recipe books or manuscripts. In practice, positive identifications point rather frequently to the use of common pine resin—the most abundant and hence the cheapest material available. It must at once be admitted that we might be unable to detect certain materials (such as the leguminous copals) that had been subjected to severe conditions to bring them into solution as varnishes, for these resins contain high proportions of polymer and the characteristic diterpenes, present in small amounts, may be destroyed by processing. This reservation applies also to the detection of fossil resins such as Baltic amber in "amber varnish" (if indeed such varnish ever was compounded using genuine amber) for although Baltic amber in lump form may be readily identified as such by GLC detection of characteristic diterpenoids which it still contains,[12] these are unlikely to be still observable after processing the resin for varnish making. The detection of pine resin and, less commonly, other conifer resins such as Venice turpentine from the European larch, is interesting enough, however, in the study of the artist or artisan's formulary. It also provides a salutary lesson as to the way in which a single material can, depending on environment, survive in several chemically changed ways, resulting in quite different chromatographic patterns.

The resins of most pines are similar in composition and contain predominantly abietadiene and pimaradiene acids of the types shown in figure 2. The abietadiene acids are interconvertible; at

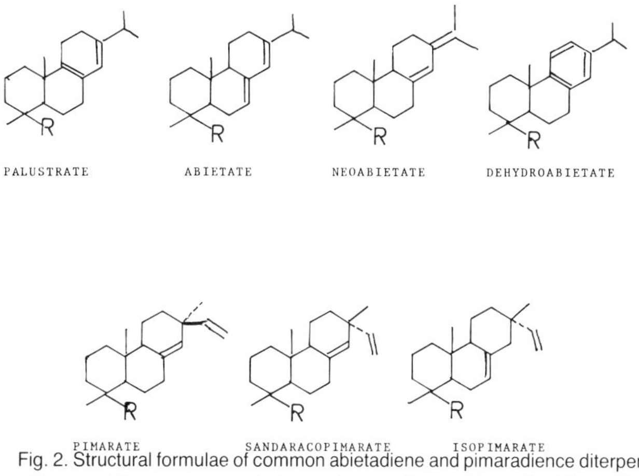

Fig. 2. Structural formulae of common abietadiene and pimaradience diterpenoid resin acids, components of pine and some other conifer resins. The methyl esters (R-COOCH$_3$) all have molecular weight 316 except dehydroabietate, which is 314.

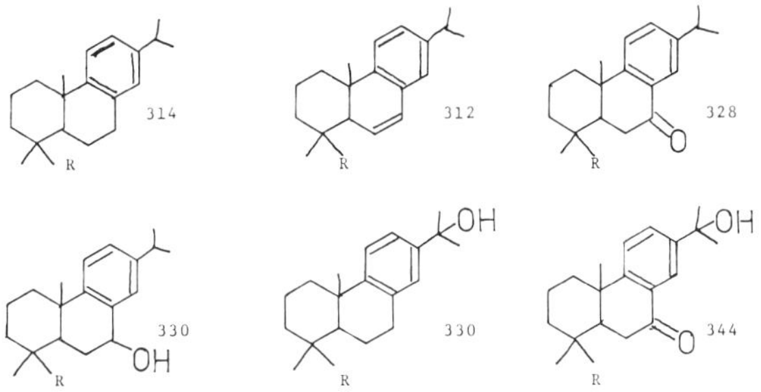

Fig. 3. Structural formulae of abietadiene acid oxidation products with their molecular weights (as methyl esters R-COOCH$_3$).

equilibrium the mixture contains mostly abietic acid. None of these acids is very stable, however, and they all tend to be oxidized either directly, by dehydrogenation, to dehydroabietic acid or with concomitant allylic oxidation to give hydroxy-or oxydehydroabietic acids (fig. 3). These compounds with the aromatic ring are relatively stable and so tend to survive, though the hydroxy compounds can be dehydrated to give compounds with extra double bonds. The pimaradiene acids do not yield aromatic compounds and simply tend to disappear on aging. The gas-chromatographic picture of composition that one can obtain with an old pine resin can therefore vary from that of relatively unchanged material (when it has been preserved in anaerobic conditions or in the presence of antioxidants) through that with both original components and a number of additional oxidation products (of slightly higher molecular weight), to one showing a single surviving identifier, dehydroabietic acid. The following results exemplify these patterns.

We have published in previous articles an example of a well preserved pine resin from a foundered Punic warship[11] and another example of similarly well preserved material from an eighteenth-century green lacquer,[2] protected in this case by the antioxidant effect of green copper pigments. Figure 4 shows the chromatogram given by a sample from a late George Stubbs panel painting, again showing resin with most of the original components detectable though most of the abietadiene acids have been converted to dehydroabietic acid.

There seems to be no particular reason for this good state of preservation here unless it be that there was no drying oil present in the medium. In admixture with drying oils, resin is less well preserved, indeed it is rarely possible to detect more than traces of dehydroabietic acid, as has been observed in samples from a number of early masters. As example we show in figure 5 a chromatogram given by green paint from Altdorfer's *Christ Taking Leave of His Mother* (National Gallery, London). The figure illustrates the use of the "mass scan" to pick out a sought-for compound even when, as in this case, it is present only as a minute peak on the total ion-current chromatogram. The mass scanned for here is 314, the molecular ion of methyl dehydroabietate. Sometimes for even greater sensitivity one scans for the more abundant mass 239, the base (largest) peak of its mass spectrum, but there is the disadvantage that this mass may also occur in the spectra of other compounds on the chromatogram, for example methyl palmitate. The very small amount of resin detected (as dehydroabietate) relative to the peaks of drying oil fatty acids raises questions as to how to interpret the finding. Is it a genuinely original component of the paint medium or simply the remains

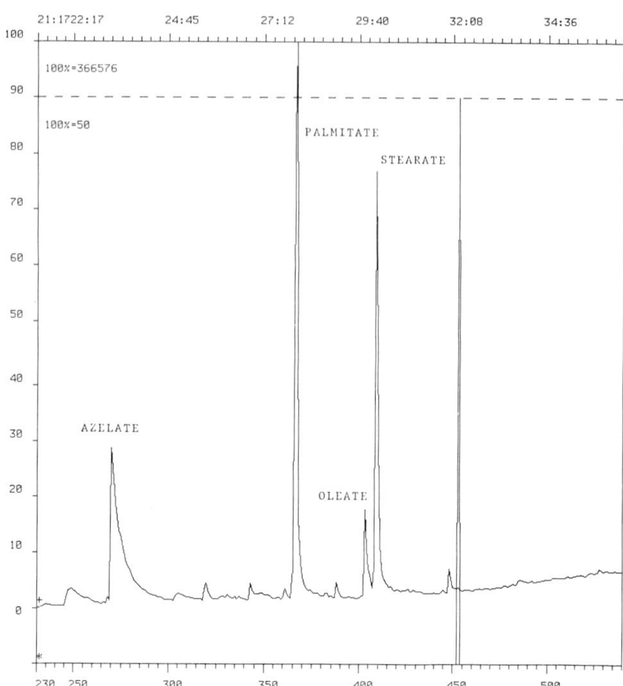

Fig. 5. Gas chromatogram of a green paint sample (after saponification and methylation) from Altdorfer's *Christ Taking Leave of His Mother* (National Gallery, London). A "mass scan" at mass 314 picks out the position of the very small dehydroabietate peak and, at about scan 455, all that remains of an original content of pine resin.

of later varnishes? We believe the former is usually the true explanation for often, with a single painting, it can only be detected (in these small quantities) in the samples where its presence might be expected, such as green paint that could include copper resinate. As an example of the third "survival mode" of pine resin, in which extra peaks of oxygenated dehydroabietic acid compounds are present, we show in figure 6 the chromatogram given (after saponification and methylation) by a nineteenth-century tracing paper. At this period numerous different natural materials, in various combinations, were used to impregnate paper to make it translucent. Many, but not all, of these impregnants have turned brown and brittle, making conservation of documents (such as ships' plans) an urgent problem.[13] A study of their composition is therefore of considerable interest. Many of the peaks on the chromatogram are of fatty acid esters and these indicate the presence of a small amount of drying oil (presence of dimethyl azelate) with a larger quantity of non-drying oil of a kind having high proportions of C-12 and C-14 fatty acids (see below). The region of the chromatogram between about scans 350 and 550, expanded in figure 7, contained the common pine resin acid esters and also large peaks of higher molecular weight compounds corresponding to hydroxydehydroabietates (330), 7-oxodehydroabietate (328), and a hydroxy-7-oxo-dehydroabietate (344). There are many other peaks in this region of the chromatogram, some of which are attribut-

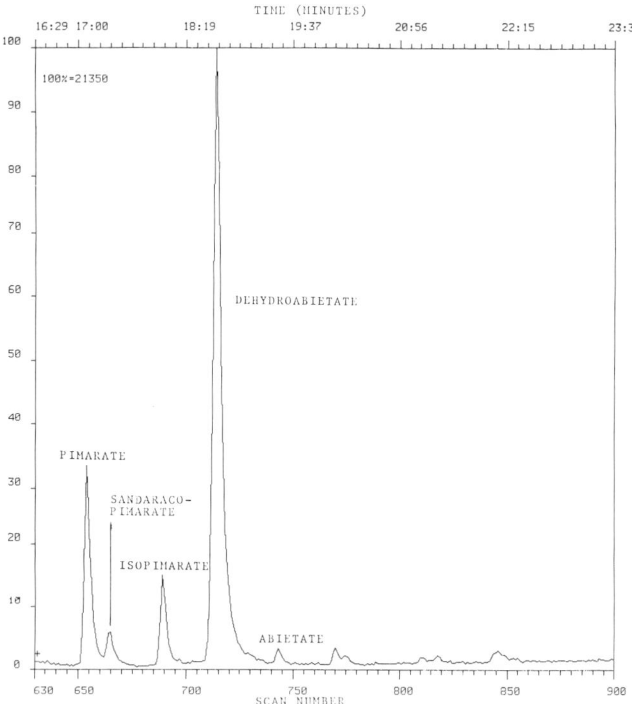

Fig. 4. Gas chromatogram yielded (after saponification and methylation) by a paint sample from a late George Stubbs panel painting, showing the presence of much pine resin.

able to the esters of higher fatty acids (C-22 and C-24 etc.) but others are still unidentified and may come from yet another natural ingredient. The pattern of oxidation products found in this sample seems to be characteristic of pine resins or mixtures containing them, kept in the dark with free access of air. It was close to that found for a sample of originally pure abietic acid kept in a bottle for thirty years in a drawer. Light, or the vigorous free radical reactions taking place in a drying oil film, seems to result in more extensive degradation leaving only the most stable compound, dehydroabietic acid.

The non-drying oil referred to in the previous paragraph was most probably coconut oil. In another sample of translucent paper, which had yellowed hardly at all, the impregnant was found to be of a simpler nature (and applied less heavily), and consisted only of this same oil admixed with a very small proportion of pine resin. The chromatogram (fig. 8) shows the regular pattern of the even-carbon number saturated fatty acid esters peaking at the C-12 acid ester (laurate) that is characteristic of this oil (and of palm oil, which is, however, a less likely contender in this instance).

Conclusions

The above survey barely hints at the range of natural organic materials that may be incorporated into museum objects but we hope that it conveys our belief that their analysis can provide useful and interesting results, both from the point of view of pure scholarship and also from that of the conservator anxious to understand the reasons for degradative change and to work with a fuller knowledge of the composition of the objects in his or her care.

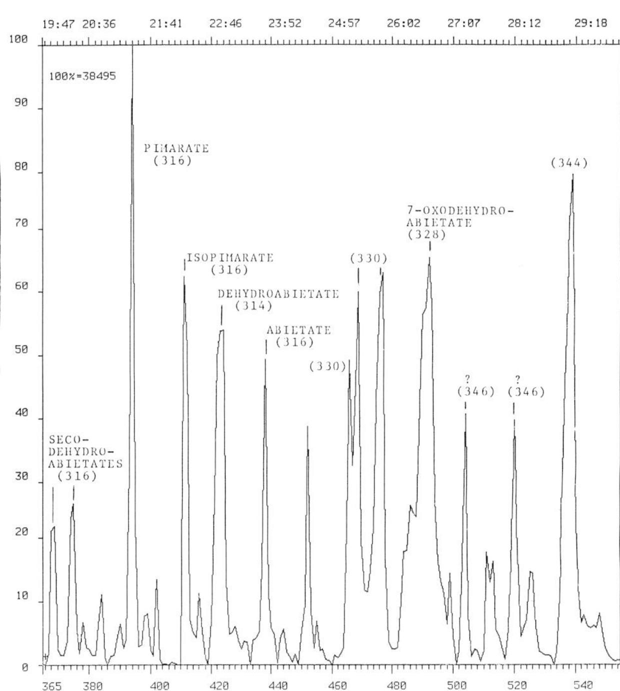

Fig. 7. Section of the chromatogram of figure 6, which includes original diterpenoid pine resin components and their oxidation products. Identifications and/or molecular weights are shown against the peaks and correspond to some of the compounds shown in figures 2 and 3.

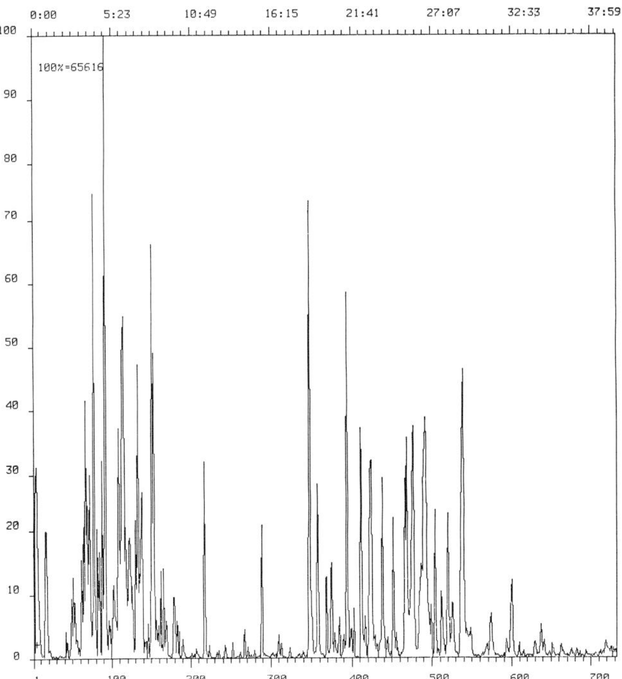

Fig. 6. Complex chromatogram yielded by the impregnating material of a tracing paper used for a ship's plan (1880). Many peaks are unidentified but those that are indicate the presence of drying oil, non-drying oil, and pine resin.

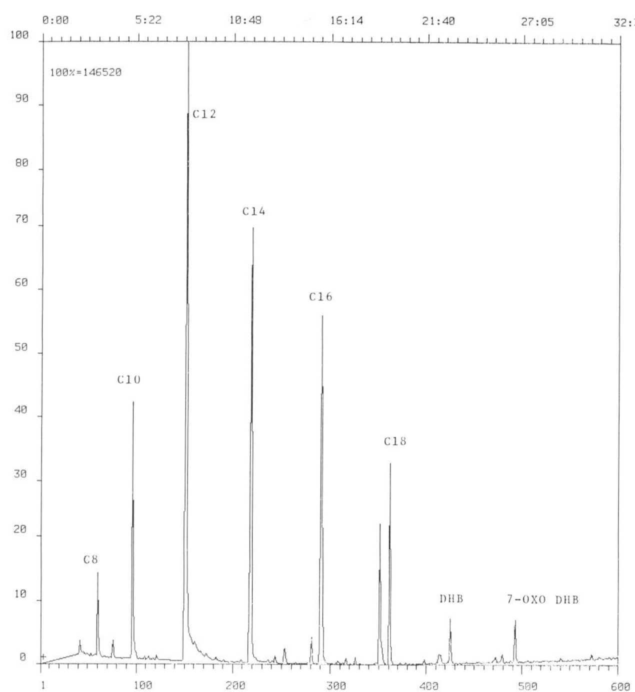

Fig. 8. Chromatogram of the impregnating material of a second tracing paper used for a ship's plan (1882). The regular pattern of saturated fatty acid esters, with a maximum at the C-12 acid ester, indicates the use of coconut oil. Small amounts of pine resin components are also present between scan 400 and 500.

References

1. J. S. Mills and R. White, "Organic Analysis in the Arts: Some Further Paint Medium Analyses," *The National Gallery Technical Bulletin*, 2 (1978): 71-76.

2. J. S. Mills, and R. White, "Organic Mass-Spectrometry of Art Materials: Work in Progress," *The National Gallery Technical Bulletin*, 6 (1982): 3-18.

3. R. White, "The Characterisation of Proteinaceous Binders in Art Objects," *The National Gallery Technical Bulletin*, 8 (1984): 5-14.

4. H. Kuhn, "Detection and Identification of Waxes, Including Punic Wax, by Infrared Spectrography," *Studies in Conservation*, 5 (1960): 71-79.

5. J. S. Mills, "Identification of Organic Materials in Museum Objects," *Conservation in the Tropics. The Proceedings of the Asia, Pacific Seminar on Conservation of Cultural Property, Feb. 7,16, 1972*, (O. P. Agrawal, ed.), New Delhi, pp. 159-170.

6. R. White, "The Application of Gas-Chromatography to the Analysis of Waxes," *Studies in Conservation*, 23 (1978): 57-68.

7. J. S. Mills, "The Gas-Chromatographic Examination of Paint Media. Part I. Fatty Acid Composition and Identification of Dried Oil Films," *Studies in Conservation*, 11 (1966): 92-106.

8. J. S. Mills and R. White, "The Gas-Chromatographic Examination of Paint Media. Some Examples of Medium Identification in Painting by Fatty Acid Analysis," in *Conservation and Restoration of Pictorial Art*, P. Smith and N. Brommelle, eds. (London: Butterworths, 1976) pp. 72-77.

9. J. S. Mills, and R. White, "Analyses of Paint Media," *The National Gallery Technical Bulletin*, 1 (1977): 57-59; 3 (1979): 66-67; 4 (1980): 65-67; 5 (1981):66-67; 7 (1983): 65-67.

10. L. E. Plahter, E. Skaug, and U. Plahter, *Gothic Painted Altar Frontals from the Church of Tingelstad* (Oslo: Universitetsforlaget, 1974) p. 91.

11. J. S. Mills and R. White, "Natural Resins of Art and Archaeology: Their Sources, Chemistry and Identification," *Studies in Conservation*, 22 (1977): 12-31.

12. J. S. Mills, R. White, and L. J. Gough, "The Chemical Composition of Baltic Amber," *Chemical Geology*, 47 (1984-85).

13. S. A. Yates, "Conservation Research into Nineteenth-Century Tracing Papers," unpublished report submitted in partial fulfillment of the requirements for the Diploma in Conservation of the National Maritime Museum, London SE10 9NF. London, 1983.

ROSALBA TARDITO AMERIO

The Study of *The Journey of Tobias's Son* by Filippino Lippi

In 1982, the Sabauda Gallery in Turin celebrated 150 years of its existence by restoring some important paintings. The restoration made possible an extensive series of scientific investigations, which has resulted in a greater understanding of individual painting techniques.

Most of the paintings in the Sabauda Gallery were from private collections, especially that of the House of Savoy. Almost all have been treated to some extent over the years.

The painting selected for this paper is *The Journey of Tobias's Son* by Filippino Lippi (1457-1504), illustrated (after restoration) in figure 1. Its condition was poor and presented a number of treatment problems. Attributed to Filippino Lippi in 1933, it is believed to have been painted in Florence between 1480 and 1485.

Condition and Treatment

The painting was in a very fragile condition owing to extensive worm damage to the panel support, which was crumbling. Because of this, the paint surface was very insecure. The worm damage had been treated in the nineteenth century when the panel, consisting of three members, had been thinned. At this time, a framework of cross-pieces was attached to the back of the panel and a gray putty of chalk and glue was spread over the wood (fig. 2). Despite or because of this, cracks have opened at the joins of the members. The central member was cut with the grain running at an angle and the tensions set up in the wood have resulted in this board cracking diagonally. This has caused severe losses in the paint along these cracks and joins (figs. 3, 4). Old restorations in those areas had severely altered the appearance of the paint, and the color balance of the work as a whole was disturbed.

The first stage of the treatment was to laboriously remove the wooden support and to transfer the gesso and paint layers onto a double canvas support. The painting was cleaned and previous restorations were removed. The paint surface was in a very poor condition (fig. 4), lacking many glazes and shadows due, not only to physical and chemical changes, but also to inappropriate treatments and incorrect restoration. The original color balance was very difficult to imagine.

Investigation of the Painting

The restorer's perception of a work of art is sometimes different from that of the art historian. The following observations are based on the report of the restorer, Pinin Brambila Barcilon, and on data obtained from the scientific examination of the painting.

Infrared reflectography has revealed some changes in the outlines of the composition, for instance, in the outline of the trees and the positioning of the right foot of the figure on the extreme left (fig. 5).

It is the belief of the author that Lippi himself elaborated on his original outline, making changes to the composition, many of which were misunderstood by early restorers and erroneously removed.

By the microscopic comparison of the color of the paint preserved beneath the frame to that on the exposed surface, it was concluded that the original tonality had been lost due to the combined effects of drastic cleaning and light action. Red and green glazes have been particularly susceptible although minute traces can be observed in the draperies.

Pigment Identification

In Tobias's cloak, the upper layer of red lake, which had been used to define the folds of the drapery, has been lost. The red paint is a mixture of lead white and vermilion, beneath which can be seen charcoal particles of the preparatory drawing. A similar layer structure is to be seen on his hose.

A cross section was made from a sample removed from the gray blue of the dress of the angel on the right. The layer structure is composed of five distinct paint layers. The lowest is gesso, over which there is a pale yellow layer that contains charcoal particles. Then there are layers of red lake with azurite. The uppermost paint is a mixture of lead white with azurite, here present in coarsely ground particles, resulting in a gray blue color. Sections made from samples removed from similar areas of drapery protected by the frame reveal the presence of a thin layer of red lake above this gray blue which would have rendered the drapery a pale violet color. Where a violet hue is still preserved, on the collar of the central angel, for example, the color was achieved by mixing lead white with azurite and red lake rather than by using a susceptible glaze layer.

The central angel's dress was painted with a copper green pigment beneath a copper resinate glaze. But very little green is now discernible although individual green particles can be seen on the paint surface when it is examined with a microscope. In addition, the copper resinate glaze has gone brown.

The red cloak of the angel with the sword has lost much of its depth of color because of the loss of red glazes.

This painting provides an example of the damage that can result from inappropriate treatment. The author stresses the importance of precise scientific investigation of a painting prior to its conservation.

Fig. 1. *The Journey of Tobias's Son* by Filippino Lippi, after restoration.

Fig. 2. The reverse of the panel before transfer, with some of the fixed cross-pieces removed.

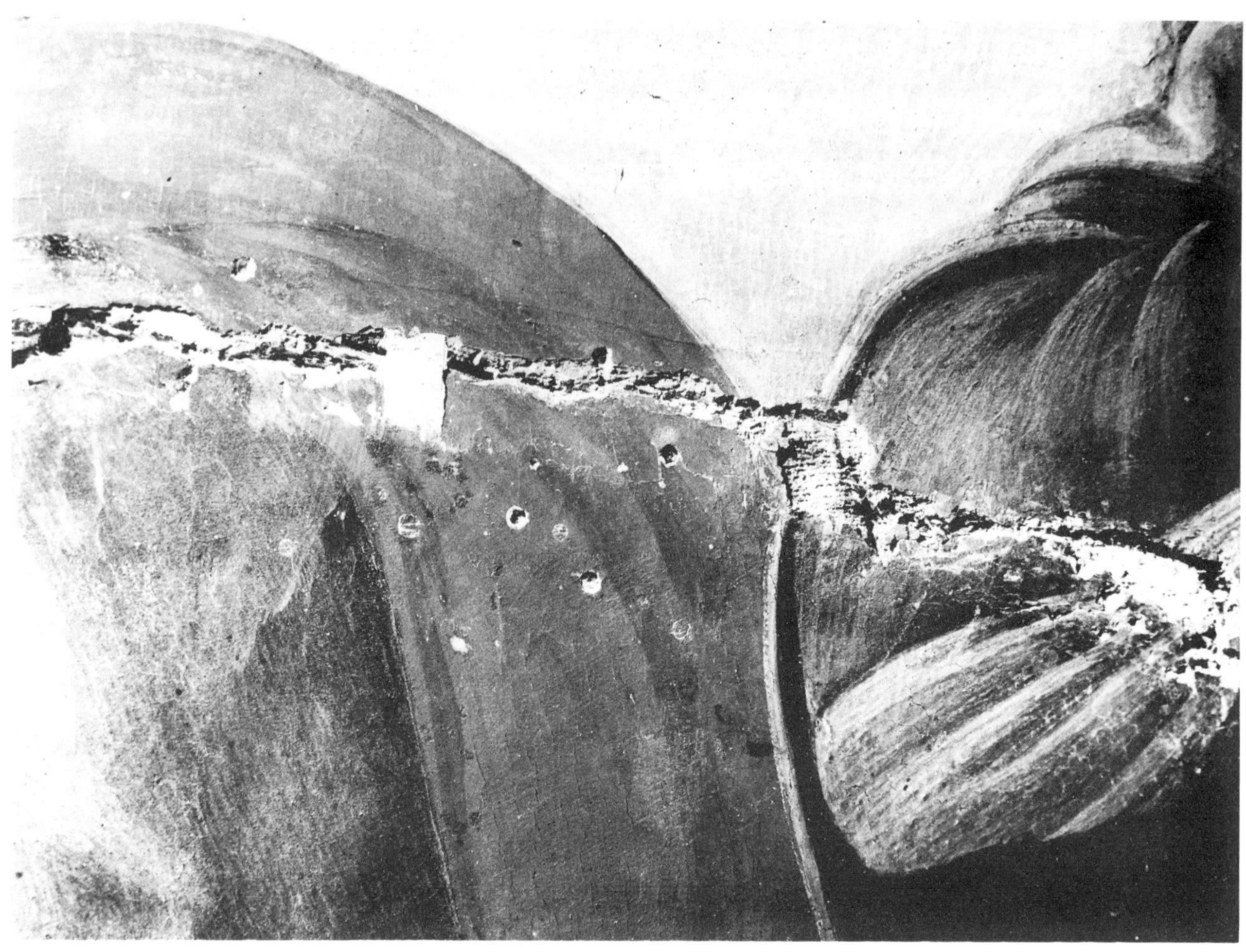

Fig. 3. Detail before restoration. Upper part of Tobias's cloak showing paint loss along a join in the upper and central members.

Fig. 4. The painting after transfer and before restoration.

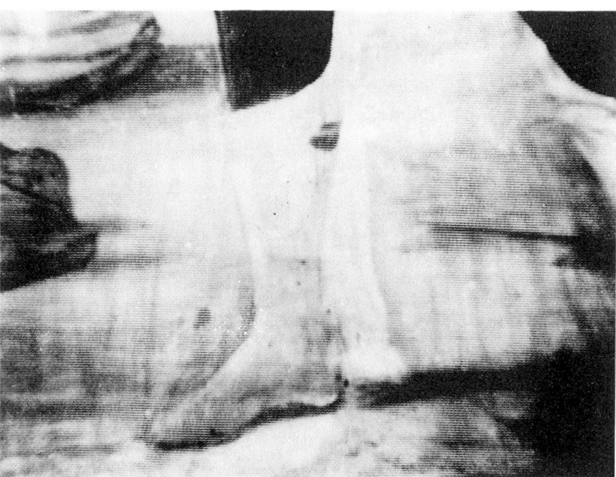

Fig. 5. Infrared photograph showing changes in the right foot of the angel with the sword.

A. VAN ROGGEN

Dimensional Analysis of Paintings and Drawings

A system that gives a quantitative description of works of art can be useful for identifying the artist, and for studying the particular habits and chronological development of any given artist. It is to be expected that such systems are complex and must entail a variety of measurements, just as experts on a given painter make their judgment by looking at style, subject, brush strokes, palette, and a variety of other aspects of the work. Nevertheless, even a simplified quantification system will make these studies easier and can advance the understanding of artists' habits and interpretations. This paper describes one facet of such a system. With dimensional analysis a direct and quantitative comparison is made between the artistic image and the real-world object that is depicted. Note that only the abstract image is analyzed. Color, medium, and other physical attributes of the work of art are disregarded for the purposes of this analysis. Both the subjective and objective aspects of an artist's expression can be quantified with these comparisons, and used to classify and identify his works. This dimensional study is demonstrated here on an image anlysis of paintings and sketches of canal scenes by John Constable.

In essence, a three-dimensional map of the area portrayed is made, together with objects displayed therein. This map is then used to make projections (calculated images) in two dimensions (the canvas or sketch paper) from any number of viewpoints (station points). Comparisons are made between each calculated image and an outline tracing of the image of an actual painting. The station point is varied systematically until correspondence has been achieved. The station point description not only includes the relative distance from the painter to the objects painted (the location on the map), but also the direction of view, the aperture (total included angle), and the horizon height on the image. A mathematical summary of the system is given in the Appendix.

The system of calculating an image from a three-dimensional map must allow for deviations from reality, whether intended by the artist for aesthetic reasons or imposed by various limitations—in sharp contrast with the usual applications of picture calculations such as computer aided design, satellite recognizance, or even cinematography. Due to artistic and other considerations an artist will change selected items observed in nature; differences from the real world can be isolated and analyzed quantitatively and their cause identified. Other complications can be incorporated, e.g. landscape painters may use standard (i.e. central) projection for normal sketches, but a differing projection for wide-angle panoramas.[1]

A mathematically perfect match between calculated and actual outlines cannot be expected for several reasons. First, small errors can occur in the measurement of the outline of remote (and thus small) items, as well as in the original placement of the paint. Some of these small deviations correspond to large errors in the location of the item on the map. Second, there is a limit to the precision with which the painting can be traced, e.g. in deep shadow areas. Third, and perhaps the most significant source of deviations, the artist may have altered the painting from a strict representation of reality. These differences, as introduced by the artist, actually can increase rather than decrease the usefulness of the technique. Such imaginative changes are characteristic of the artist, at least for a given period and medium, and thus will help in characterizing his works. While this system of analysis cannot identify a copy forged directly from an

Table 1
Index to Flatford Map

Code	Description
BARN	One of the outbuildings near the Valley Farm
BTH	Boathouse
DOCK	Dock buildings near the Mill
FLM	Flatford Mill buildings
FNCn	Fence n
FORD	The ford near Flatford
HWW	Wagon from *The Hay Wain*
LOCK	Flatford Lock
MOAT	The moat near Flatford
RBn	Rowboat n
SPONG	Island separating the Mill stream from the Stour
TRn	Tree n
WLH	Willy Lott's house (cottage)

The following are the station points for the works indicated:

SP40	*The White Horse*, 1819 (Frick Collection, New York)
SP41	*The Mill Stream*, 1813-14 (Tate Gallery, London)
SP42	*Flatford Lock and Mill*, 1812 (Corcoran Gallery, Washington, D.C.)
SP57	*View on the Stour*, 1813 (Victoria and Albert Museum, London
SP69	*Lock on the Stour*, 1827 (Victoria and Albert Museum, London)
SP70	*The Farmhouse*, 1832 (British Museum)
SP72	*Flatford Mill on the Stour*, 1817 (Tate Gallery, London)
SP73	*Boatbuilding near Flatford Mill*, 1817 (Victoria and Albert Museum, London)
SP85	*View on the Stour*, 1813 (Victoria and Albert Museum, London)
SP87	*Landscape*, 1814 (Victoria and Albert Museum, London)
SP93	*The Valley Farm*, 1835 (Tate Gallery, London)
SP104	*The Hay Wain*, 1821 (National Gallery, London)
SP129	*The Lock*, 1813 (Victoria and Albert Museum, London)
SP132	*Track by a Pool*, 1804 (see Gadney p. 43)

original (nor from a photograph taken of either one), it can detect forgeries wherein various pieces of differing original paintings are placed together to form a new image "in the style of" the original artist. The reason is that these individual elements will have differing aspects on the originals, which are very difficult to remedy without having access to the original items or scene (see the comments on *The Hay Wain* below).

Application to Constable's Canal Scenes

In John Constable's works there is a remarkable interplay between topographical accuracy and artistic invention. His earlier works are more objectively accurate than his later ones, which are known to have become more creative. This has been noted in many publications on Constable[2] and now can be quantified directly. Constable had a great admiration for nature and the "natural landscape," and thus in his compositions, rivers and land formations are usually constant, while buildings are frequently more variable. On close examination, the changes he made can be detected. As an example, the tracing for *The Hay Wain*, when compared with the calculated projection from the best station point on the map, shows clearly that the wagon itself was drawn with an aspect differing from that of the landscape, proving that the wagon was sketched separately and then added to the composition, rather than having been painted *in situ*

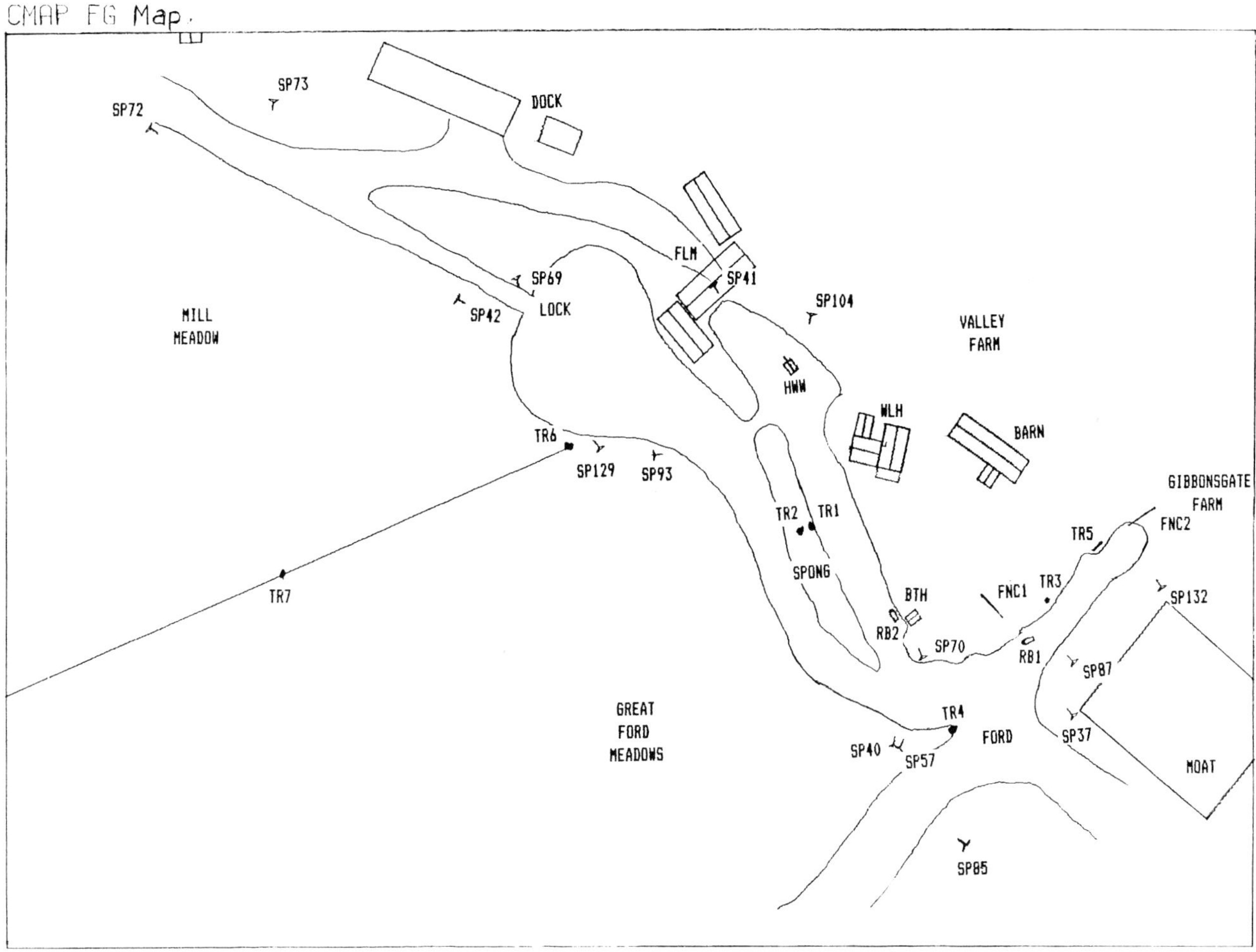

Fig. 1. Composite map of the Flatford Mill area, ca. 1800.

with the rest of the composition. Even more obvious changes noted in the canal scenes are modifications in Willy Lott's cottage, which was altered (usually only when it was a main point in the composition) by enlargement, rotation, or displacement.

The concentration of Constable's art on subjects taken from a very limited area near Flatford has provided enough individual views so that a reasonably accurate map could be extracted. The correlation between the various paintings and sketches has permitted the fullest use of dimensional analysis as a quantitative means for a scientific study of his works, because the larger the number of correlations, the more precisely the station points can be determined. Under these circumstances, even a slight change in the station point can cause a large mismatch between traced and calculated image.

As indicated in the Appendix, a visual criterion of outline matching, e.g. the use of overlays, is currently more accurate than computing the difference between traced and calculated outlines. For purposes of visual comparisons between images, it is convenient to make all items appear transparent. In other words, in the calculated image remote features are not obscured by nearby objects. Furthermore, the

outline should be sufficiently simple to avoid unnecessary crowding; usually the minimum number of objects required for the purpose is shown. Finally, there is no need to make elaborate models representing the finest detail, because only the location and size on the projection are significant. A tree can be represented only by its main trunk, a house usually need not require such details as windows or doors. The human eye and brain can fill in these details without difficulty, whereas there are no computer algorithms to solve such problems.

To apply image analysis to the series of canal scenes by John Constable, all located near Flatford, Suffolk, England, we begin with the composite map of the Flatford area as it existed between 1731 and 1840.[3] For each of the paintings and sketches analyzed, a preliminary station point is selected and the image calculated. The projection variables are changed systematically until a reasonable likeness is achieved. Then these final projection variables are used to make a reversed projection from selected items on the tracing, and these extracted items are added to the map description. The whole process is iterated using the improved map and other paintings while corrections are made to reduce matching errors.

Fig. 2. *The White Horse* and its calculated outline. Photograph copyright of the Frick Collection: New York.

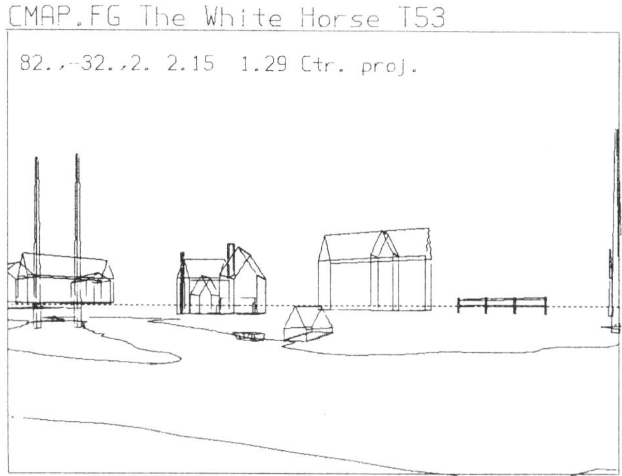

The resulting map of Flatford is shown in figure 1. The map contains not only the banks of the river Stour and the Mill Stream, with some of the main buildings painted by Constable, but also several marked items used in one or more of his works: the wagon from *The Hay Wain*, some rowboats, trees, etc. The locations marked "SP" are the station points for some of the paintings or sketches by Constable. For an index to the marked items, see table 1.

The utility of the described system of image analysis can best be shown by means of a few specific cases. For example, the literature gives several possibilities for the location represented in *The White Horse* (Frick Collection, New York). Most authors[4-6] place it upstream from Flatford Mill, towards Dedham where the towpath changed from the right bank to the left one and the horse had to leap on the barge to cross the river.[7] However, Parris says that the view is "probably from downstream at Flatford" and gives the angle of view as being almost opposite that of *The Mill Stream* (Tate Gallery, London).[8] Figure 2 shows *The White Horse* and the outline calculated from station point SP40 on the map. The outline does not show the boat with the

horse, and has added a fence not used in Constable's final rendition, but shown on the full-sized oil sketch (National Gallery, Washington, D.C.). The large trees on the Spong (among others TR1 and TR2) obscure the mill and the upper end of the Mill Stream, visible in the outline. There is no question that the downstream location fits the painting very closely. Upstream there is no location that matches. Further evidence of the correct fit between tracing and outline comes from a sketch made at a station point (SP57), approximately one meter removed from that of *The White Horse*, and turned slightly to the right. The outline of this sketch is shown in figure 3 and shows to the far right a willow tree (TR4), while on the left the Spong and the partial view of the Mill Stream have disappeared from view. Willy Lott's cottage on the far left has been rotated as seen by the relative locations of the chimneys. The tree on the right of this sketch is found in at least two other paintings, *Flatford Lock and Mill* (Corcoran Gallery, Washington, D.C.) and *The Mill Stream*, with station points SP42 and SP41, respectively, and shown in figures 4 and 5.

Note that in figure 4 the mill buildings appear to fit better with the original, higher roofline shown by pentimenti in the painting. The only item that is out of place is the boathouse, which on the painting is near Willy Lott's cottage, while on the outline it is given in the location shown in *The White Horse*, i.e. at the downstream end of the Spong, where it is not visible from locations near the lock. In *The Mill Stream*, again there is no boathouse in the upstream location, and, according to the calculated outline, the trees on the bank near Willy Lott's cottage (not shown on the outline) would obscure the boathouse in the location of *The White Horse*. Other markers, such as the two tall poplars (TR1 and TR2) on the Spong, as well as the willow (TR4) mentioned above, fit perfectly on all these images. It is the consideration of such observations that leads to the conclusion that Constable often drew items from nature more faithfully than he drew human artifacts.

A second example of applied image analysis, mentioned above, is a case where more than one sketch was used to create the final work of art: *The Hay Wain* (National Gallery, London) (fig. 6a). The calculated outline of the cottage, banks, etc. fits as accurately as the examples above, and a model of the logging wagon[9] can be placed in the same location where Constable painted it. For clarity, rectangular boxes are used to represent the wagon body and the driver; the horses are not shown. Note that, due to the high station point on the bank, one looks down into the wagon. However, Constable drew this wagon with his eyelevel at almost the same height as the top of the wagon body (the painting is well known and not reproduced here). Figure 6b shows the same view and wagon when seen from a station point corresponding to that of the wagon sketch; the landscape has changed drastically. The wagon may have been sketched from a sitting position on the lid of his paintbox, as he made many of his sketches.[10] In transposing the wagon to the scene of *The Hay Wain*, the difference in height due to the bank of the mill pool was not taken into account. It is curious that this deviation from reality not only does not harm the artistic impact of the work, but apparently never has been noted before this quantitative analysis was made, even though *The Hay Wain* is the most famous of Constable's paintings.

Many other Constable paintings and sketches have been matched. The more significant ones are listed in table 1; their station points and directions are indicated in figure 1.

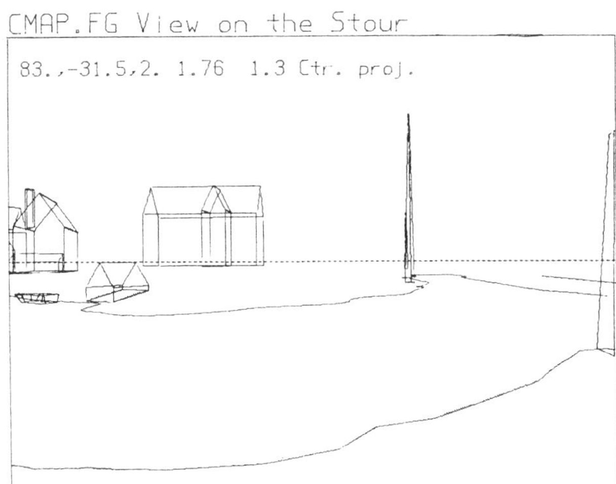

Fig. 3. *View on the Stour,*[12] and its calculated outline. Photograph from the Victoria and Albert Museum, London.

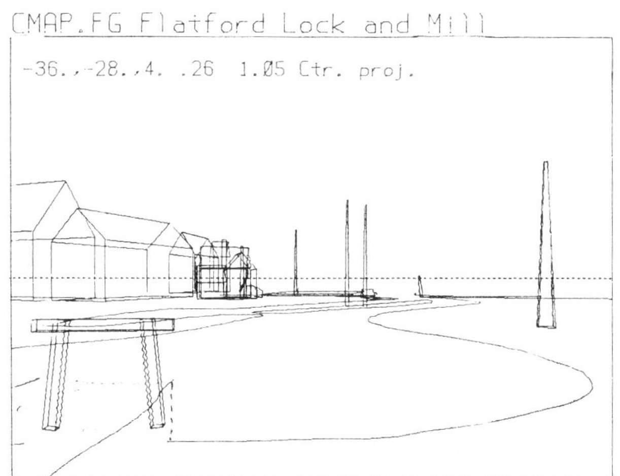

Fig. 4. *Flatford Lock and Mill*, and its calculated outline. Photograph from the collection of the Corcoran Gallery of Art, anonymous loan, Washington, DC.

Conclusion

It is shown that dimensional analysis as applied to works of an individual artist can produce a better understanding of his creative habits and special characteristics. This type of study, especially when expanded with fast video hardware and with new algorithms for image analysis, can give a practical and useful quantitative system for the characterization of the work of artists. The data so generated are completely similar to the concordances used in the analysis of works of written art, and are essential in bringing complete image analysis and style descriptions to a quantitative science.

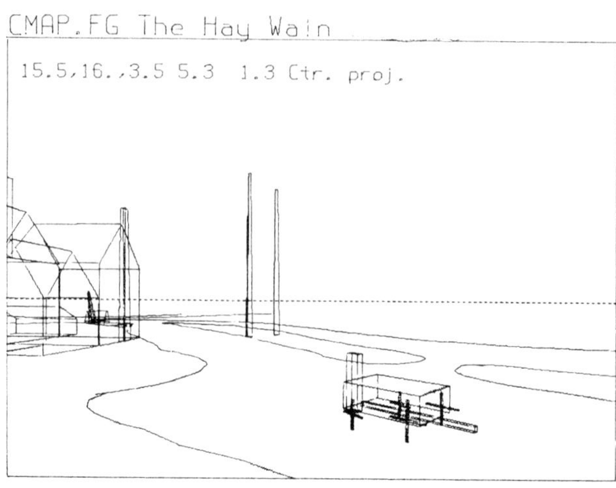

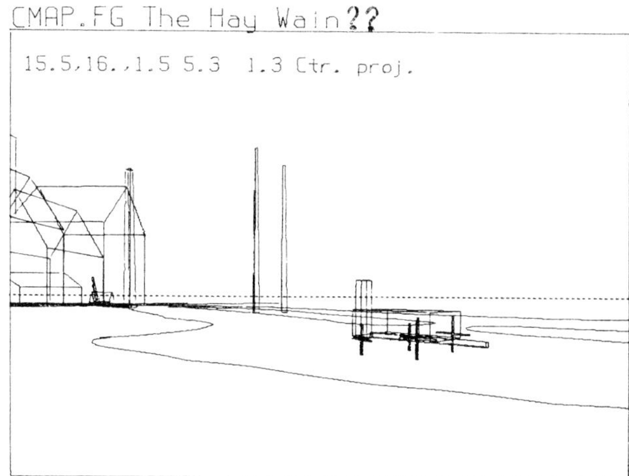

Fig. 5. *The Mill Stream*, and its calculated outline. Photograph from the Tate Gallery, London.

Fig. 6. *The Hay Wain*, calculated outline: (a) with wagon in correct location, (b) with wagon in the same location, but seen from a lower station point to make the wagon outline match that of Constable's sketch.

Appendix

Projection

The transformation from the three-dimensional world to a two-dimensional plane can be described most concisely by using matrix notation in homogeneous coordinates.[11] A point in PC = PW $\star$ XF Euclidian space, described by the vector **PW**, is mapped to a vector **PC** on the 2-d canvas by the matrix multiplication

$$\mathbf{PC} = \mathbf{PW} \star \mathbf{XF} \tag{1}$$

where **XF** is the 4x4 transformation matrix, built up from submatrices **SP** (the vector shift to the station point), **RT** (from the rotation of the view angle), and **PR** (which depends on the aperture):

$$\mathbf{XF} = \left[\begin{array}{c|c} \mathbf{RT} & \mathbf{PR} \\ \hline \mathbf{SP} & 1 \end{array} \right] \tag{2}$$

For a given station point location, angle, and aperture, all elements of **XF** are known and the calculated projection **PC** can be obtained for any point **PW** on the map.

Reverse Projection

The reverse projection, i.e. obtaining the 3-d map description from a point on the 2-d canvas cannot be calculated from the normal inverse, $\mathbf{PW} = \mathbf{PC} \star \mathbf{XF}^{-1}$, because **XF** is singular. This is to be expected, because an infinity of 3-d shapes can correspond with a single 2-d projection. However, given a set of 4 independent points PC1...PC4 of an object tracing, and an approximation to the corresponding values **PW** (e.g. the size and shape are known), one can form

$$\begin{bmatrix} PC1 \\ \overline{} \\ PC4 \end{bmatrix} = \begin{bmatrix} PW1 \\ \overline{} \\ PW4 \end{bmatrix} \star XF \tag{3}$$

or

$$C = W \star XF \tag{4}$$

which can be solved by

$$XF = W \star C \tag{5}$$

With this **XF**, the complete set of points defined in the map can be transformed with Eq. (1) to a calculated image **PC** for comparison with the traced image.

Convergence

The projection variables, i.e. the elements of **XF**, are first estimated by calculating one or more reverse projections using equation (5). From these matrices the weighted elements form the first approximation to **XF**. This **XF** is tested by applying equation (1) to the whole set of map points for the given tracing, and then refined iteratively by varying individual elements of **XF**. Using a different tracing, i.e. one from a painting with a differing station point, this procedure is repeated, giving a set **PW'**. The deviation between the two sets, **PW-PW'**, ideally to be zero, is minimized by systematically varying the station point elements.

In general, there is no problem of converging to an accurate transform **XF** when a visual criterion is used to indicate the amount of deviation between tracing and computed outline. This differs drastically for automatic, algorithmic convergence, because there is not yet a generalized algorithm applicable to this minimization problem. In contrast to the usual image comparison techniques such as used in space satellite surveyance and mapping, artists' renditions of the world make comparisons more complex. One noted complication is the trend of artists to switch from the classical 'central' projection to a cylindrical projection when the aperture is large. However, once it is known at what aperture a given artist makes this change, it can be incorporated in the calculations and quantitatively expressed. In fact, this aperture range is one more numerical datum useful in establishing differences among artists.

References

1. H.E. Malde. *Am. Scientist* 71. pp. 132-140. 1983.

2. A. Smart and A. Brooks. *Constable and His Country* (London: Elek. 1976) p. 72.

3. *A Survey of the Parish of East Bergholt*. Brasier. 1731.

4. R. Gadney. *Constable and His World*. (New York: W. W. Norton Co., 1976) p. 52.

5. J. Walker. *John Constable*. (New York: Henry Abrams Inc.. 1978) p.96

6. A. Smart and A. Brooks, *Constable and His Country*. p. 39.

7. A. J. R. Waller, *The Suffolk Stour*. (Ipswich: N. Adlard Co. 1957) p. 20.

8. L. Parris, I. Fleming-Williams, and C. Shields; *Constable Paintings, Watercolours and Drawings*. (London: Tate Gallery. 1976) p. 108.

9. J. G. Jenkins. *The English Farm Wagon* (Oakwood Press. 1961).

10. G. Reynolds. *Constable's England*. (New York: The Metropolitan Museum of Art, 1983) p. 25.

11. W. K. Giloi, *Interactive Computer Graphics* (Englewood Cliffs: Prentice Hall, 1978) pp. 105-107.

12. G. Reynolds, *Constable's England*, p. 70.

JOHN WINTER

Gold in Japanese Paintings: A Case History Involving the *Kirikane* Technique

Gold has been used in several ways in Japanese paintings. *Kimpaku* (literally "gold leaf") is the technique used to apply gold leaf over sizable areas of a painting, rather than to delineate a design. In the *sunago* method, gold particles or flakes are sprinkled on the support, or some part of it, before execution of the painting or calligraphy. For *kirikane* or *kirigane* ("cut gold"), into gold leaf cut strips or small geometrical shapes is arranged into linear designs within the overall composition. *Kindei* ("gold mud") uses pulverized gold mixed with a paint vehicle and painted on like any other pigment. Other metals, particularly silver, have also been used, and the above techniques, or variants of them, may be found on sculpture and in lacquer work. Here we are concerned with the last two methods: *kirikane* and *kindei*.

The *Kirikane* Technique

The creation of design elements with cut gold leaf is believed to have originated in China[1-3] and to have been introduced at an early date into Japan, where it underwent a period of technical and stylistic development culminating in the latter Heian period (approximately the tenth to twelfth centuries). However, its use continued thereafter, and good examples are seen from the fourteenth and fifteenth centuries. Eventually, the use of *kirikane* became less common, and by the seventeenth century (the date of the paintings described here) it was unusual, and perhaps quite rare. It has always been considered an extremely difficult technique to do well, and was apparently executed largely by specialists. With paintings and sculpture, it is generally found on religious works, where a conservative tradition and a predetermined approach allowed the incorporation of such methods. Art historical studies of *kirikane* development in Japan have been written by Noma.[4-6]

Kirikane has been revived in Japan in the twentieth century, and the technique has been described in detail.[2,4,7] According to these descriptions, standard gold leaf is too thin to be cut successfully, and is first laminated to give thicker leaf of, typically, two or three layers. Two leaves are held together, stroked carefully over hot charcoal embedded in ash (which gives a uniform, controlled source of heat) and then pressed firmly between sheets of paper; a third leaf is added in the same way if desired. The laminated leaf is cut using a bamboo knife on a block covered with deerskin, the cutting surface and tools being rubbed with talc to prevent the leaf from sticking to them.[7] Strips are first cut: where these are to be applied as such they are extremely narrow, usually a fraction of a millimeter, but wider strips may be cut further into small geometrical shapes. The cut leaf is applied to the painting support with an adhesive, often seaweed glue; a strip is picked up from one end with the tip of a moistened brush and laid down along an adhesive line applied from a second brush.

Once applied, the leaf may be gently burnished to improve the gloss,[4] and the greater glossiness of gold leaf is commonly cited as typical of *kirikane*, as opposed to the easier *kindei* method using gold-powder paint.

Description of Paintings

Scroll 1, No. 62.27 in the Freer Gallery collection, is a handscroll consisting of an ornate frontispiece painting (fig. 1) followed by a chapter of the *Lotus Sutra* executed in gold paint on dark blue paper.

The columns of characters in the sutra are separated by triple lines, of which the outer two are in gold paint and the middle line is in a silver-colored metallic pigment. The frontispiece painting is on a different piece of paper joined to the calligraphy section and apparently originally dyed purple, but now faded. It is executed in gold and a silver-colored powder (now tarnishing to brown), with outlines in black and touches of blue and red. The painting end of the scroll is backed with a heavy brocade. In 1982, the roller was detached from this scroll to reveal an inscription that could be dated in accordance with 1666.[8]

Scrolls 2 and 3 are Nos. 85-213-TA-1971.47 and 1961.141a in the collection of the Fogg Art Museum of Harvard University, and belong to the same set as does Scroll 1, being other chapters of the *Lotus Sutra*. Scroll 3 has been described by Rosenfield;[9] the general descriptions are similar to that given above for Scroll 1.

Two more handscrolls of this set have been described by Murase.[10] The complete set of thirty-two scrolls, forming a copy of the entire *Lotus Sutra*, is reputed to have been commissioned by the retired Emperor Go-Mizunoo as a presentation, in 1666, on the fiftieth anniversary of the death of Tokugawa Ieyasu. However accurate, this story is supported by the inscription on the roller of Scroll 1 and by such circumstantial evidence as Tokugawa family crests on various parts of the mountings. A dating to the mid seventeenth century appears to be generally accepted for the works.

Observations and Results

Visual Observations

On all three paintings, the gold work is extensive and falls into two types (fig. 1). Clouds, mountains, faces and clothing of the figures, and similar areas are clearly executed with gold-powder paint applied with a brush in the *kindei* technique. Geometrical patterns based on a repeating lattice, generally in the lower parts of the paintings, are of the second type and are referred to as "lattice-pattern gold." Scrolls 2 and 3 have different lattice patterns from that seen on Scroll 1, but observations pertaining to the technique were identical for all three.

The lattice-pattern gold has a noticeably matte appearance by comparison with the painted gold. This is seen especially when the paintings are viewed obliquely, and also when observed at low power through a microscope (fig. 2). Use of an infrared viewer shows the lattice-pattern gold to have a higher infrared reflectivity than painted gold, though that of the latter shows a strong dependence on the angle of viewing.

The lattice-pattern areas are obvious candidates for *kirikane*, or cut gold-leaf decoration, except that the matte texture seems illogical, or "wrong," for this type of gilding, which is usually more glossy. The point is reinforced by microscopic examination, which gives images such as figure 3, where the gold appears almost particulate in nature, even though the parallel bands with squared or angled ends suggest a cut-leaf technique. Further microscopic examination reveals that, where the gold is missing (possible having flaked off or worn away), a layer of white particulate material is present underneath. This is seen in figure 4, and may be observed in many places on all three paintings. The layer of white material is exactly co-extensive with the gold: nowhere was it found extending beyond the over-

Fig. 1. Frontispiece painting to chapter 17 of a copy of the *Lotus Sutra* dated to 1666 (Scroll 1). Freer Gallery of Art No. 62.27.

lying gold and, so far as its presence could be observed, it appears to be present everywhere beneath the gold.

The white particulate material was identified by x-ray diffraction as pulverized talc. Order-of-magnitude estimates of the thickness of the talc layer were made by two methods. First, small samples of lattice-pattern gold with the talc still adhering were examined with a scanning electron microscope. Thickness measurements were in the range 10-14μm; since it is difficult to guarantee that the whole of the talc layer remained in place during sampling and mounting, this may be a minimum estimate. The vertical edge of the painting of Scroll 1 that joins and overlaps the continuation of the scroll was originally formed by cutting through a lattice-pattern gold area. The resulting series of cross-sections enabled the thickness of the talc to be measured by photomicrography. A silvered square microscope coverslip was used as a 45° mirror to obtain images from which the talc layer thickness was estimated at 15-23μm.

The thickness of the talc layer, which is of course unlikely to be constant from place to place, appears therefore to be of the order of 10-25μm (0.010-0.025mm).

Nature of the Lattice-Pattern Gold

Samples of gold from lattice pattern areas on all three paintings were examined in the scanning electron microsocope (SEM) to give images such as figures 5, 6, and 7. These made clear that the gold is a continuous leaf, rather than discrete particles, though it is also markedly crinkled, evidently from the talc layer having impressed a corrugated microtopography upon it.

X-radiographs were done of the Scrolls 1 and 3 paintings (fig. 8). Inspection of these showed that the narrow bands of gold were laid down as individual strips that pass one over another at crossing points, the radiographic density dropping by increments corresponding to the multiplicity of strips.

The preceding observations by microscopy, by SEM, and by x-radiography lead to the conclusion that the lattice-pattern gold is gold leaf with a layer of talc applied to one side, which has been cut into strips (and diamond-shapes for Scrolls 2 and 3), and then applied to the painting support as *kirikane*. The unusual matte texture of the gold results from the talc layer, which has impressed a crinkled microtopography on the leaf.

Lamination of Leaf

The technique, in modern *kirikane* practice, of laminating individual gold leaves to give a correspondingly thicker leaf has been previously described. SEM images of samples from these seventeenth-century paintings afford evidence that laminated leaf was also being used then. Figure 6 clearly shows two laminae at the edge of the leaf. Figure 7 is from a sample that was washed free of talc before mounting and is consequently rather distressed, but shows separating laminae at the points marked. Study of micrographs such as these sometimes shows the presence of three laminae in one place; no unambiguous evidence of more than three laminae was discovered. The softness of gold and the fragility of gold leaf mean that the failure to observe a certain number of laminae at the few locations sampled does not definitely eliminate the possibility of their presence: a lamina may have been worn away or become mechanically detached. However, the SEM evidence does suggest a triply-laminated leaf, with additional laminae not being completely ruled out.

Fig. 2. Two types of gilding on Scroll 1 painting, 5X. Upper area is painted gold; lower area is lattice-pattern gold.

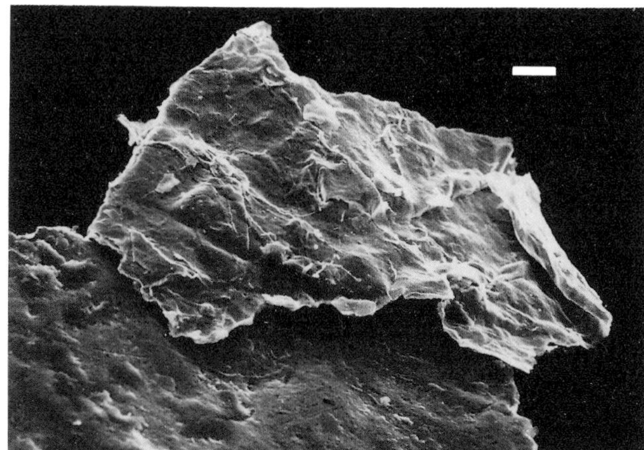

Fig. 5. Scanning electron micrograph of lattice-pattern gold sample from Scroll 1. Scale bar = 10μm. The lower part shows the talc layer with the gold leaf face down; in the upper part the leaf is folded back to reveal its upper surface.

Fig. 3. Lattice-pattern gold on Scroll 1, 25X.

Fig. 4. Lattice-pattern gold on Scroll 3, 40X. Arrows show areas where the gold has been lost, revealing the talc layer.

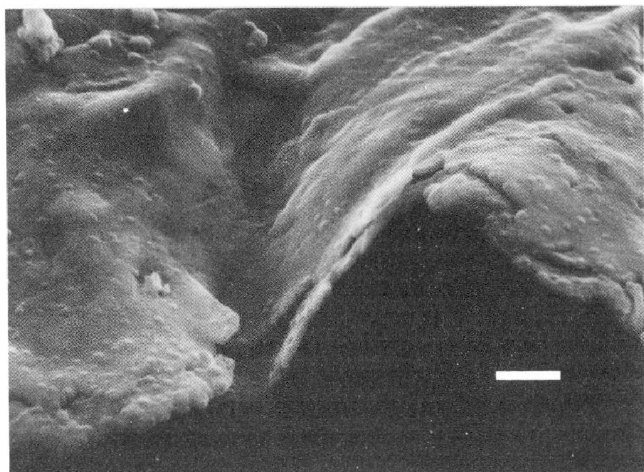

Fig. 6. Scanning electron micrograph of lattice-pattern gold leaf from Scroll 1, showing lamination of leaf. Scale bar = 1μm.

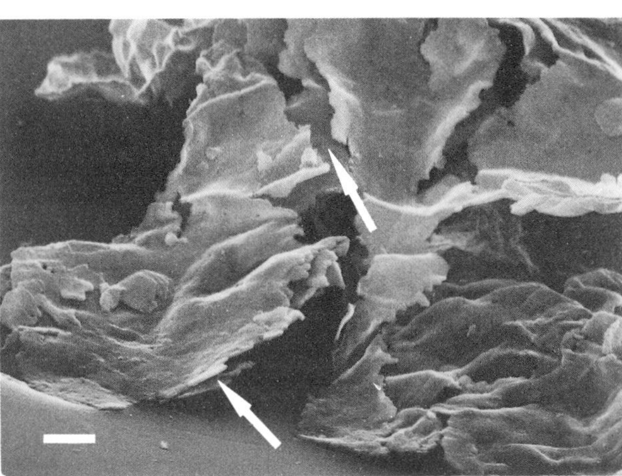

Fig. 7. Scanning electron micrograph of gold leaf from Scroll 1. Scale bar = 2μm. Arrows show laminae becoming separated.

Fig. 8. Detail of x-radiograph of Scroll 3. The broken line shows a typical densitometer scanning track.

Thickness of Gold Leaf: SEM

A sample of modern Japanese gold leaf was investigated for comparison and for calibration purposes using x-ray densitometry. A piece of modern leaf measuring 126 x 126mm weighed 30.4mg. By assuming the specific gravity of pure gold, 19.32, an average thickness of 0.099μm is found. Measurements on scanning electron micrographs of the cut edges of pieces of gold leaf gave a mean thickness of 0.08μm (table 1). An overall mean of 0.09μm was therefore assumed for modern Japanese gold leaf.

Table 1 also shows spot measurements of leaf thickness (single lamina) from SEM images of painting samples. The mean of 0.10μm is close to that found for modern leaf, but two points should be remembered. The first is that these measurements were made on free edges seen in the very small SEM samples originally removed to solve the problems of leaf versus particulate gold and of leaf lamination, and can only be very approximate. The second is that gold leaf is not expected to have a completely constant thickness, and indeed the SEM images do suggest variations. To deal further with such problems would require more extensive sampling from the paintings, as well as more elaborate sample preparation. It was not felt that the additional sampling of these works was justified at this stage.

Thickness of Gold Leaf: Radiography

An alternative approach to the problems of leaf thickness and number of laminae was to use densitometry of x-radiographs. This was done on the Scroll 3 painting while it was removed from its mount, thus avoiding interference from the heavy brocades that contain gold-wrapped threads and are normally mounted behind the paintings. Figure 8 shows a detail of one of the radiographs. A step wedge consisting of one, two, three, and four layers of modern Japanese gold leaf on a thin paper support was placed close to the areas of interest in the painting during radiography, and was used as a calibration. The radiographs were scanned through various locations showing single and double thicknesses of gold, and figure 9 shows a typical scan, along a path such as that marked in figure 8. Other scans (with an appropriate distance ratio) were taken through the step-wedge image to give a series of calibration steps; the results are in table 2.

The total optical density range scanned in these radiographs is small; consequently there is considerable random variation in the background, in the peaks themselves, and in the step-wedge scans. This factor, with others, seriously limits the attainable accuracy. However, the small radiographic density range (the heights of most peaks appeared to be less than 0.1 optical density units) meant that a straight line calibration was justified in producing results of this modest attainable accuracy, even though the appropriate characteristic curve is not in fact linear.[11]

The crinkled nature of the gold leaf on the paintings is expected to introduce positive systematic errors, which will be partly, but probably not completely, offset by any lack of flatness in the step-wedge gold leaf. Further, the presence of a layer of talc (a magnesium silicate) under the gold also introduces a positive error, though one that is probably negligible. The latter point was checked by scanning radiographs taken at 25 and 40kV in corresponding areas; the results are compared in table 3. Since the relative x-ray absorptions of gold and magnesium silicate will change on going to a higher tube potential, the fact that the results overlap well within experimental errors suggests that the talc contribution is indeed negligible.

Both peaks in figure 9 show a double maximum, a feature present in many, though not all, of the peaks scanned. This is possibly the result of a microscopic turning or crumpling of the edge of the gold strip as it was cut. If so, the *minimum* within the peaks should be most representative of the actual gold thickness. Heights designated as mean were estimated graphically for a hypothetical squared-off peak of the same area. These, along with the maximum height of each peak, are tabulated in table 3 and summarized in table 4. The results were calculated initially as τ-units, where τ is the thickness of the step-wedge leaf, then converted to absolute thickness in table 4 using $\tau = 0.09\mu$m.

Since the gold leaf is known, from SEM evidence, to be laminated, table 4 also lists the mean thicknesses of individual laminae on the assumption that two, three, and four such laminae are present. Comparison of these results with the estimate of 0.10μm from SEM measurements confirms the observation of laminated leaf having been used, and is consistent with three laminae being generally present. The result of 0.11μm for three laminae when using the "minimum" estimate of the peak heights is in remarkably good agreement. However, it should be remembered that considerable errors are present in the thickness estimates of both methods used, and that there is consequently still some room for ambiguity in the number of layers.

Table 1
Thickness of Gold Leaf (Scanning Electron Microscopy)

	Range	Mean
Paintings leaf	0.04 - 0.14μm	0.10 ± 0.03μm
Modern leaf	0.06 - 0.11μm	0.08 ± 0.02μm

(9 laminae were measured in each case, the range indicating the smallest and largest results. Errors in this and subsequent tables express the standard deviation of the sample)

Table 2
Densitometric Scans of Step Wedge (Millimeters along Density Axis)

	25kV		40kV
Step	Scan 1	Scan 2	
1st	—	22	41
2nd	22	21	35
3rd	27	27	21
4th	28	26	44
Mean	24.7 ± 2.9mm		35.2 ± 10.2mm

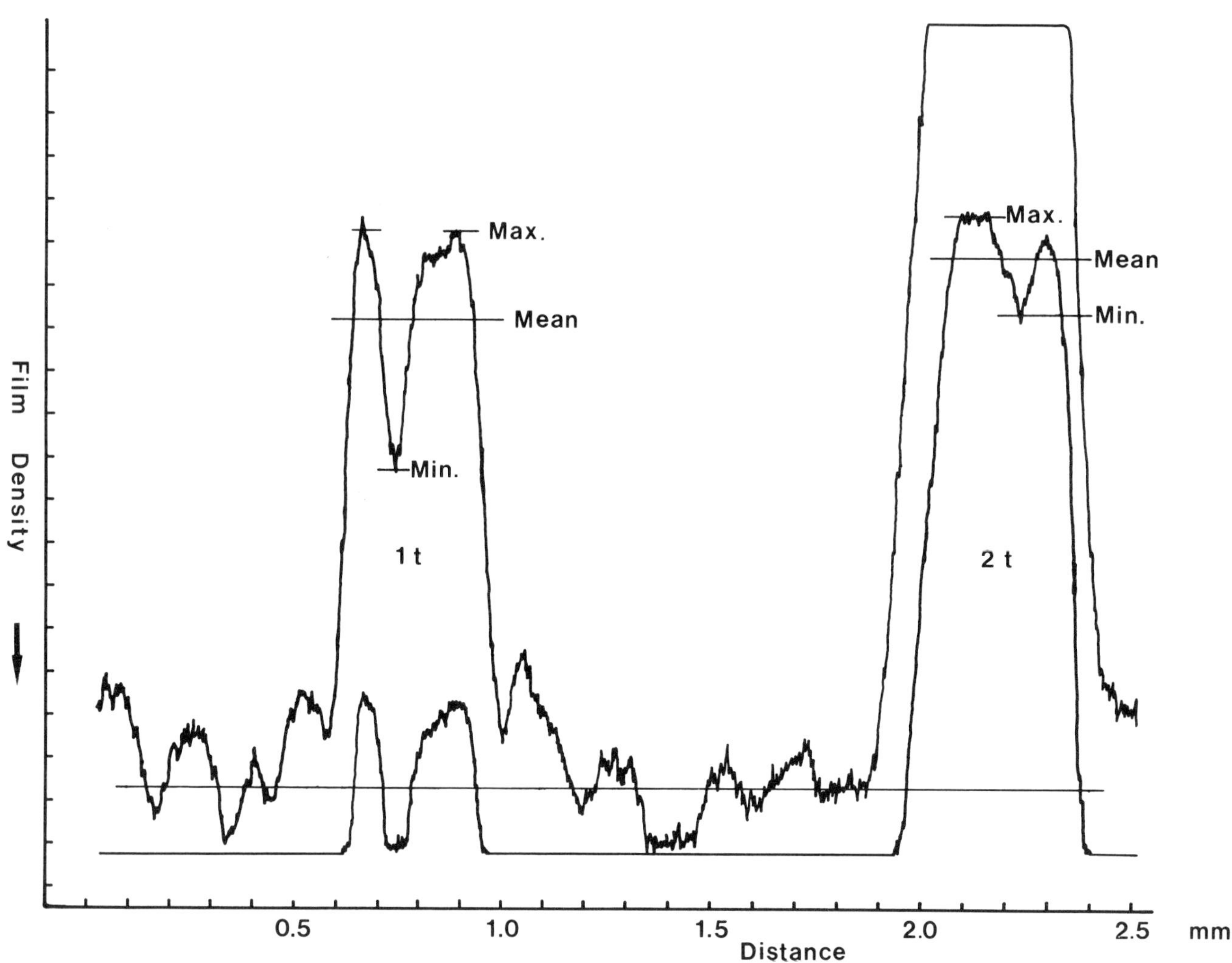

Fig. 9. Typical densitometer scan. *1t* is the peak at a single gold strip; *2t* is the peak at two superposed gold strips. Scans (a) and (b) were taken down the same track with the chart-recorder zero shifted to bring the peaks successively on scale.

Table 4
Thickness of Leaf from Paintings (Densitometry on X-radiographs)

τ = mean thickness of modern leaf in step wedge, taken as 0.09μm

	Min.	Mean	Max.
Overall Thickness			
τ-units	3.59 ± 1.37	4.07 ± 1.11	4.97 ± 1.10
Micrometers	0.32 ± 0.12	0.37 ± 0.10	0.45 ± 0.10
Mean Thickness of 1 lamina (μm)			
2 laminae present	0.16	0.185	0.225
3 " "	0.11	0.12	0.15
4 " "	0.08	0.09	0.11

Table 5
Composition of Silver-Gold Alloy (Electron Microprobe)

	Painting	Calligraphy
	Ag/Au	Ag/Au
	0.409	0.556
	0.423	0.579
	0.431	0.536
	0.482	0.552
	0.500	0.708
	0.463	0.636
	0.438	
Mean	0.449 ± 0.033	0.595 ± 0.066
Propn. Ag	31% (± 1.5)	37% (± 2.6)

Table 3
Thickness of Leaf on Scroll 3 Painting
(Densitometry on X-Radiographs)

τ = height above background relative to mean step height for step wedge. Scans went through either a single strip of gold leaf or two superposed strips at a crossing point; in the latter case the tabulated result has been divided by two to give the mean thickness for a single strip.

Location of Scan	No. of strips	25kV τ/No. strips			40kV τ/No. strips		
		Min.	Mean	Max.	Min.	Mean	Max.
1	1	2.47	2.91	3.60			
	2	1.86	2.39	3.40			
2	1	2.15	2.71	3.52	2.41	2.81	3.21
	2	1.64	2.715	4.27	—	2.27	3.69
3	1	3.00	4.25	5.26	3.01	3.41	4.03
	2	4.455	4.735	4.92	—	2.24	4.02
4	1	—	2.31	2.71			
	2	4.15	4.29	4.455			
5	1	—	4.82	5.51			
	2	4.82	5.10	5.65			
6	1	—	3.16	4.82			
	2	—	4.29	5.345			
7	1	—	4.98	5.75			
	2	—	3.545	4.435			
8	1	—	5.02	6.84			
	2	—	3.685	5.63			
9	1	4.17	4.53	5.38			
	1	5.30	5.47	5.59			
	1	3.44	4.17	5.22			
	1	9.80*	10.28*	10.81*			
	1	5.67	6.32	7.00			
	Mean	3.59	4.07	4.97			
	Standard deviation	1.37	1.11	1.10			

*These results were omitted from the calculation as probably representing a double thickness, or other fortuitous thickening, of the leaf.

Identity of Other Pigments

On the Scroll 1 painting, the red pigment used in the lips of the figures was identified as vermilion and the blue pigment used in the hair of some of the figures was identified as azurite, both by x-ray diffraction.

A sample of the silver-colored metallic powder present on the same painting was analyzed in the SEM; an x-ray spectrum for a mixture of silver and gold was obtained. Since gold powder is also present on the painting, analysis of this pigment, and also of the silver-colored pigment in the rulings on the sutra portion of the scroll, was done by the electron microprobe, in which single particles or small clumps of particles could be analyzed individually. The results, in table 5, show that both pigments are in fact silver-gold alloys, with about 31% silver for that on the painting and 37% silver for that on the sutra itself.

Conclusions

These paintings are of a late period for an occurrence of the *kirikane* technique. Nevertheless, the lattice-pattern gold is executed in cut gold leaf, which carries a layer of talc on the underside. The talc must have been present on the gold at the time it was cut, and has impressed a microscopically crinkled texture upon the leaf, resulting in a matte appearance quite different from that usually seen in *kirikane* work.

The reasons why this was done are not entirely clear. Since talc may be used on the leather cushion used to cut the leaf, it is perhaps arguable that the layer was picked up, more or less fortuitously, at this stage. However, while the textural effects of talc on gold leaf may have first been noticed in some such way, the presence of the talc layer in all lattice-pattern areas on at least three paintings suggests strongly that it is there intentionally. Thus, it seems probable that a layer of talc was applied to the leaf, presumably using some binding medium, before it was cut for formation of the design elements, the talc layer having impressed the unusual matte texture upon the gold.

The gold leaf appears to have a thickness comparable to that of modern Japanese leaf, roughly $0.1\mu m$, and has been laminated much like that for modern *kirikane*. Probably (but not certainly) three laminae are present; it is not known whether the lamination technique was the same as that used in modern times.

The "silver" pigment used on both the painting and calligraphy parts of Scroll 1 is actually a silver-gold alloy containing between 30 and 40% silver. The use of this seems fairly logical. Silver powder is well known as a pigment on Japanese paintings, but its tendency to tarnish and discolor must surely have been noticed at an early date. The use of a silver-gold alloy would reduce the rate of discoloration and this would be a natural material to adopt in paintings of value. It is interesting that the proportion of gold is about as high as can be tolerated while still retaining a silver color. The result of 31% silver found here for the painting may have rather too high a proportion of gold (assuming that a silver color was in fact desired) but since the alloy has now become tarnished, it may have suffered depletion of the original silver content.

The author is aware of no previous report of the use of a silver-gold alloy on a Japanese painting where silver was expected. Thus, the interesting problem of when this alloy was introduced remains uninvestigated at present.

Experimental Notes

X-Ray Diffraction

The Debye-Scherrer technique was used with 114.3mm diameter Gandolfi-pattern cameras and CuKα radiation (λ = 1.542Å) generated at 40kV and 20 mA. The talc samples gave diffraction patterns identical with JCPDS pattern 19-770 and with that from a sample of known material.

Scanning Electron Microscopy

The Cambridge Stereoscan 250 Mk2 owned by the U.S. National Museum of Natural History was used for both the secondary electron images shown in figures 5, 6, and 7, and for the elemental x-ray spectra.

X-Ray Densitometry

X-radiographs of the demounted painting from Scroll 3 were taken at 25 and at 40kV using Kodak Industrex Type M film, and incorporating a four-level gold-leaf step wedge as described above. The films were scanned on a Joyce-Loebl Model 3CS Microdensitometer owned by the National Bureau of Standards (Reactor Building). For most scans, slit dimensions of 0.06 x 0.06mm were used, but scans with slits down to 0.03 x 0.03mm were run to confirm the absence of peak truncation across the very narrow gold lines (0.2 - 0.4mm wide). The densitometer had a standard density wedge of 0.2 d-units fitted for these measurements.

Electron Microprobe

Samples of the silver-colored pigment from the painting consisted of particles adhering to paper fibers; samples from the sutra portion of the scroll were paint flakes detached from the surface. The microprobe owned by the Department of Mineral Sciences, Smithsonian Institution, was operated at 15kV and the count rates for the $AgL\alpha$ and $AuM\alpha$ emissions determined for a series of spots. The instrument software reduced these to proportions of elements; in table 5 silver-gold ratios only are reported to avoid possible complications from minor and trace elements in the paper fibers or the binding medium. The proportions of silver and gold are derived from these ratios. One result, on a painting sample, showed a greatly elevated gold content, and was discarded as probably involving a pure gold particle.

Acknowledgments

Mr. Yasushi Egami (Tokyo National Research Institute of Cultural Properties), while doing research at the Freer Gallery of Art in 1981, first pointed out the differences between the two types of gold on Freer Scroll 62.27 when seen with an infrared viewer; this observation catalyzed the work described here. Professor John M. Rosenfield, acting director of the Fogg Art Museum, Harvard University, kindly permitted the studies on Scrolls 2 and 3 while they were at the Freer Gallery for remounting. Mr. Walter R. Brown (U.S. National Museum of Natural History) operated the scanning electron microscope. Mr. Martin Ganoczy and Mr. Tony Cheng provided services on the microdensitometer at the National Bureau of Standards. Dr. Eugene Jarosewitch and Mr. Joseph Nelen did likewise for the electron microprobe in the Smithsonian Institution. Mr. R. Nishiumi (Freer Gallery of Art) provided samples of modern Japanese gold leaf and made up the densitometric step wedge. All are thanked profusely.

References

1. S. F. Moran, "The *Kirikane* Decoration of the Statue of Fugen Bosatsu," *Oriental Art*, 6 (1960): 125-129.

2. D. Seckel, "Kirikane: Die Schnittgold-Dekoration in der Japanischen Kunst, ihre Technik und ihre Geschichte," *Oriens Extremus*, 1 (1954): 71-88.

3. D. Seckel, "Kirikane in der Kunst Chinas und des Nahen Ostens," *Oriens Extremus*, 11 (1964): 149-161.

4. S. Noma, "Kirikane zatsuwa (Chats on *Kirikane*)," *Urushi to Kogei*, (1931), No. 363, 1-18 .

5. S. Noma, "Kirikane aya-ko (Design Considerations in Kirikane)," *Bukkyo Bijutsu* No. 17, (1930): 57-66.

6. S. Noma, "Kirikane no jidaisei ni tsuite (On the Historical Development of *Kirikane*)," *Kokka* (1937), No. 560, 193-199 (Part 1), and No. 561, 227-233 (Part 2).

7. K. Ishiodori, S. Takasaki, *Nihonga no Hyogen Giho* (Expressive Techniques of Japanese Painting), (Tokyo: Bijutsu Shuppansha, 1978) pp. 45-48.

8. Y. Shimizu, entry, dated 1982, in the Freer Gallery of Art Object Notes for 62.27 (Scroll 1).

9. J. M. Rosenfield in J. M. Rosenfield, F. E. Cranston, E. A. Cranston, (eds.), *The Courtly Tradition in Japanese Art and Literature: Selections from the Hofer and Hyde collections*, (Cambridge, Mass.: Fogg Art Museum, 1973) pp. 82-83.

10. M. Murase, "Japanese Art," *Selections from the Mary and Jackson Burke Collection*, (New York: Metropolitan Museum of Art, 1975) pp. 47-51.

11. Eastman Kodak Co., *Radiography in Modern Industry*, (Rochester, N.Y., 3rd edition, 1969).

C. EUGENE CAIN and VICTOR F. KALASINSKY

Characterization of Eighteenth- and Nineteenth-Century Paper Using Fourier Transform Infrared Spectroscopy and Thin-Layer Chromatography

Fourier Transform Infrared Spectroscopy

Theoretical Background

The traditional method of recording infrared data involves optically separating the various infrared wavelengths using a prism or a grating and sequentially determining which wavelengths are absorbed by a sample. Fourier transform infrared (FT-IR) spectroscopy, however, provides an interesting alternative to this traditional method.[1-3]

In a Michelson interferometer, as shown in figure 1, the source beam is divided into two parts by a partially reflecting "beamsplitter," the two parts reflect off mirrors M_1 and M_2, and a recombined beam exits the interferometer. If the distances from the beamsplitter to the two mirrors are the same, then the recombined beam is undistorted and there is constructive interference of the light rays. If, however, M_1 is movable and its distance from the beamsplitter is changed, then there is no longer a condition of constructive interference in the recombined beam. At a given position of M_1, some wavelengths will interfere constructively while others will interfere destructively, and by continuously changing the mirror position, each wavelength from the source can be brought into and out of constructive interference. If a detector is responsive to this changing interference pattern, plots similar to those shown in figure 2 can be obtained. The ordinate represents the extent of constructive interference and the abscissa represents the position of the movable mirror. Since the relationship between mirror position (distance, x) and the wavelengths of the light determines the intensity $(I(x))$ of the interference pattern, it is possible to extract wavelength or frequency information (a spectrum, $B(\upsilon)$) from these "interferograms" (fig. 2). The Fourier transformation relates the two quantities of interest, $B(\upsilon)$ and $I(x)$, in the following way:

$$B(\upsilon) = \int_{-\infty}^{\infty} I(x) \exp(-2\pi i \upsilon x)\, dx$$

The complicated nature of this mathematical transformation requires that it be done by a computer, and modern FT-IR instruments are generally equipped with dedicated minicomputers and magnetic disk or tape storage facilities.

This sophisticated method of obtaining an infrared spectrum has a number of inherent advantages over the more traditional dispersive spectrophotometers. These advantages are generally known as (1) Fellget's advantage, (2) Jacquinot's advantage, (3) Connes's advantage, and (4) the stray-light advantage. The first two are of primary concern and require brief elaboration.

Fellget's advantage refers to the fact that all the wavelengths are "viewed" by the detector at the same time so the *entire* spectrum can be obtained from a single scan of the movable mirror. The distance traveled by the movable mirror determines the ultimate spectral resolution, and for routine sampling a single scan requires a time of one second or less. In both dispersive and FT-IR spectroscopy the signal-to-noise ratio (S/N) in a spectrum can be enhanced (and the noise level thereby reduced) by co-adding successive scans in a computer, so the rapid scanning capability of the FT-IR instruments provides a substantial advantage.

Jacquinot's advantage (the "throughput" advantage) concerns the amount of source energy that actually falls on the detector. In a dispersive instrument, a slit is used to isolate a particular wave-length of light after the light has been dispersed by a prism or grating. There-

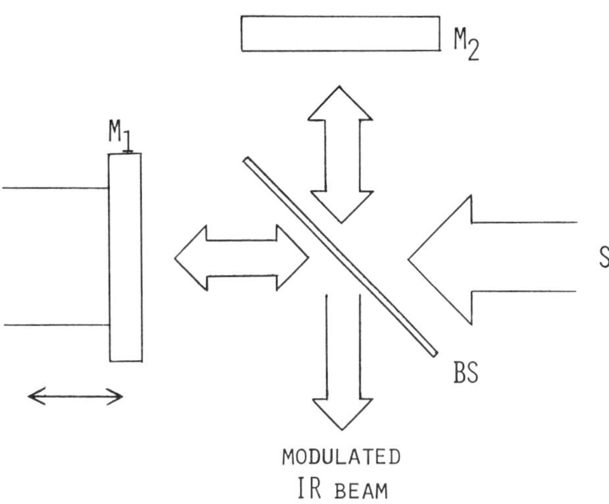

Fig. 1. Diagram of a simple Michelson interferometer; S, source; BS, beamsplitter; M1, movable mirror; M2, fixed mirror.

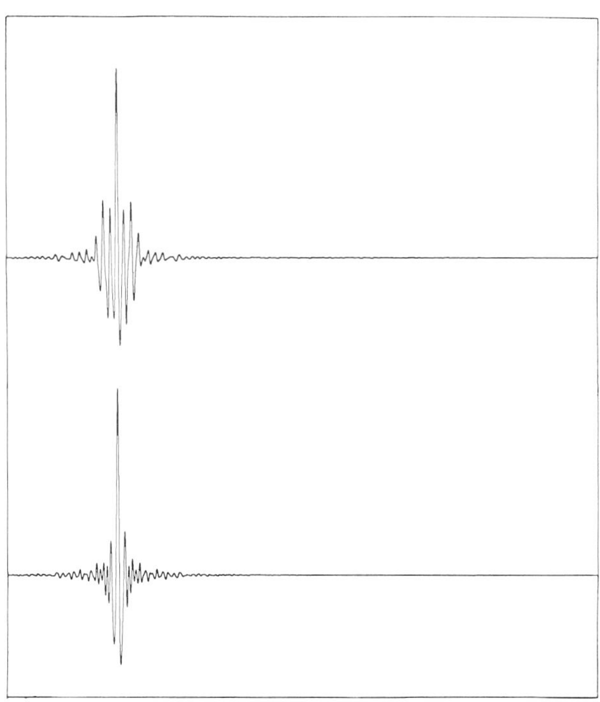

Fig. 2. Infrared interferograms without sample (lower trace) and with sample (upper trace).

fore only a very small portion of the source radiation reaches the detector, and since the ability of infrared detectors to respond accurately to low light levels is limited, considerable photometric error is possible. Since no slits are necessary in FT-IR instruments, the light flux reaching the detector is relatively large and spectral data can be measured more accurately than for the dispersive case. Jacquinot's advantage is particularly important when the analyst is faced with an "energy-limited" situation; that is, when the sampling method itself reduces the amount of available source radiation.

Sampling Methods

All the methods of handling samples that are standard for dispersive infrared spectroscopy are equally applicable to FT-IR instruments. Because of the advantages of the latter, however, there are additional sampling methods that can be used with advantage for infrared spectroscopy studies. Transmittance, reflectance, and emission spectra can be obtained, but the first two are more important in the identification of unknown chemical compounds.

Advantages of FT-IR over dispersive IR were cited in a recent paper[4] describing how a core sample comprising four paint layers taken from a painting known as the *Virgin and Child*, provided sufficient material for the FT-IR method even though the sample mass of each layer was less than $0.5\mu g$.

Similarly in the analysis of paper, interest is generally focused on the trace organic components, and sample limitations are common. Furthermore, some method of separating the components of a mixture must be used in conjunction with the infrared analysis. Combinations of this type require the various advantages of FT-IR systems.

Chromatography Interfaced with FT-IR

Gas chromatographic separations have been combined with FT-IR instruments (GC/FT-IR) for rapid analyses of microgram (and sometimes nanogram) quantities of sample. The infrared beam is directed into a gold-lined sample holder (2mm ID; 40cm length) through which the effluent of a gas chromatograph is allowed to flow. The rapid-scanning capabilities of the FT-IR permit low-resolution infrared spectra of the individual components to be recorded as they flow through the cell. While most spectrometers can collect between two and seven scans per second, one type of instrument can collect up to fifty scans per second. Again the throughput advantage is very important, but the ability to rapidly collect the entire spectrum (multiplex) is essential.

High-pressure liquid chromatography FT-IR (HPLC/FT-IR) interfaces are not as highly developed as GC/FT-IR but progress in this area should result in commercially available systems in the near future.

Thin-layer chromatography-FT-IR (TLC/FT-IR) interfaces are the most complicated because the silica gel of a TLC plate can absorb atmospheric water in large enough quantities to obscure the analyte spectrum. Consequently, the bands of separated compounds on a TLC plate are scraped off and extracted. Because of the difficulties in obtaining pure compounds from TLC separations, it is useful to be able to obtain high-quality spectra from very small sample quantities. The FT-IR method, however, is well suited for such studies, and results of this type are presented in a later section.

Reflectance Spectroscopy

Reflectance spectroscopy can be useful in infrared analyses because under certain conditions, reflection can be incomplete as a result of sample absorption. This is especially true if the reflection occurs at the interface between the sample and a crystal having a large refractive index. The term "attenuated total reflection" (ATR) refers to the case where there is absorption by the sample that "attenuates" the infrared beam. Ordinarily a KRS-5 crystal (thalium iodide-bromide mixed crystal) is used as the internal reflection element, and any number of sampling geometries can be employed.[5]

An advantage of ATR infrared spectroscopy in the study of paper is that the attenuation of the beam (absorption by the sample) takes place near the surface of the sample. The "depth of penetration" varies from one to five microns and depends upon the angle of incidence and the refractive indices of the crystal and sample. By varying the angle of incidence, depth profiling can be accomplished, and in favorable cases the thicknesses of surface layers may be determined.

A potential drawback to ATR spectroscopy is that the amount of absorption by the sample is generally small. Coupled with this is the problem of surface contact between the ATR crystal and the sample. For samples with a rough surface, such as old paper, the lack of uniform surface contact reduces the intensity of the reflected light. Therefore the analysis of paper requires measuring a small absorption in a very weak (and possibly very noisy) spectrum. The advantages of FT-IR spectroscopy help alleviate these problems.

A relatively new FT-IR technique that holds the prospect of non-destructive analysis, has recently been described.[6] In "photothermal beam deflection" spectroscopy (PBDS), a low-powered laser is directed across the surface of an area that is to be analyzed, and an infrared beam from an FT-IR instrument impinges on the sample in the same area. If the sample absorbs the infrared light and subsequently releases that energy as heat, the temperature gradient produced by the release of the heat can cause the laser beam to be deflected. The magnitude of the deflection depends upon the amount of infrared energy absorbed by the sample. An interferogram is generated by measuring the extent of the beam deflection, and the Fourier transform can convert this interferogram into a spectrum. This technique holds promise for the non-destructive analysis of relatively large samples without having to perform any sample preparation.

The applications of FT-IR spectroscopy to the analysis of paper and other art objects are manifold and have only recently been explored with enthusiasm. The analysis of the eighteenth- and nineteenth-century papers that follows provides but one example.

The Analysis of Degradation Products of Selected Papers

TLC of Paper Extracts

Methanol extracts of twenty nineteenth-century papers were concentrated and applied as spots to silica gel TLC plates and the resulting separation patterns were observed.[7-9] On comparison the TLC patterns were found to have considerable resemblance. The remarkable similarity of the twelve nineteenth-century papers shown in the photograph of TLC plate 51C (fig. 3) demonstrates that many papers from the same period have similar degradation products. This relationship holds even though some of the papers appear to be much more degraded than others. The paper represented by the TLC pattern from spot 4 in figure 3 is from a well-preserved. unused full sheet produced in the Wilcox Mill near Philadelphia between 1825 and 1846. Extracts of this sheet were also used in the separation of major degradation products for identification by IR and FT-IR.

The similarity in TLC patterns found in these preliminary investigations appeared promising and indicated that this approach should be pursued further.

TLC of Foxed Spot Extracts

When foxed spots from a sheet of paper were excised and extracted. and the products separated by TLC. comparison of separation patterns for the foxed spots and the associated paper indicated little difference. Of course similarities are to be expected since the excised foxed spots are on the paper. Even when paper sheets with maximum foxing were chosen and care was used in excision to include as little unfoxed paper as possible. results indicated little or no differences in the TLC patterns of the papers and of their associated foxed spots. The results suggest that foxing and aging have the same degradation products and, therefore, possibly the same degradation mechanisms. It is possible then that the difference between overall degradation of paper and foxing may be a matter of degree. If so, foxing could be considered to be localized, accelerated aging catalyzed by some agent at the foxing site.[10]

Separation of Major Degradation Products

In order to have larger samples for use in the identification by FT-IR and IR, the isolation of major decomposition products of Wilcox paper was accomplished by applying extract across the full width of a 20 cm plate and carrying out the TLC separation in the same manner as before. The individual horizontal bands, some of which are shown in figure 4, were carefully scraped from the plate and eluted. The extracts of the isolated bands were used in the IR and FT-IR spectral analysis. Or, as in the case of the major band indicated in figure 4, the extract could be reapplied to a fresh plate for a second TLC separation to ensure a pure sample for IR and FT-IR analysis.

Identification of Major Degradation Products

The purified bands were extracted from the silica gel absorbant and spectra taken by FT-IR and IR were used in determining the chemical nature of the products. The spectra obtained were compared with standard spectra of known compounds. For example, comparison of the spectra showed a close match between certain compounds such as: (1) Band 7(51) and a standard, palmitic acid; (2) Band N-1-B-9 and a standard, dibutyl tin dilaurate; (3) Band NC-2-B-1 and a standard *cis*-9-octadecene-1,12-diol diacetate. It was sometimes helpful to convert the product into a derivative and to match the derivative with standard spectra. For example a close match was found for: (1) the methyl ester of Band 6(51) and a standard, methyl tridecanoate, and (2) the salt of band 9(37) and a standard, the potassium salt of stearic acid. Additional checks of sample purity (after TLC) were performed using the GC/FT-IR system. The methyl esters were prepared from various of the TLC fractions, and the FT-IR spectra were recorded as the ester eluted from the GC. Each sample was determined to be an ester of a fatty acid, and the similarities of the spectra make unequivocal identification difficult without the aid of standard

spectra. Further work in this area will be pursued.

It is interesting that the compounds identified so far all seem to be related to naturally occurring fatty acids. The significance of this is not yet clear and it may be that the pattern will not persist. Bands analyzed thus far are the less polar bands. Analysis of the more polar bands is underway at present. It should be mentioned that in later. rosin-sized papers there also occur abietic acid type compounds. as expected.

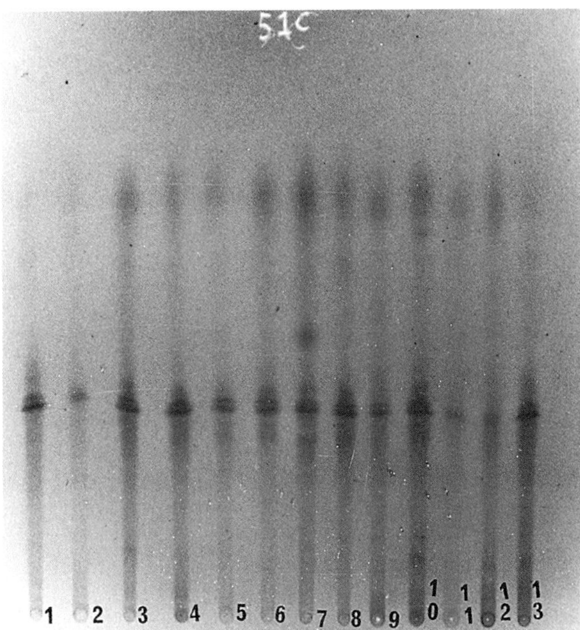

Fig. 3. TLC Plate 51C. Extracts of twelve nineteenth-century papers. Spots 1 and 13. Book, New Orleans, 1856; 2. Book, New York, 1854; 3. Confederate Records, 1863; 4. Paper, J. M. Wilcox, 1825-46; 5. Book, New York, 1893; 6. Book, New York, 1880; 7. Book, Chicago, 1875; 8. Book, New York, 1862; ° Book, Philadelphia, 1857; 10. Book, London, 1853; 11.Book, London, 1⌂ 12. Book, London, 1822. Moving phase; benzene, methanol, acetic acid; 110:15:0.5.

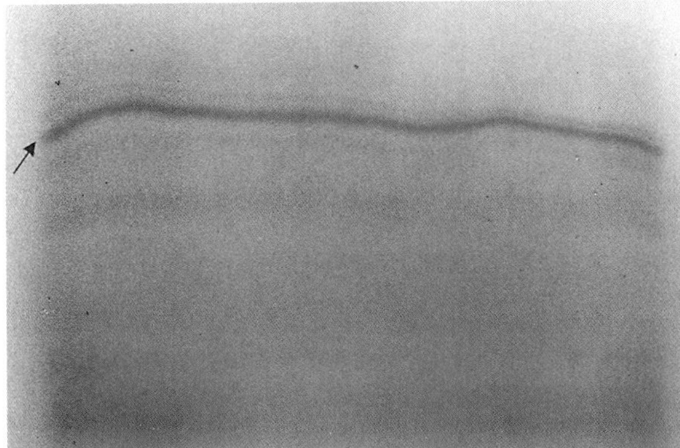

Fig. 4. TLC Plate 52S. Some bands from separation of Wilcox paper extract for FT-IR and IR spectral analyses. Moving phase; benzene, methanol, acetic acid; 120:10:0.5.

Reflectance Spectrometry

Another application of FT-IR in paper studies is surface analysis using reflectance attachments. The detection of gelatin sizing was carried out using this technique. The individual ATR spectra of the sized and unsized cotton papers were recorded, and the former exhibited bands attributable to the cellulosic material in the paper along with additional bands. By using the computer to compare the two spectra it was possible to generate a "difference spectrum" that clearly showed evidence for the gelatin sizing. Infrared bands at 1630 and 1540cm^{-1} in the difference spectrum were similar in relative intensity and shape to those in an authentic gelatin sample. Other relatively intense bands in the difference spectrum have not been identified as yet, and continued work is necessary.

The identification of degradation products is a continuing project. Great care must always be used to ensure that the compound identified is not some artifact of the method or solvents used. It is hoped that additional definitive results will be available before long.

Acknowledgments

The authors would like to acknowledge the findings of hard-working student researchers. Appreciation is due to Frank Wade, Bert Tagert, John Turner, Janet Hamilton, Stuart Simon, Wendy Perry, and John Johnson. Appreciation is also extended to the Analytical Laboratory of Winterthur Museum, to the Chemistry Department of Millsaps College for use of their facilities, and to the Office of Research at Mississippi State University.

References

1. G. Horlick, *Appl. Spectrosc.*, 22, (1968): 617 .

2. P. R. Griffiths, C. T. Foskett, and R. Curbelo, *Appl. Spectrosc. Rev.*, 6, (1972): 31.

3. P. R. Griffiths, *Chemical Infrared Fourier Transform Spectroscopy*, (New York: Wiley, 1975).

4. J. C. Shearer, D. C. Peters, G. Hoepfner, T. Newton, *Anal. Chem.*, 55, (1983): 874A.

5. N. J. Harrick, *Internal Reflection Spectroscopy*, (New York: Wiley, 1967).

6. M. J. D. Low, "Some Uses of Infrared Photothermal Beam Deflection Spectroscopy," Twenty-Fifth Rocky Mountain Conference, Denver, Colorado, August 14-17, 1983, paper 175.

7. K. Randerath, *Thin-Layer Chromatography*, 2nd Ed. (New York and London: Academic Press, 1966).

8. R.M. Scott, *Clinical Analysis by Thin-Layer Chromatography Techniques* (Ann Arbor: Hunphrey Science Publishers, 1969).

9. S. Simon and C. E. Cain, "Analysis of 19th Century Papers Using Thin-Layer Chromatography," *J. Miss. Acad. Sci.*, 27, (1982): 9 .

10. C. E. Cain and B. A. Miller, "Photographic, Spectral, and Chromatographic Searches into the Nature of Foxing," *AIC Preprints* (1982), p. 34.

PAUL T. CRADDOCK

Three Thousand Years of Copper Alloys: From the Bronze Age to the Industrial Revolution*

*The analytical data referred to in this paper are to be found in an envelope in the back of this volume.

Most programs of copper alloy analysis have concentrated on material from Bronze Age cultures, yet it is only at the end of this period that an interesting range of alloys, including for the first time ternary and quaternary alloys, was in regular use. Attention has focused on the earlier periods because most projects until recently have centered on the problems of artifact typology and provenancing of the metal sources, based on the trace elements in the copper. From the end of the Bronze Age the problems of multiplicity of sources and mixing of scrap metal have been deemed too daunting for this class of study and consequently relatively few analyses have been published from which information on the alloying practice might be obtained.

That situation is now being transformed by the publication of important analytical projects based on existing museum collections coinciding with the as yet unpublished reports of the analysis of large numbers of the many thousands of copper alloy small finds dating to Roman times and onward, from the major urban excavations that are now in progress over much of Europe. Thus it is now possible to begin coherently tracing the overall development of copper alloys through the last 3,000 years, although some large gaps still exist, notably in the European Iron Age and post-Han China. This paper draws on more than 5,000 analyses of material from classical antiquity, prehistoric Britain and Spain, medieval Islam, India and Tibet, and West Africa (see microfiche). These analyses have been performed at the British Museum Research Laboratory over a number of years, the majority were previously unpublished and they have never before been considered together or with other related major analytical projects. Among these, the most useful are those of Moorey on Luristan bronzes; Gettens, Barnard, and Chuang and Kwong on early Chinese bronzes; Tanabe on Japanese bronzes; Riederer, Werner and Liszak-Hours on Indian and Tibetan bronzes.[1-8] In Europe and the Middle East, Roman bronzes have been extensively analyzed by Picon, Condamin and Boucher, den Boesterd and Hoekstra (vessels) and Cope and Billingham (late Roman coins).[9-11] For the Dark Ages and medieval Europe, the work of Werner is preeminent and he has also published analyses of many hundreds of West African bronzes.[12-13] There are as yet relatively few major analytical publications devoted to the last two millennia, although there is an enormous quantity of unpublished analytical data principally on material from recent excavations.[14] More generally, the Laboratory of the Berlin Museum has produced a large number of analyses on a very wide range of objects.[15]

The Trace Elements

With this great and ever-expanding wealth of data appearing it is necessary to give some thought to its proper use. Analytical studies of Bronze Age material have been dominated by attempts to correlate the trace element composition with the typology of the artifacts or with the source of the copper. Research continues even now with hopes centered on more data, more sophisticated analytical techniques or statistical treatments, but the real problems lie not in the data or its treatment, but much more fundamentally in the almost total lack of information on the chemical processes and compositional changes between the ore source and the finished metal of the analyzed artifact which can only be bridged by often untenable assumptions.

The first such assumption is that the ore used from one locality had a consistent composition. This may have been true in the earliest Bronze Age when only the leached oxide ores are believed to have been exploited, but certainly by the full Bronze Age all the main copper-bearing horizons within a mine would have been simultaneously exploited giving a wide range of composition.

The second untenable assumption is that the smelting processes had at least a consistent effect on the trace elements within the copper. The smelting experiments carried out by Tylecote and his students showed that minor changes in the reduction potential or temperature within the simple clay furnaces used produced a dramatic change in composition.[16] Purification of the copper in clay crucibles also gave very variable compositions from experiment to experiment.

The problems of working with metal that is a mixture of metals from several sources are self-evident, especially when long-distance trade in metal is well attested.

Lead Isotopes

Recent work has been done to establish the provenance of copper by determination of the isotopic ratio of the lead within the copper.[17] Small amounts, below 1%, may reasonably be supposed to come from the copper ore itself, and the isotope ratio should survive chemical alteration within the ore body and furnace intact, thus surmounting the above problems. Hence one should be able at least to rule out many potential copper sources, but even here, lack of knowledge of the chain of events between ore and metal can lead to potentially disastrous mistakes.

Lead isotope determinations at Timna, southern Israel, one of the very few ancient metal mine-smelter complexes to have been properly investigated, provide an excellent salutary warning. The lead isotope ratios of the copper ores, and both the iron and manganese fluxes varied quite widely both internally and in relationship to one another. This spread was reflected in the copper metal from the site. Attribution of metal to Timna without the understanding that much of the lead in the metal originated from the flux rather than from the ore itself must, therefore, be treated with caution.[18] The main point here, though, is not the specific problem of lead in fluxes, but the dangers inherent in working with materials where the processing stages are now unknowable. From the Late Bronze Age varying amounts of lead were often added to copper alloys, and from Roman times onward, brass became common, with the zinc acting as a further source of lead. Thus in most alloys produced over the past two millennia, there is the possibility of lead from the copper ore, the flux, the zinc, as well as the possibility of deliberate addition of lead, probably itself a mixture. These possibilities coupled with the problems of mixed copper, generally make it really impossible to disentangle meaningful lead isotope data on the copper ore sources. This is reflected in the lack of success that has attended lead isotope analyses of more recent copper alloys.[19]

Is there, then, any point in trace element analysis? The answer is a resounding "yes," as these data provide an invaluable indication of the technical history of the metal, especially if the bulk analyses can be coupled with an investigation by scanning electron microscopy of the inclusions in the metal. Information on the ore type, and on the smelting process, is given by the trace metals. Even the subsequent degree of refining can be inferred; reuse is indicated by the prevalence of sulphides or oxides in the inclusions.

Iron

In terms of the bulk analyses, iron is perhaps the most neglected but useful of the common trace elements. The following section, on its value as a process indicator, will serve as an example of the usefulness of trace element content, and also of a more general approach to trace element studies in metal.

It is believed that the earliest smelting involved the reduction of very high grade copper oxide or carbonate ores directly to the metal, either without the formation of a slag, or by producing globules of copper in a matrix of slag and a cinder of unreduced ores and fuel.[20] The process was quick but only a small proportion of the available copper was recovered. There was little iron in the ore, or if present, little opportunity for it to enter the copper. By the Late Bronze Age, much more efficient processes were in use whereby much lower grade ore could be used and recovery was much higher.[21] This was done by removing the waste material as a highly ferruginous liquid slag periodically tapped from the furnace. If the ore was sulphidic then it usually contained iron, if a carbonate then iron oxides were added as a flux; either way iron minerals were present in the furnace and some were reduced to metallic iron in the slag. The process ran for many hours to allow the copper to drain through the iron rich slag. Whilst this was happening some of the iron could and did become incorporated in the copper.[22] Thus for example the 300 or so spills and artifacts of copper from Late Bronze Age deposits at Timna regularly contained several percent of iron (table 3). Iron is only barely soluble in copper and most could be easily removed by melting in a crucible and skimming the surface or adding clean sand to form a slag. Experiments showed that it became progressively more difficult to remove the iron below 0.5% in the metal.[16] Comparing analyses of Early Bronze Age and much later bronze, there is a significant difference in the average iron content. Generally, the simple process gives an average iron content well below 0.1%, the more sophisticated well over that figure (fig. 1). The date at which the sophisticated process was adopted varies tremendously. One observes a sharp rise in the average iron content of bronze in Third Dynasty Egypt in the mid third millennium B.C. (table 1) whereas the changeover does not seem to have occurred in much of Western Europe until two millennia later.

So far, iron has only been considered as a trace metal that the smiths were at pains to minimize, but in a very few instances iron-rich copper was deliberately produced. An excellent example are the *Ramo Secco* bars, (table 65) found in Italy in association with early coin hoards of the sixth through third centuries B.C. They are assumed to be a primitive form of currency, probably precursors of the more familiar Roman *Bar* Aes.[23] Typically the *Ramo Secco* bars have in excess of 20% iron. At this sort of concentration quantitative analysis becomes rather meaningless as the iron is distributed very unevenly as large dendrites in the copper-rich copper-iron matrix, or as cast iron globules.

The consistently high iron contents restricted to one class of artifact show that they are intentional, leading to questions of how the iron content was achieved, and why. Fortunately the smelting experiments carried out by Merkel provide the answer to the first question.[24] By increasing the proportion of charcoal in the furnace charge,

	Sample size	Average Fe content	% below 0.05% Fe
British, Welsh and Spanish Bronze Age Metal (tables 4 - 17)	936	0.05	79
Greek, Etruscan and Roman Metal (tables 20 - 67)	3062	0.23	26

Fig. 1. Comparison of iron content of Bronze Age and classical copper brought about by different smelting practices.

he was able to make the furnace much hotter and more reducing, reproducing the conditions for iron smelting. The copper so produced had between 20 and 50% iron and appeared metallographically similar to the *Ramo Secco* copper. The metal was extremely difficult to cast and impossible to work.

It now remains to explain why such a useless material should have been deliberately produced. The answer lies in its role as currency, where the value was related just to the weight, not to its utility. It was in the producers' interest to make as much metal as cheaply as possible, and this they certainly achieved! Slightly later, when lead became very common, the ferruginous copper of the *Ramo Secco* bars was replaced by the heavily leaded Roman *Bar Aes* (table 65), similarly useless for ordinary purposes, although uncommon iron-rich coins have been found elsewhere. There are the magnetic Chinese *cash*, although in this case the iron content (usually below 5%) is typical of ordinary unrefined copper. Some Indian coins contain between 20 and 30% of iron and the metal must have been produced in a similar fashion to that of the *Ramo Secco* bars.[25]

The only ancient literary reference to ferruginous copper appears in Pliny's *Natural History* where the Hellenistic sculptor Aristonidas is said to have "made a blend of copper and iron in order that the blush of shame should be represented by the rust of the iron shining through the brilliant surface of the copper."[26]

The Alloys

By the end of the second millennium B.C., the use of tin bronze was almost universal throughout the Old World after over a millennium of coexistence with arsenical copper. Items of arsenical copper such as the Late Bronze Age Atlantic culture axehead from Spain (PRB Reg 1926.10-17.1) with 1.1% of arsenic (tables 13-17) or the sub-Minoan female statuette (CAT. 30) with 2.2% of arsenic (table 21) from Crete are rarities, although the alloy was to reappear much later amongst the alchemists (see below). Given the upheavals in the Mediterranean at the end of the Bronze Age, and the enormous distances over

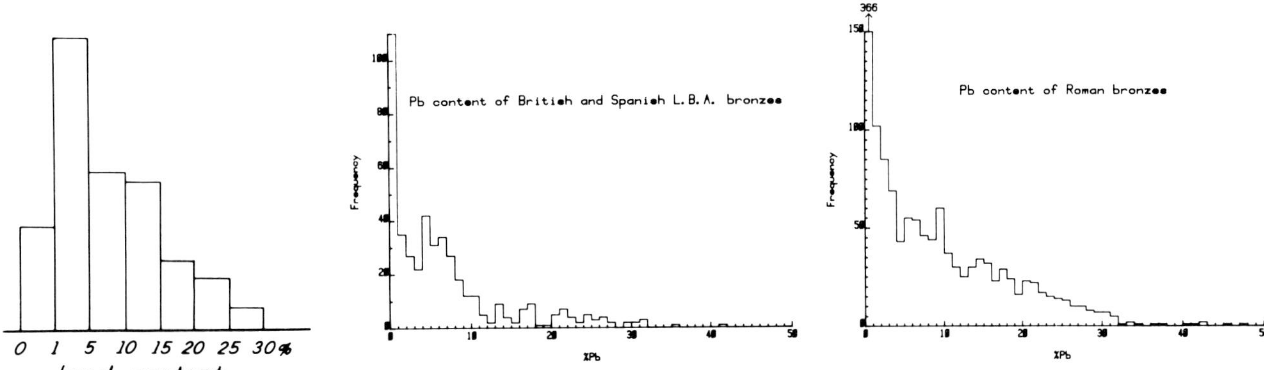

Fig. 2. Comparison of lead content of bronze from China. Atlantic Bronze Age Europe. and the Roman Empire showing the basic similarity of pattern. (The Chinese analyses are only from heavy cast vessels resulting in the absence of unleaded bronze in the table. Reproduced from *Freer Chinese Bronzes. Vol II* with permission.)

which the tin must perforce have been distributed, the ubiquity of tin bronze and the uniformity of its composition is surprising. There is little evidence from the bronzes themselves of the scarcity of tin in the Near and Middle East, which has been proposed as one of the main causes of the rapid rise in the popularity of iron.[27]

Leaded bronze

From the middle of the second millennium B.C., leaded bronze was used with increasing frequency for cast metal work. The practice seems to have developed independently in Shang China,[2] Egypt[28] (table 1), and in Mycenaean Greece[29] (table 21). The range of use however was very different. In China leaded bronze was used for large cast vessels, which in the West would have been made of hammered bronze. In contrast, in the West leaded bronze was generally used in small castings, with lead probably used as a cheap additive. By the early first millennium B.C., leaded bronzes were widely used over much of the Old World, with some cultures, such as the Atlantic Bronze Age, frequently producing heavily leaded bronze,[30-32] (table 4).

The overall impressions gained from the analyses is that the occurrence of leaded bronzes in any given group of bronzes, and the actual percentage of lead in the individual bronzes, is highly erratic. There are a few classes of artifact, notably currency, that regularly have high lead contents, but in general it does not seem possible to correlate lead content with object type, beyond the fact that leaded bronzes are restricted to castings (metallurgical reason given below), and large castings tend to contain more lead. It must however be stressed again that random lead content does seem to be a general feature of bronzes through the period under review. This is exemplified by the very similar pattern of lead content shown in the histograms of bronzes from the British and Spanish Late Bronze Age, Roman, and Chinese vessels shown in figure 2. The frequency and content of lead in bronze does seem to increase through the first millennium B.C. reaching in the West a plateau by the latter part of the millennium that was maintained through the Hellenistic, Roman, By-

zantine, Islamic, and medieval periods. Occasional reversals of this pattern, such as the sharp decrease in the use of leaded bronze in the British Bronze Age from the Wilburton to the Ewart Park phase, most probably merely reflect the local availability of lead.

In the Far East the picture is confused at present by the incomplete state of the analytical record. Thus up to the fifth century A.D., there are abundant analyses of Chinese and Japanese material but little from India, and afterward the situation is reversed with a great deal of analytical data for Indian and South East Asian material generally (tables 79, 82), but little for China or Japan (tables 80, 81). The early Chinese vessels and coins are usually heavily leaded, whereas the few contemporary Indian bronzes that have been analyzed have a much lower and more erratic lead content,[33] although it should be borne in mind that they are much smaller in size and thus not strictly comparable.

However, the hundreds of analyzed statuettes, many of which are quite substantial castings from the fifth century onward, also demonstrate that Indian and Tibetan brass and bronze rarely contains more than 10% of lead (table 82).[6-8,34] In contrast, unpublished analyses of some hundreds of Chinese coins in the British Museum suggest that heavily leaded alloys persisted there.[35]

The properties of lead in copper alloys and the possible reasons for its addition must now be discussed. Lead is a low melting point metal, which is insoluble in copper. Up to 2% lead in the alloy significantly increases the mobility of the molten metal. Increasing the lead content further has little effect on the mobility but does lower the melting point of the alloy. Increasing the mobility of the molten metal and decreasing the melting point clearly aids casting and, in addition, leaded bronzes are relatively easy to drill, file, or grind.

Finally lead has always been cheap, an important consideration if large castings or large numbers of casts were produced. However, the disadvantages are severe. Ideally the lead remains dispersed throughout the copper as minute globules, but at concentrations of more than a few percent, there is an increasing tendency for the

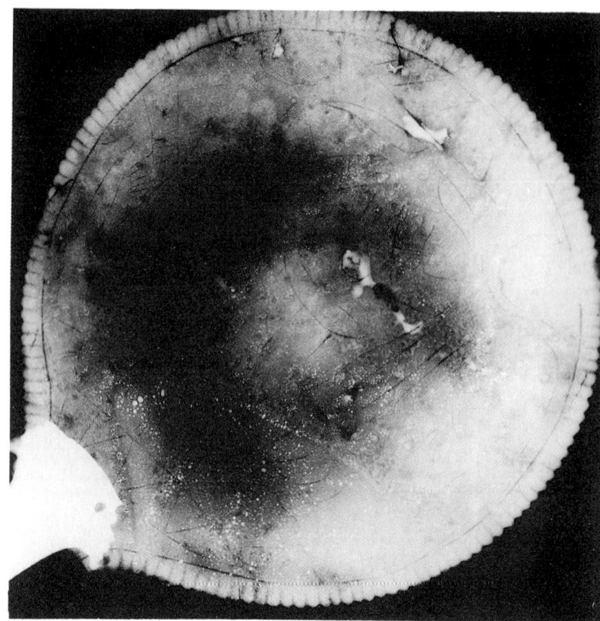

Fig. 3. Radiograph of Late Etruscan Mirror (cat. 702, table 47) showing macroscopic "lakes" of lead in the bronze.

globules to link up to form more macroscopic "lakes" of lead with the lead-bronze interface forming a serious weakness that would tend to crack open if any deformation was attempted (fig. 3). For this reason hammered metal never contains more than small quantities of lead. The tendency of lead to concentrate on the surface was a problem for certain classes of objects such as mirrors or metal intended to be mercury gilded.[36] On mirrors the lead would tend to smear during polishing and so Greek and Etruscan mirrors are rarely leaded. (The mirror in figure 3 is a piece of burial chamber furniture, and was not intended for use). Chinese mirrors,[3] and the Roman mirrors, which possibly consciously use the same alloy, are of leaded, high-tin bronzes (table 57). The Roman mirrors were often tin plated after fabrication. The nature of the surface treatment of the Chinese mirrors is far more problematic.[37]

If the alloy was to be mercury gilded then lead, because of its solubility in mercury, had to be excluded. As Theophilus stated in the twelfth century, "if you see white spots emerging on the gilding so that it refuses to dry evenly this is the fault...of lead because the copper was not purged and refined free of it."[38]

The usual alloy for castings to be mercury gilded was copper with just one or two percent of tin, while sheet metal was of unalloyed copper. Perhaps the most famous mercury gilded bronzes are the horses of St. Mark's in Venice. These are copper with about 1% tin.[39] The metal must have been extremely difficult to cast, but it did allow the statues to be mercury gilded. This information has allowed the horses to be firmly dated, thus ending at least two centuries of speculation on their origins. Oddy had shown that mercury gilded statuary was rare before Late Imperial times in the West. Thus, as the horses were made to be gilt, they must be Roman rather than Greek.[40] Mer-

cury gilding of small objects is now well attested from Hellenistic Greece[36,41] and early Han China.[42]

Thus, although it is possible to explain on technical grounds the presence of small quantities of lead in copper alloys, it is very difficult to find any convincing metallurgical or aesthetic reason for the large quantities randomly encountered in large castings or more regularly in currency. This very randomness, and the lack of association with any particular type of artifact (currency apart) suggests that, when available, it was added merely as a cheap substitute. Lead has always been much cheaper than tin or copper through the last 2,000 years and this is likely to have been true at earlier periods. However there still does not seem to be any convincing explanation of why lead, which was known and used from very early times, should have

Fig. 4. Sectioned retort for producing zinc, from Zawar, Rajasthan.

only begun to be regularly added to bronze as late as the second millennium B.C., and then over such a wide area. Lead is a by-product of silver production and one could argue that increased silver production during the second millennium B.C. lay behind the greater availability of lead. Unfortunately, although this theory may hold for the West, it falls down completely for the East, where the regular use of leaded bronzes precedes the earliest silver by at least a millennium. However, much later in the first millennium B.C., the great increase in the use of lead in copper alloys in the West from Hellenistic times onwards is almost certainly due to the prodigious increase in silver production from argentiferous lead at centers such as Laurion. The conclusion that lead served merely as a cheap additive is strengthened by its more regular occurrence in alloys used for currency. As discussed above in the case of the *Ramo Secco* bars, the objective there was to produce the maximum amount of metal as cheaply as possible, regardless of its properties, while still retaining something of the expected appearance and color of the original metal. With leaded currency all pretence was sometimes abandoned and one gets items of pure lead as well as heavily leaded alloys. Examples of this are many and include the hoards of unused (and unusable) axeheads from the Atlantic Bronze Age,[30] the Etruscan and Roman *Bar Aes*,[23] the Chinese *cash* (heavily leaded from their origin until the sixteenth century),[35] and the trade manillas made in Europe for use in West Africa (table 92). It should be noted that these items were regarded as currency, and they were no more consciously fraudulent than modern "silver" coinage. Furthermore, it is most unlikely that they were ever regarded as usable ingots of copper. In most cases just a tap with a hammer, if not their actual appearance, would have revealed their true nature.

Brass

While leaded bronze was becoming established in the first millennium B.C., a new alloy of great potential was developing somewhere between Asia Minor and India. This was brass, the alloy of copper and zinc. Literary references to, and artifacts of, brass appear with increasing frequency through the first millennium B.C.[43-45] Although zinc ores are common and easy to reduce, the low boiling point of zinc means that it is produced as a gas and, without special condensing apparatus, it would be lost. There are two solutions, and it is possible that both were developed early on. The zinc reduction could be carried on in a retort and the zinc separately condensed (fig. 4). Zinc was certainly produced in this way on an industrial scale in medieval India, although recent radiocarbon dates of about 100 B.C. from the zinc production site of Zawar, Rajasthan suggest a much earlier origin.[46]

From much further west there is a description of what sounds like zinc distillation in the fourth century B.C. by the geographer Theopompus, preserved in Strabo.[47] This refers to a stone found at Andeira in ancient Phrygia in west Asia Minor, that, when smelted with a certain earth, dripped down "mocksilver." When this mocksilver was mixed with copper it gave brass. The interpretation that this is a description of zinc production is strengthened by the discovery of metallic zinc in contemporary levels of the Athenian Agora.[48] The process at Andeira is described as an evident curiosity and brass was certainly rare at this period. However this was changed by the beginning of the first century B.C. at the latest, when the process by which brass was produced directly from copper and zinc ore in a closed crucible at 1,000°C was introduced. Significantly, the earliest evidence comes from brass coins minted in Phrygia (table 66).[49] In the West, brass made by this cementation process rapidly became popular, and, with little technical change, it remained the usual way to produce brass until the nineteenth century. The zinc and iron contents often enable cementation brass to be recognized. Inspection of the tables in the fiche shows that none of the brasses from the West prior to about 1600 contains more than 28% zinc. In the eighteenth century, Champion, one of the famous brass smelters of Bristol, stated that using the cementation process he could make a maximum of fifty-six pounds of brass from each forty pounds of copper; this of course represents just over 28% of zinc.[50] Further proof of the significance of the 28% zinc level was shown by Haedecke and Werner.[51] In separate experiments they heated zinc oxide, charcoal and

first copper, and secondly a 40% zinc brass at 1,000°C in a closed system. The result in each case was a brass with 28% of zinc. It should be noted however that some postmedieval cementation brasses contain up to 33% of zinc (table 78).[52]

Also of diagnostic interest is iron. If brass is made directly from calcined zinc carbonate then any iron oxide in the ore is reduced to metal and becomes incorporated in the copper. Thus, for example, the early Roman brass *Dupondii* and *Sestertii* tend to have a higher iron content than the contemporary bronzes (table 69). This increased iron content was recognized in the past. Thus, when the first brass was made in Europe by mixing copper and zinc metals in the late eighteenth century, one writer remarked, "The brass is quite free of knots or hard places arising from iron to which other brass [i.e. cementation] is subject and this quality as it respects compass needles renders it of great import in making compasses."[53]

Large deposits of zinc carbonate were available to the European smelters, but in the Middle East the sulphide was much more prevalent. This had to be converted to the oxide before it could be used. This was done by roasting a heap of ore and collecting the clouds of dense white fumes of zinc oxide sublimate. These were condensed on an iron frame such as Marco Polo witnessed at Kerman in central Persia[54] or on a framework of ceramic pegs,[55] large numbers of which can still be found on old smelting sites in Iran.[56] Galen seems to describe a similar process performed in Cyprus in the third century A.D.[57] A feature of this process is, of course, that any iron present in the ore would be left behind, and the zinc oxide would be relatively pure. It can be seen from tables 66 and 72 that the brass made in the Near and Middle East from the Phrygian coins right through the Islamic period do not seem to have the high iron characteristic of the European brass.

Further east in India the situation is completely different. There is little documentary or analytical evidence for the cementation process. However, it seems possible that brass was always made from copper and metallic zinc, produced by reduction and condensation in retorts such as those found at Zawar, Rajasthan (fig. 4).[58] The Indian iatro chemists give detailed descriptions of the process from the thirteenth century A.D. onward, and ascribe its origins to the seventh century A.D.[59] The radiocarbon dates from Zawar of the first century B.C., and examples of brass excavated from Taxila (an ancient trading center lying in present-day Pakistan that flourished between the fourth century B.C. and the second century A.D.) suggest an earlier date. One of these brasses contains 34.4% zinc and can surely only have been produced from copper and zinc metals.[60]

Rather surprisingly, there is little evidence of early brass in China. The few extant analyses of medieval Chinese material show little brass, and the earliest brass coinage dates from only the sixteenth century.[36] Primitive zinc distillation is now practiced in Yunnan, using a varient of the Mongolian type of still, which is very different from that used at Zawar.[61] The origins of this Chinese zinc industry are at present very obscure, but probably quite late.

The use of brass spread rapidly over the rest of the Old World, and after the collapse of the Roman Empire in the West, with the attendant loss of the two main tin producing provinces, Britannia and Iberia, brass largely supplanted bronze for most purposes.[62] The prevalence of brass in the art objects that now adorn our museums and galleries (tables 72 and 77), is fully reinforced by the analyses of everyday medieval copper alloys. Among the thousands of pins, buckles, hinges, strap ends, needles, etc, from recent urban excavations, approximately 75% are of brass, 25% of copper, with just an occasional bronze (although most of the brasses have a little tin).[14] The reason for this transformation must lie solely in the relative difficulty of obtaining supplies of tin (significantly, English alloys tend to have more tin and less zinc).[63] Even so, the widespread use of tin continued for tinning vessels, in soft solders, and of course as the main constituent of pewter. Bronzes as such, however, seem to have become restricted to two main classes of objects. The first includes high-tin alloy specialties, such as bells, mirrors (OA 66.12-9 75 and 76, table 73), and the hot-forged high-tin vessels that probably originated in the Iron Age of South East Asia,[64] but spread through India (table 79) and finally developed into the superb Islamic *Haft Jush* vessels (OA 1950.7.251 and 1891.6-23.4, table 73). The second class of bronzes are the heavy castings, used in Europe especially for mortars and vessels. These typically contain under 5% of tin, and around 20% or more of lead (remainder copper), but little or no zinc (tables 77 and 78). This alloy is called *"Caldarium"* in the ninth-century Mappae Clavicula, and described as white metal of one part lead to four parts copper or bronze.[65] This use of the term *Caldarium* seems to hark back to the *Caldarium* described by Pliny as being a brittle alloy only suitable for castings.[66] In the Middle East leaded brass was preferred for mortars and other heavy castings, presumably because being that much further removed from supplies of tin, bronze was even more expensive to produce than in the West (table 74).[62]

Special Alloys

As well as the mainstream alloys discussed above there were other alloys of copper, much rarer, but not without interest. These include the beginnings of alloys which, although initially very unusual, were eventually to become commonplace, alloys made to imitate other more precious metals such as the arsenical copper so popular with the alchemists, and finally alloys made by special treatments to produce attractive sheens exemplified by the copper-gold-silver alloys of the Japanese *Shakudo* and possibly the Corinthian bronze of classical antiquity.

The surviving ancient texts from the Near and Middle East of the first millennium B.C. mention materials such as "copper of the mountain" and "white copper" as rare and often expensive commodities. From late in the first millennium B.C., there are several Greek references to copper alloys that are particularly bright, shiny, and corrosion-free but are apparently not bronze.[46,47]

These alloys were nearly all described as being made to the east in what had been the Persian Empire, and were often used for vessels. In this context should be noted the Biblical reference to "shiny bronze vessels precious as gold," brought back from Babylon by the Jews to the Temple in Jerusalem after their return from captivity in the mid fifth century B.C.[67]

The term "copper of the mountain" appears in both Assyrian and Archaic Greek contexts, and latterly becomes synonymous with brass in Greece.[44,45] The interpretation of the term "white" or shiny copper has far more possibilities. It could refer to high-tin bronze, cupro-nickel, or arsenical copper. The term occurs several times in Pliny's *Natural History* and is now usually translated as brass for no

obvious reason. Examples of the hot-forged high-tin bronze vessels (see above), are known from late in the first millennium at Taxila,[56] and further south in a whole series of bowls from the Nilgiri Hills in southern India, which date to the end of the first millennium B.C. (table 79).

The second possible interpretation, cupro-nickel, is one that has been unduly neglected. Copper alloys with nickel and zinc were imported into Europe during the seventeenth century A.D. from China, where the alloy was known as Bai-Tong. Needham traces the term back to the Han period,[68] but due to the lack of analyses, the real history of Bai-Tong in China is not known. One Chinese source of the fourteenth century A.D., quoted by Needham, states that Bai-Tong was originally imported from the island of Kish, in the Persian Gulf. It has long been known that the copper ores from Oman on the Persian Gulf are rich in nickel and several Sumerian copper objects contain up to 8% of the metal,[69] although this does not qualify them for classification as cupro-nickel. However, Seeley has recently examined the head of a bull from Anatolia or Mesopotamia that is not later than the second millennium B.C., and that contains copper with 20% nickel.[70] Toward the end of the first millennium B.C., several artifacts of cupro-nickel (21.35% and 19.0%) were found in Taxila,[33] and at the same date, but further to the north, in Bactria, cupro-nickel was regularly used for coinage.[71] In the frequently acrimonious discussion of these coins it has always been assumed, even by Needham, that the metal came from China, but it now seems at least as likely that it came from the Middle East, and provides another possible interpretation for "white copper."

A third possible interpretation is that of arsenical copper. During the Early Bronze Age arsenical copper was widely used, but apparently fell out of use in the face of competition from tin bronze. However, by Roman times, arsenical copper appears frequently in the recipes of the alchemists and others wishing to imitate gold or silver. The manner in which arsenical copper was prepared and the compositions were controlled is described in detail in *The Book of Minerals* by Albert Magnus, and follows older descriptions. As the method is likely to derive from those used in still earlier times it is given here in full:

> And *arsenicum* [realgar, As_2S_2] when calcined changes from red to black but afterwards if sublimed in an *aludel* [a covered vessel with a long neck], as we have often said it again becomes as white as snow[1] [i.e. oxidized to arsenous oxide As_2O_3] and if such calcination and sublimation are repeated a number of times it becomes extremely white and sharp. And because of its sharpness ... if added during the fusion of copper, penetrates into it and changes it to a shining white. But if the copper stands a long time on the fire, the *arsenicum* evaporates and then the copper returns to its original colours.[72]

Despite the frequency with which the alloys are mentioned, very few have ever been identified. One example is the base and escutcheon of a small fourteenth-century Tibetan statuette of Manjusri, (OA Reg 1905.5-19.4, table 83). This has 7% of arsenic in the body rising to about 15% on the surface, giving an attractive soft steel gray texture. Similar alloys were used by the Japanese (*Shirome*) at least from the seventeenth century.

After the rare alloys, and alloys made to imitate precious metals, it now remains to discuss the alloys designed to produce a fine surface after chemical treatment. The most familiar of these are the Japanese *Shakudo* alloys of copper with gold and silver. The alloys were treated with a variety of chemicals. typically metal salts. and the juice of the bitter plum as a source of malic. acetic. and oxalic acids. These alloys were produced during the Edo period. but probably have a much longer history.

Needham equates the purple-sheen gold of the Chinese Taoists with *Shakudo* type alloys.[68] Once again. in the absence of work on more recent Chinese material none has yet been identified. However. the Tibetan alloy *Dzne-Ksim* was said to contain copper. gold. and silver together with other metals including nickel.

An example of a Tibetan copper-gold-silver alloy is in the British Museum collection (OA Reg 1880-4072. table 82). There is. however. now no trace of surface treatment on this object. although examples of surface-treated objects were seen by Turner on his travels through Tibet in 1800 when he reports seeing at the Tashilumpo monastary "some of those images composed of that metallic mixture which in appearance resembles Wedgwood's black ware."[34]

In classical antiquity. there are references to the highly prized alloy of copper-gold-silver, "Corinthian bronze" from the third century B.C.[73] Pliny describes several varieties.[74] The Alexandrian alchemists do not mention the term Corinthian bronze but pay great attention to the surface treatments of copper alloys, sometimes containing gold and silver.[26] The most unambiguous description is given by Zosimus who specifically states that the rusting or purpling process (*iosis*) was to be carried out with pimpernel and rhubarb, a good source of acetic, malic, and oxalic acids.[75]

A small Roman plaque (G&R Reg 1979.12-13.1, table 60) with a composition similar to that of the Japanese *Shakudo* has been recently examined, and shown to have an identical surface treatment.[76] The link between Corinthian bronze of Pliny and *iosis* of the alchemists is provided by Pausanias in his description of Corinth, where he states that the fountains of Peirene were used in the coloring of Corinthian bronze.[77]

The Organization of Metal Production

Analytical data can provide indirect evidence on the sequence of smelting, purification, alloying, and fabrication. In almost every instance, metal was smelted near the ore source. Even the richest copper ore usually contained only about 20% copper and it can have made little sense to transport ore, containing 80% waste material, for any distance. The mines tended to be in remote upland areas where communications were difficult and timber for charcoal abundant. The massive heaps of tap slag associated with primary smelting are almost always found at mine sites rather than in urban metalworking centers. In the rare instance where substantial quantities of metal have been recovered by excavation from ancient mine-smelting sites, the composition has been shown to be very different from that usually encountered in finished metal objects. Timna, in the south of Israel, is a classic example. More than 300 items ranging from copper prills embedded in slag to finished mining tools were analyzed. The unusually high iron content of the copper (table 3) gives a rare glimpse of the raw product of antiquity and shows that it was not usual to purify the metal at the smelter. The different ores and smelting techniques used throughout the Old World produced copper of very variable composition before purification. This was appreciated by the ancients and thus Pliny in the opening chapters of Book 34 of

the *Natural History* spoke of the natural "types" of copper from various localities and listed them in order of excellence. Clearly Pliny believed. for example. that the copper from Gaul was intrinsically different and better than that from Cordova. To modern minds brought up on the concept of the ninety-two immutable natural elements and the Law of Constant Composition. this seems strange. However. the theoretical treatment of the origin of matter. as stated by Aristotle and based on the four elements of fire. air. water and earth. held that metals were a "vaporous exhalation" formed by the sun's rays upon water. The exhalation was entrapped in the earth. the dryness of which gradually converted it to metal.[78] Thus there was no intrinsic theoretical reason in the time of Pliny why a metal should have constant properties. Indeed common experience would suggest that the very varied environmental and geological circumstances in which the metal "grew" should produce a varying product. Pliny says of iron. "There are numerous varieties of iron the first difference depending on the kind of soil or climate."[79]

Sometimes. however. Pliny realized that the smelting and purification could be held responsible. Thus in the case of Cyprus copper he states:

> Bar copper also is produced in other mines. likewise fused copper. The difference between them is that the latter can only be fused as it breaks under the hammer. whereas bar copper otherwise called ductile copper is malleable, which is the case with all Cyprus copper. But also in the other mines these differences of bar copper from fused copper is produced by treatment: for all copper after impurities have been rather carefully removed by fire and melted out of it becomes bar copper.[80]

This must be a reference to the presence of sulphides or iron in the metal which would certainly have to be removed before the copper became ductile. Ingots from the Roman mining area of Be'Ora, near Timna, contain about 5% iron (table 3). If the metal from some other production centers was of as poor a quality as this, it is small wonder that these coppers were recognized as different from the copper from centers such as Cyprus with more rigorous quality control. There was little incentive for the smelter to purify the metal as all this did was reduce the quantity of salable metal at the expense of large quantities of fuel. It was up to the smith who was to alloy and fashion the metal to purify it. This too was expedient as, in the absence of any analytical facility, the only way the smith could rely on the quality of the metal was to purify it himself. The smiths' workshops were normally in settlements and the surveys of archaeological sites through the Middle East carried out by Berthoud et. al[81] found that the slags from urban sites tended to be rich in arsenic and tin as well as copper, suggesting that they were the waste product of alloying rather than being associated with the primary smelting.

There is evidence that in medieval Europe the craftsmen alloyed the metals they fabricated. The authorative descriptions of metalworking processes given by Theophilus include good detailed descriptions of how to purify copper, how to alloy it with the tin, and how to make brass by the cementation process.[39] The preceding section in Theophilus on copper smelting is much more sketchy and full of error, to quote from C. S. Smith's notes, "as with all the Theophilus sections describing the preparation of raw materials with which he had no direct contact."[82] The implication is that Theophilus was familiar with and expected to perform all the operations from the purification stage on to the completed object. Recent analysis by the author of some Romanesque metalwork has confirmed this view. Analysis of all the copper-based components of the eleventh-century shrine of St. Walburg, now in the Gronigen Museum, Holland showed that just one source of copper had been used throughout to make the cast alloys, the copper for repoussé panels and the copper trays to hold enamels inset into the castings (table 77). The only explanation is that the craftsmen started off with one piece of copper and purified or alloyed it to suit their particular requirements. Another example is given by the twelfth-century dishes excavated at Bugle Street, Southampton, England and examined by the author (table 77). One dish is of bronze and the other of brass, but the trace element patterns are identical and sufficiently unusual to demonstrate that the different alloys had been made from one piece of copper. Once again the only explanation is that both alloys and vessels were made in the same workshop.

Specialized mass production was a distinctive feature of the Industrial Revolution contrasting with the individual craft production of earlier times. There are hints of this specialization in Pliny's *Natural History*, which states that Aegina had formerly specialized in the production of the upper parts of candelabra, whereas the stems had been fabricated in Tarranto. Although it is impossible to recognize candelabra specifically from these two towns one can test the general statement. Candelabra of fairly standard design were produced in large numbers by the Etruscans and Romans. They usually consisted of a number of cast components (base, stem, crown, etc.) which were rivetted or soldered together. When analyzed (table 51) the trace element patterns of the various components making up a given candelabra show a marked similarity (pastiches apart), and demonstrate that the components were cast from the same supply of copper (i.e., in the same workshop) and could not have been produced in specialist workshops many miles apart. Thus the analysis of the surviving objects does not support Pliny's statement and suggests the candelabra were the product of a craft industry rather than representative of incipient mass production.

References

1. P. R. S. Moorey, *Catalogue of the Ancient Persian Bronzes in the Ashmolean Museum*, (Oxford: Oxford University Press, 1971).

2. R. J. Gettens, *The Freer Chinese Bronzes, Vol. II Technical Studies* (Washington: Smithsonian Institution, 1969).

3. N. Barnard, *Bronze Casting and Bronze Alloys in Ancient China* (Canberra: Monumenta Serica, 1961).

4. L. S. Chuang and L. S. Kwong: Appendix to the *Catalogue of Bronze Seals of the Art Gallery of the Chinese University of Hong Kong* (Hong Kong: Chinese University, 1980).

5. G. Tanabe, "A Study of the Chemical Compositions of Ancient Bronze Artifacts Excavated in Japan," *Journal of the Faculty of Science University of Tokyo,* Section V, Vol. II, pt. 3 (1962), pp. 263-317.

6. J. Riederer, Zur Metallanalyse der Statuetten in *Das Bild des Buddha*, ed. H. Uhlig (Berlin: Safari Verlag, 1979).

7. O. Werner, *Spektralanalische und Metallurgische Untersuchungen an Indischen Bronzen* (Leiden: Brill, 1972).

8. J. Liszak-Hours, "Stude en Laboratoire d'Objects Himalayens en Metal du Muse Guimet," *Annales du Laboratoire de Recherche des Musees de France* (1982): 28-79.

9. M. Picon, J. Condamin, and S. Boucher, "Recherches techniques sur des bronzes de Gaule I-IV," *Gallia 24, 25, 26 & 31* (1966, 67, 68 and 71), pp. 189-225, 153-168, 245-278 and 157-178.

10. M. H. P. Den Boesterd and E. Hoekstra. *Spectrochemical analyses of The Bronze Vessels in the Rijksmuseum G M Kam at Nijmergen* (Leiden: Brill, 1966).

11. L. H. Cope and H. N. Billingham, Various reports and notes on 3rd-4th century coinage appeared in the *Bulletin of the Historical Metallurgy Group* between 1967 and 1971.

12. O. Werner, "Analysen mittelalterlicher Bronzen und Messinge I and II-III," *Archäologie und Naturwissenschaften I and II* (1977 and 1981): 144-221, 106-171. "Analysen mittelalterlicher Bronzen und Messinge IV," *Berliner Beiträge zur Archäometrie 7* (1982): 144-221.

13. O. Werner, "Westafrikanische Manillas aus deutschen Metallhütten," *Erzmetall 29* (1976): 447-453.

14. These are as yet unpublished but exist as student dissertations, or as reports awaiting publication. The work of C. Caple on medieval pins, and the work of D. A. Brinklow on copper alloys from York are available at the School of Archaeological Science, University of Bradford BD7 10P. The work of R. White on copper alloys from Flaxengate, Lincoln is available at the Department of Conservation and Material Science, Institute of Archaeology, University of London. The Ancient Monuments Laboratory Fortress House, 23 Saville Row, London W1, has also been active in this field, and acts as a clearing house of information (contact J. Bayley).

15. *Berlinger Beiträge zur Archäometrie Vol. 1-7* (Berlin: Rathgen Forshungslabor, n.d.)

16. R. F. Tylecote, H. G. Ghaznavi, and P. J. Boydell, "Partitioning of Trace Elements, During the Smelting of Copper," *Journal of Archaeological Science 4* (1977): 305-333.

17. N. H. Gale and Z. A. Stos-Gale, "Bronze Age Copper Sources in the Mediterranean, a New Approach," *Science 216* (1982): 11-19. It should be pointed out that the work of the Gales includes careful investigation of the production sites as well as the artifacts, and the comments on lead isotope analysis carried out in isolation do not apply to this work.

18. N. H. Gale, pers. comm.

19. N. C. Goucher, J. H. Teihet, K. R. Wilson, and T. J. Chow, "Lead Isotope Studies in Metal Sources from Ancient Nigerian Bronzes," *Nature 262, 5564* (1976): 130-131.

20. B. Rothenberg, R. F. Tylecote and P. J. Boydell, *Chalcolithic Copper Smelting*, I A M S monograph no. 1 (London, 1978).

21. H. G. Conrad and B. Rothenberg, *Antiques Kupfer im Timna-Tal.* (Bochum: Der Anschnitt Beiheft, 1980).

22. S. R. B. Cooke and S. Aschenbrenner, "The Occurrence of Metallic Iron in Ancient Copper," *Journal of Field Archaeology 2* (1975) 251-266.

23. A. Burnett and P. T. Craddock, "Italian Currency Bars," in J. Swaddling ed. *Italian Iron Age Artefacts in the British Museum* (London: British Museum Publications, 1985).

24. J. F. Merkel, *Bronze Age Copper Smelting Experiments based on Timna, Israel* (PhD thesis, Institute of Archaeology, London: 1982).

25. K. T. M. Hegde, "Analytical Study of Paunar Coins," *Journal of the Numismatic Society of India 37,* (1975): 180-183.

26. Pliny, *Natural History* (Book 34.40).

27. A. M. Snodgrass, *The Dark Age of Greece* (Edinburgh: Edinburgh University Press, 1971).

28. J. Riederer, "Die Naturwissenschaftliche Untersuchung der Bronzen der Staatlichen Sammlung," *Berliner Beiträge zur Archäometrie 7* (1982): 5-34.

29. P. T. Craddock, "Copper Alloys Used by The Greeks," *Journal of Archaeological Science 3* (1976): 93-113.

30. J. R. Bourhis and J. Briard, *Analyses Spectrographiques D'objects Pré-historiques et Antiques* I-IV (Rennes: Universite de Rennes, 1961, 1972, 1975 and 1979).

31. P. T. Craddock, "Deliberate Alloying the Atlantic Bronze Age," in M. Ryan ed. *The Origins of Metallurgy in Atlantic Europe.* (Dublin: Stationery Office, 1980), pp. 369-385.

32. M. A. Smith and A. E. Blin-Stoyle, "A Sample Analysis of British Middle and Late Bronze Age Materials using Optical Spectrometry," Proceedings of the Prehistoric Society 25 (1959): 188-208.

33. S. Prakash and N. S. Rawat, *Chemical Study of Some Indian Archaological Antiquities* (Bombay: Asia Publishing, 1965).

34. P. T. Craddock, "The Copper Alloys of Tibet and their Background," in ed. W. A. Oddy and W. Zwalf, *Aspects of Tibetan Metallurgy* (London: British Museum Occasional Paper, 1981), pp. 1-33.

35. S. Bowman *pers. comm.*

36. P. T. Craddock, "Copper Alloys Used by the Greeks II," *Journal of Archaeological Science 4* (1977): 103-123.

37. W. T. Chase and U. M. Franklin, "Early Chinese Black Mirrors and Pattern Etched Weapons," *Art Orientalis* 11 (1979): 215-258.

38. Theophilus, *On Divers Arts.*

39. M. Leoni, "Metallographic Investigation of the Horses of San Marco" in ed. G. Perocco *The Horses of San Marco* (London: Thames and Hudson, 1979), pp. 191-200.

40. W. A. Oddy, L. Borrelli and N. D. Meeks, "The Gilding of Bronze Statues in the Greek and Roman World," in ed. G. Perocco *The Horses of San Marco* (London: Thames and Hudson, 1979), pp. 182-190.

41. K. Assimenos, "Technological and Analytical Research on Precious Metals from the Chamber Tomb of Phillip II (Vergina)," in preprints of *The 2nd Internationales Symposium Historische Technologie der Edelmetalle",* (Meersburg, 1983).

42. W. A. Oddy, "Gold in Antiquity," *Journal of the Royal Society of Arts* (October 1982): 1-14.

43. P. T. Craddock, "Composition of Copper Alloy Used by the Greeks III," *Journal of Archaeological Science 5* (1978): 1-16.

44. F. R. Caley, *Orichalcum and Related Ancient Alloys,* (New York: American Numismatic Society Notes and Monographs no. 151, 1964).

45. R. Halleux, "L'orichalque et le Laiton," *Antique Classique 42* (1973): 64-81.

46. P. T. Craddock, L. K. Gurjar, and K. T. M. Hegde, "Zinc Production in Medieval India," *World Archaeology 15* no. 2 (1983): 211-217.

47. Strabo, *Geography Book XIII, 56.*

48. M. Farnsworth, C. S. Smith, and J. L. Rodda, "Metallographic Examination of a Sample of Metallic Zinc from Ancient Athens," *Hesperia (Supplement 8)* (1949): 126-9.

49. P. T. Craddock, A. M. Burnett, and K. Preston, "Hellenistic Copper Base Coinage and the Origins of Brass," in ed. W. A. Oddy *Scientific Studies in Numismatics* (London, British Museum Occasional Paper 18, 1980).

50. J. Day, *Bristol Brass* (Newton Abbot, England: David and Charles, 1973).

51. O. Werner, "Uber das Vorkommen von Zinc und Messing im Altertum und im mittelalter," *Erzmetall 23* (1970): 259-269.

52. J. Percy, *Metallurgy* (London: Macmillan, 1861).

53. R. Watson, *Chemical Essays 4* (1786): 45-8.

54. Marco Polo, *The Travels.*

55. A good description of Islamic brass making is given in J. W. Allan, *Persian Metal Technology, 700-1300 A.D.* (London: Ithaca Press, 1979).

56. J. W. Barnes, "Ancient Clay Furnace Bars from Iran," *Bulletin of the Historical Metallurgy Group 7.2* (1973): 8-17.

57. J. Walsh, "Galen Visits the Dead Sea and the Copper Mines of Cyprus," *Geographical Society of Philadelphia Bulletin 29* (1927): 93-111.

58. P. T. Craddock, I. C. Freestone, L. K. Gurjar, K. T. M. Hegde, I. V. Sonawane, "Early Zinc Production in India," *Mining Magazine* (January 1985): 45-52.

59. P. Ray, *History of Chemistry in Ancient and Medieval India* (Calcutta: Indian Chemical Society, 1956).

60. J. Marshall, *Taxila.* (Cambridge: Cambridge University Press, 1951).

61. R. F. Tylecote, "Ancient Metallurgy in China," *Metallurgist and Materials Technologist 15 no. 9* (September 1983): 435-439.

62. P. T. Craddock, "The Copper Alloys of the Medieval Islamic," *World Archaeology* 11 no. 1 (1979): 68-79.

63. R. Brownsword and E. F. H. Pitt, "Alloy Composition of Some Cast 'Latten' Objects of the 15th-16th Centuries," *Journal of the Historical Metallurgy Society, 17,1* (1983): 44-49.

64. W. Rajpitak and N. J. Seeley, "The Bronze Bowls from Ban Don Ta Phet, Thailand," *World Archaeology* 11 no. 1 (1979): 26-31.

65. Anon., *Mappae Clavicula,* Chapter 79.

66. Pliny, *Natural History,* Book 40.94.

67. *The Bible,* Book of Ezra 8, line 27.

68. J. Needham, *Science and Civilisation in China Vol. 5 Part 2.,* (Cambridge: Cambridge University Press, 1974) pp. 225-236 and 252-268.

69. M. E. L. Mallowan, *The Development of Cities,* fascicle from Vol. I Chapter 8 Part 1 of the revised Cambridge Ancient History. (Cambridge: Cambridge University Press 1967) esp. p. 66.

70. N. Seeley pers. com.

71. C. F. Cheng and C. M. Schwitter, "Nickel in Ancient Bronzes," *American Journal of Archaeology* 61 (1957): 351.

72. Albert Magnus, *The Book of Minerals,* Book 3 Tractate 2 Chapter 3.

73. J. Murphy-O'Conner, "Corinthian Bronze," *Revue Biblique 90 no. 1* (1983): 80-93.

74. Pliny, *Natural History,* Book 34.3.

75. *Zosimus III,* 16.5.

76. P. T. Craddock, "Gold in Antique Copper Alloys," *Gold Bulletin 15.2* (1982): 69-72.

77. The best discussion of the relevant passage in Pausanias is in E. R. Caley, "The Corroded Bronze of Corinth," *Proceedings of the American Philosophical Society 8.5* (1941), pp. 689-761, esp. pp. 749-751.

78. E. R. Holmyard, *Alchemy* (Harmondsworth: Pelican Books 1957).

79. Pliny, *Natural History,* Book 34.41.

80. Pliny, *Natural History,* Book 35.20.

81. Th. Berthoud, S. Bonnefous, M. Dechoux, and J. Franciox, "Data Analysis: Towards a Model of Chemical Modification of Copper from Ores to Metal" in P. T. Craddock, ed., *Scientific Studies in Early Mining and Extractive Metallurgy (London: British Museum Occasional Paper 20, 1980), pp. 87-102*

82. *Theophilus, On Divers Arts,* C. S. Smith, trans., (Chicago: University of Chicago Press) p. 139.

83. Pliny, *Natural History,* Book 34.6.

I. R. SELIMKHANOV

Arsenical Copper: Some Results from Optical Emission Spectrochemical Studies

A large number of ancient bronze artifacts has been excavated in the territory of Azerbaijan SSR and from analytical study of these it is the intention to establish the date of origin of the bronzes and further to determine what metals and alloys preceded the use of tin-bronze.

It is worthy of note that a number of publications[1] report local sources of tin in the Caucasus, but our studies have demonstrated such claims to be without scientific foundation.[2]

For the study of ancient metallurgy involving systematic historico-analytical investigations we have found the use of optical emission spectrochemistry invaluable. Our work commenced in 1952-53 with the development of analytical procedures for quantitative spectroscopy reported by the author in 1960.[3] It should be mentioned that previous examinations of ancient metal artifacts of Azerbaijani origin were conducted by wet chemical procedures with limited success and indeed, no clear characterization of ancient Azerbaijani metals resulted.

Notes on Spectrochemical Analysis

Using optical emission spectroscopy quantitative determinations of the principal elements of copper and copper alloys were made. When, however, the alloying ingredients totaled more than 10%, the sampled material was diluted with pure carbon powder. For minor and trace metal ingredients semi-quantitative studies prevailed.

The spectrochemical procedure used 10mg specimens with excitation produced by an AC arc discharge. The lower concentration limits (i.e. detectability) in percentile values (%) for impurity elements are:

Sn	Pb	As	Sb	Ag	Au	Bi	Ni	Co
0.001	0.001	0.02	0.005	0.0005	0.0005	0.005	0.0005	0.002

It is to be noted that care and circumspection are required in the interpretation of data obtained from highly oxidized metals derived from ancient settlements or burial grounds by reason of the exchange of oxidation states with the ambient surroundings. Such a circumstance could prevent the detection of arsenic or antimony even though originally such metals were present. To take into account such possibilities it is essential that conditions under which each artifact was found be duly recorded, including: (1) weight; (2) discovery site (e.g. settlement or burial); (3) condition of metal artifact (slight, partial, or complete oxidation); and (4) sample analyzed (metal, oxide or patina).

By reason of such considerations, it is valid to ask whether the copper beads and pendants discovered in the historically important neolithic settlement at Catal Hujük in Anatolia are really of pure copper.[4]

To further illustrate this point we cite the analysis of an arrowhead discovered in the Kiultepe Hill which was in better condition than some of the Catal Hujük artifacts, and observe that the green patina of the arrowhead contained 0.2% arsenic whereas the base metal of the object had 1.4% arsenic. Other metal impurities in the patina and base metal were:

	Sn	Pb	As	Ag	Bi	Ni	Fe
Patina %	0	0.05	0.2	0.002	0.003	0.003	1.0
Base metal %	0.003	0.15	1.4	0.1	0.02	0.005	0.15

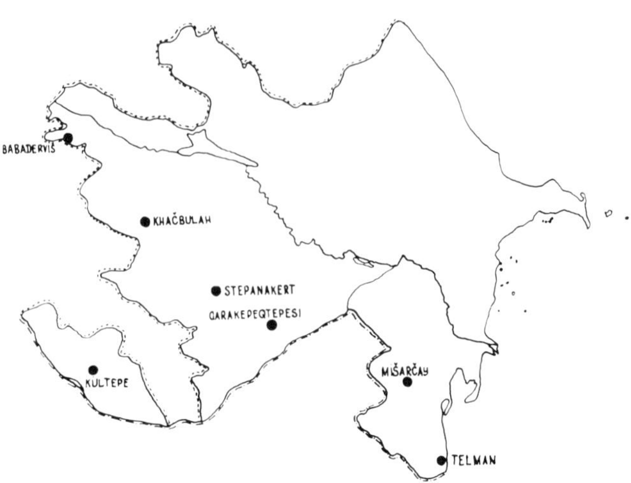

Fig. 1. Sites of most early metals in Azerbaijan SSR.

It is important to note that one ancient copper bead weighing 1.2g was totally oxidized and no impurity elements were detected except for iron. Presumably, other impurity elements were washed out by ground water. The conclusion that the bead was made from native copper would be uncertain.

Investigation: Stage 1

Our first studies in 1953 were of the oldest Azerbaijani artifacts recovered from monuments dating to the third millennium B.C. Archaeologists accepted these as belonging to the Copper Age. The monuments, namely two kurgans in Stepanakert and one on the right bank of the Khachinchay River, are in the Nagorno-Karabakh Autonomous Region. A third monument on Kiultepe Hill is in the Nakhichevan Autonomous Region in Azerbaijan.

On examination, the objects—eight from Kiultepe and two from Stepanakert—proved to be arsenical copper. The arsenic concentration ranged from 1.74 to 6.09% for the Kiultepe artifacts and from 1.64 to 2.11% for those from Stepanakert. Tin was absent or of insignificant concentration in all objects as were other trace elements in the copper base.

An artifact in the form of a copper blade from Khachinchay, on the other hand, was of complex composition containing 1.82% Sn, 2.26% As, 1.22% Pb, and 1.38% Sb. Accordingly this object was reclassified as dating from the Bronze Age and not the Copper Age, as formerly accepted. The examinations thus established (1) there is no evidence for a pure Copper Age in Azerbaijan; (2) the most ancient metals yet discovered are of arsenical copper; and (3) the arsenical "Copper Age" in Azerbaijan dates to the third millennium B.C.

Investigation: Stage 2

More recently, continued archaeological excavations have produced artifacts dating to early in the fourth millennium B.C. as well as others dating to the third millennium B.C. The older objects came from a site on Kiultepe Hill 18.2 meters from the crown and ^{14}C dating procedures at that stratum level yield a date 3807 ± 90 B.C. Six artifacts were obtained from that stratum whereas a seventh object, retrieved from a higher stratum, is most probably of later origin. The analysis of this particular object yielded 1.15% As, 1.6% Ni, and 0.2% Sb and since it is known that there are no copper-nickel deposits in Transcaucasia, it is supposed that the object in question was imported.

Of the six objects from the fourth millennium, four were excluded from examination on the basis of small size (from 0.3 to 2.5g) and because they were totally oxidized. The remaining two artifacts, one an arrowhead and the other unidentifiable, were in a better condition and analyses revealed copper with 0.4 to 1.4% arsenic. Since high-grade deposits of arsenical ores including realgar and orpiment exist in the close vicinity of Kiultepe Hill, it may be presumed that the objects were of local origin. Nonetheless no crucibles, molds, or smelting artifacts have been found.

Data from Other Locales

The earliest artifacts in Armenia were excavated from Tekhuta Hill near Echmiadzin dating to 3800-3600 B.C. The objects, in the form of a knifeblade or spearhead, a drill, and two awl fragments, were subjected to spectrochemical analyses and were shown to be arsenical copper alloys with concentrations of 3.6 to 5.4% arsenic. Given their advanced state of corrosion it is possible that the arsenic concentration was originally higher.[7]

In Iran early use of arsenical copper alloys is reflected in the composition of tools excavated from a hill settlement at Tal-i-Yahya near Kerman. The artifacts, a chisel and an awl, were examined by Tylecot and McKerrel,[7] who report the alloys contained 0.3 to 3.7% arsenic and levels of tin and other metals not considered as significant to the alloy identification. The artifacts dated to the fourth millennium B.C.

Two spatulas were recovered from the same locale; one of them dated to ca. 3500 B.C. and the other to ca. 3000 B.C. Notably there is a recognizable change in alloy composition since the older object contains 1.7% arsenic with negligible tin whereas the more recent artifact contained 1.19% arsenic and 3% tin. H. H. Coghlan[8] considers this period to be transitional between the use of arsenical copper and arsenical-tin bronze.

Metals of the Third Millennium B.C.

Excavations at several Azerbaijani sites in the regions of Fizuli, Astara, Kazakh, Astrakhanbazar, Kedabek, and Dashkesan, and at sites in Nagorno-Karabakh Autonomous Region and in Nakhichevan Autonomous Region, have produced metal artifacts that date throughout the third millennium B.C. Nevertheless, only at sites in the Astara region were fragments of crucibles and casting molds excavated.

Spectrochemical analysis revealed that all metal artifacts were of arsenical copper although some of them also contained 0.6 to 1.6% nickel and two objects also contained lead. At sites that dated to the end of the third millennium some gold and silver artifacts were also excavated.

Conclusion

Analytical investigations of ancient Azerbaijani metals recovered from archaeological excavations have shown that the oldest metals

in the form of small tools or ornaments, dating to the first half of the fourth millennium, were arsenical copper alloys. At the beginning of the third millennium arsenical copper alloys with lead and other copper-arsenic alloys with up to 1.6% nickel appeared. Gold and silver ornaments appeared as the third millenium drew to a close.

Arsenical copper tools and objects were produced from local raw materials, as indicated by the excavation of fragments of crucibles and casting molds.

It is to be noted that our investigations yield no evidence to support medieval claims for the use of elemental arsenic in Transcaucasia. These written claims are most likely to refer to the use of the arsenical minerals orpiment or realgar.

Fig. 2. Chronology of metals used in ancient Azerbaijan.

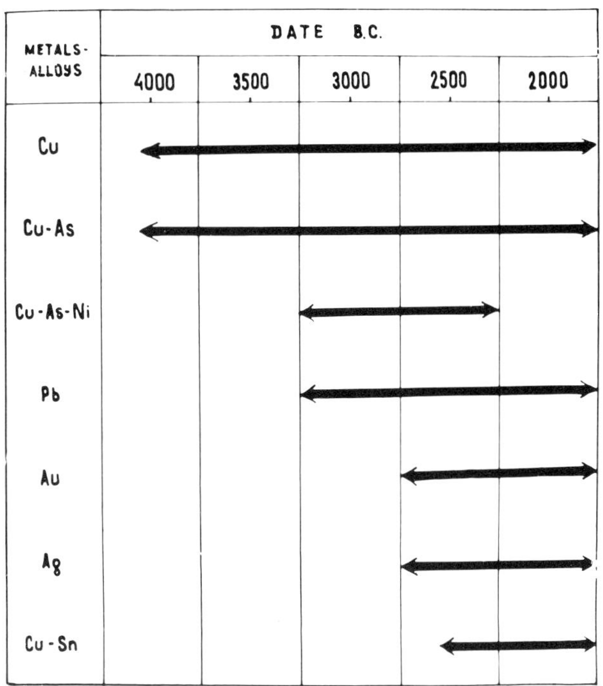

References

1. P. Knaurth et al., *The Metalsmiths* (New York: Little Brown and Co., 1974) p. 57.

2. M.-A. Kashkay, I. R. Selimkhanov, "O Himicheskoy Chrakeristike Nekotorih Predmetov iz Kuvshinnih Pogrebeniy Drevnego Mingechaura," *Izvestiya Akademii nauk Azerb. SSR*, No. 9 (Baku: 1954) pp. 21-38.

3. I. R. Selimkhanov, *Istorico-himicheskiye i Analiticheskiye Issledovaniya Drevnih Predmetov iz Mednih Splavov* (Baku: 1960).

4. H. Neuninger, R. Pittioni, W. Siegl, "Frühkeramische Kupferlegierungen," *Archaeologia Austriaca*, Bd. 35 (Vienna: 1964) pp. 98-100.

5. I.R. Selimkhanov, "K Istorii Razvitiya Matalloobrabotki i Gornorudnogo dela v Azerbaijane, *Vestnik Akademii nauk SSSR*," No. 9, (Moscow: 1958) pp. 56-57.

6. I. R. Selimkhanov, R. M. Torosian, "Metallograficheskiy analiz drevneyshih metallov v Zakavkazie," *Sovetskaya arheologiya*, No. 3 (Moscow: 1969) pp. 229-234.

7. R. F. Tylecot and McKerrel, "An Examination of Copper Alloy Tools from Tal-i-Yahya, Iran," *Bulletin of the Historical Metallurgy Group*, 5, No. 1 (London: 1971) pp. 37-38.

8. H. H. Coghlan, "Notes on The Prehistoric Metallurgy of Copper and Bronze in the Old World," *Occasional Papers on Technology*, 4, Second Edition (Oxford: 1975).

W. T. CHASE

Compositions of Oriental Bronzes: Another Look

Much data has recently become available from excavations in the
People's Republic of China and from publications about the compositions of early bronze objects in the Asian area. This data bears on
questions such as the early history of zinc. colors and properties of
the bronzes as they were originally made. groupings of bronzes by
composition, ritual urns of bronze versus objects made to function.
etc.

Using ternary diagrams. the available data was presented and
comparisons made. The author hopes to publish a definitive compi-
lation of analyses of early Oriental bronzes. Computer techniques for
graphic plotting of three-and four-dimensional data were also discussed.

JACK OGDEN

Potentials and Problems in the Scientific Study of Ancient Gold Artifacts

The scientific study of ancient jewelry has been prompted largely in order to establish criteria by which we can determine provenance and age of the objects concerned.

With gold objects, the use of analysis of major or trace elements to investigate their provenance is hampered by the fact that gold has always been traded, mixed, alloyed, refined, and reused. Only in the case of an isolated primitive society (or a modern industrialized one!) can we expect any consistency in the composition and trace element ratios of goldwork. Some grouping by trace element and lead isotope ratios might be possible with, say, Bronze Age societies with limited sources of supply, but with the major trading civilizations all we can hope for is some unity of overall purity due to state control. Thus the gold-silver-copper ratios of Roman and post-Roman goldwork can be used to support other evidence for their dating but any groupings on the basis of trace elements would be unexpected except in the most limited of cases.

Additionally, there is a dearth of relevant geological and mineralogical data about the sources themselves—many of the sources of archaeological significance are, of course, only of minimal importance to economic geologists today. As an example, ancient goldwork frequently contains small visible gray metallic inclusions that are essentially alloys of the platinum group metals: osmium, iridium, ruthenium, and platinum (fig. 1). Initial perusal of mineralogical textbooks suggested a limited number of possible sources of gold with such platinoid content and thus it seemed feasible to draw trade routes on maps. A thorough investigation, however, eventually showed that there were few, if any, Old World placer gold sources that had not at some time or other been found to contain platinoid grains.[1] The only conclusion that could be drawn was the surprisingly great extent to which gold, even in Roman and Byzantine times, was derived from placer rather than vein deposits. The inclusions also turn up in fakes.

It is usually possible to expose fakes of ancient jewelry on purely scientific grounds.[2] Unfortunately the current state of our knowledge —and the skill of forgers—means that we can seldom scientifically *prove* a gold item to be ancient. Examination of materials and techniques can enable us to state that an object is genuine to a very high level of probability; but the final leap to asserting its genuineness is still usually a subjective one often left to the art historian with the necessary stylistic knowledge. The presence of this "objectivity gap" is more obvious with gold jewelry than with many other categories of antiquity, particularly those where scientific dating is possible.

The scientist, looking at possible fake jewelry, often works from a mental checklist of a dozen or so criteria that indicate spuriousness. In many cases a single "positive" identification can prove the piece to be fake, as, for example, the presence of more than a minute trace of cadmium in a gold alloy[3] or the use of electroplating to give a matt, pure surface, often revealed by the presence of gilding blisters (fig. 2).

The researcher can thus set about demonstrating an object to be a fake in a systematic and logical way. Unfortunately, in most cases, the best he can do to establish genuineness is to get negative results on his tests for forgery! The problem is that whilst there are techniques used today that were not used in antiquity there are few, if any, procedures used in ancient times that have not been at least attempted by recent forgers. In the case of alloy composition the com-

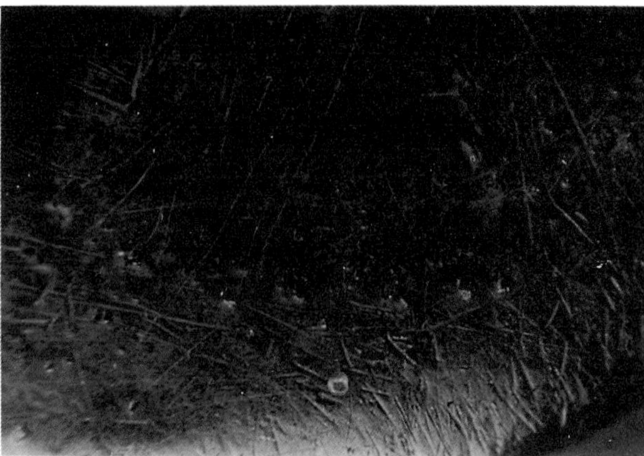

Fig. 1. Detail of the bezel of a Hellenistic gold ring showing small silver gray inclusions of osmium-iridium alloy grains.

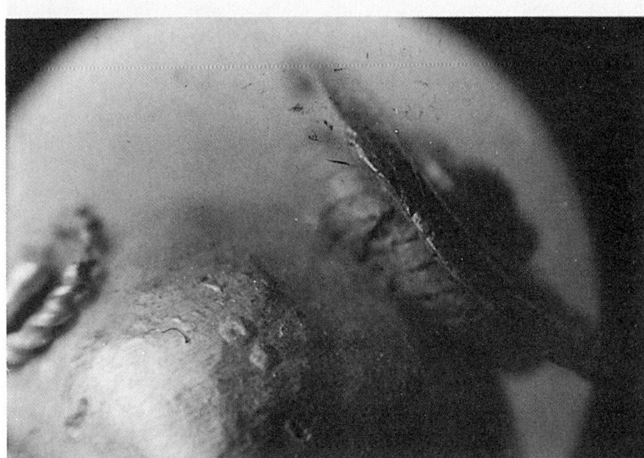

Fig. 2. Detail of the pendant from a recent fake "Greek" gold earring showing small blisters resulting from the use of electro-gilding to give a rich matt color to the gold.

mon re-use of ancient damaged or valueless items means that fakes can have plausible compositions without any deliberate matching of alloys on behalf of the forger.

The manufacture of ancient wire is a good example of the potential dangers of using isolated techniques to establish authenticity. Wire drawing probably does not predate the Late Roman period and so its use in gold jewelry (usually identified by the longitudinal striations along the wire) is frequently an indication of spuriousness. It is general knowledge today (and well published) that gold wires in antiquity were usually made by processes involving twisting a strip of gold; these wires have characteristic spiral "seam lines" (fig. 3). Curators and others still take the presence of such "spiral seams" as proof of age but in fact forgers have been copying such handmade wires for at least twenty-five years (fig. 4).

As forgers improve their skills, it becomes necessary to increase the number of criteria by which we make our judgments. Thus we should consider not only the materials and techniques but also the object's historical, stylistic, and craftsmanship qualities.

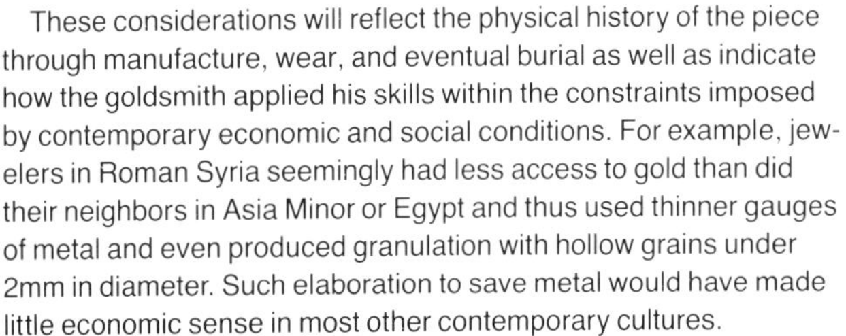

Fig. 3. A Roman earring pendant shows the spiral "seams" caused by the twisting manufacturing process.

Fig. 4. Three different attempts at producing ancient-style wires with spiral "seams," (from the same item as figure 2).

These considerations will reflect the physical history of the piece through manufacture, wear, and eventual burial as well as indicate how the goldsmith applied his skills within the constraints imposed by contemporary economic and social conditions. For example, jewelers in Roman Syria seemingly had less access to gold than did their neighbors in Asia Minor or Egypt and thus used thinner gauges of metal and even produced granulation with hollow grains under 2mm in diameter. Such elaboration to save metal would have made little economic sense in most other contemporary cultures.

A preliminary study of the materials and techniques of ancient jewelry has been published by the author,[4] and is in the process of being extended to include some of the aspects of early goldwork discussed above. A starting point with relevance and potential for forgery detection is an examination of the possible changes in appearance of ancient gold caused by burial and time. The following is a brief summary of the writer's research on the changes in buried gold objects.

Diffusion to the Alloy Surface

Dry environments (such as many Egyptian tombs)

Gold and silver form single-phase alloys of high chemical stability. Surface films of silver chloride are usually only to be found on very silver-rich alloys in the presence of salt (or other chlorides) in the ground. For example some Egyptian electrum items have a visibly crystalline silver chloride patina.[5] Silver sulphide films are usually the result of recent, rather than ancient, history.[6]

Iron and copper can migrate to the surface of ancient gold to form reddish iron oxides and mixed iron-copper oxides.[7] These can give an attractive patina and can act as barriers retarding further corrosion.

Wet Environments

When the gold item is in occasional or constant contact with an aqueous solution (such as ground water), attack can continue indefinitely when corrosion products are removed in solution as they are formed.

Corrosion in aqueous solutions is an electrochemical reaction, so in general the baser metals will be leached from the surface of the gold item when it is buried in moist soils. This process, which will give a purer, slightly matt, surface to the gold, is traditionally termed "surface enrichment" by archaeologists, though the term is used differently by surface scientists.[8]

Deliberate ancient pickling to improve color could match the effects of surface enrichment, but the existence of the natural processes is indicated by the similar surfaces on different parts of an item subjected to extensive wear and abrasion in ancient times, and the depletion of silver and copper from the rims of placer gold nuggets.

A study of small placer grains in cross-section with an electron microprobe has showed them to have porous rims of a very low silver content, even on grains with interior silver contents of up to 46%.[9] Copper was similarly depleted—sometimes none remained on the grain surface. The thickness of the rims depended on grain and source but most were between ten and twenty microns. This depletion process could be presumed to occur in any environment where gold is exposed to aqueous solutions not already saturated in silver or copper.

The steep gradients of the enriched rims show that any tendency to even out the composition by diffusion must be very slow compared to the rate of surface leaching. (With even a slight increase in temperatures, diffusion in gold-silver alloys becomes rapid. A simple illustration of this is the almost instant paling of much Roman gold when it is heated during repair; the silver present in the bulk of the gold diffuses to the silver-depleted surface.)

Surface enrichment has been detected on various ancient gold items using cross sections of damaged items, analysis before and after localized abrading of the surface, and analytical techniques that give differing degrees of penetration into the surface of the item.[10] The enriched rims seem to be in the order of ten microns thick. So far surface enrichment effects cannot be quantified, but recent analyses have detected differences in surface compositions between coins from the same mint but different burial environments.[11] On the other hand differing enrichment gradients have been found on gold items from the same burial.

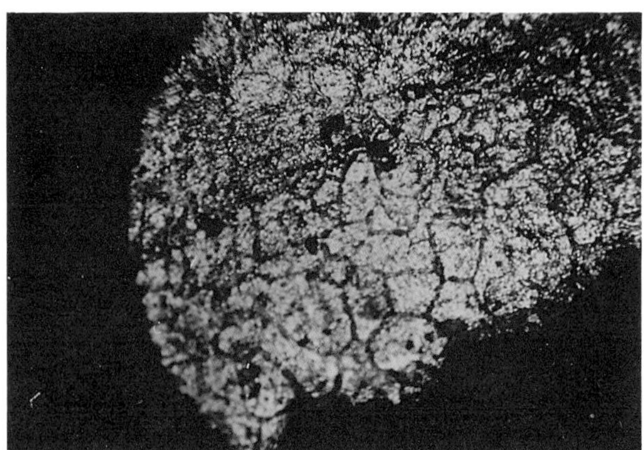

Fig. 5. The surface of an Anglo Saxon gold pendant showing grain boundary etching probably due to burial effects.

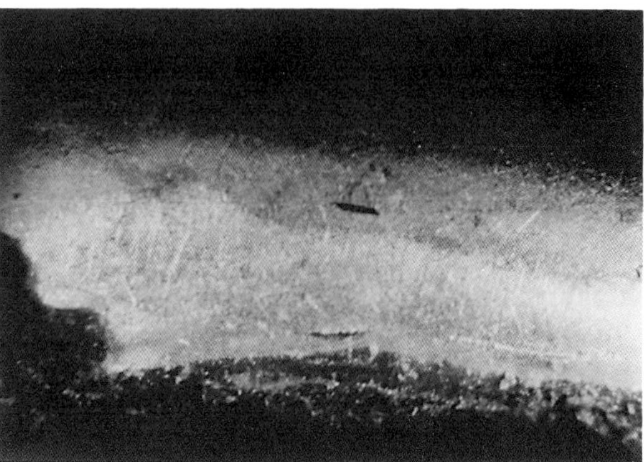

Fig. 6. Detail of ancient gold showing the paler color where copper corrosion products have been recently removed.

Copper in gold diffuses preferentially to and within the grain boundaries resulting in the grain boundary etching on many ancient gold items (fig. 5). Some items with relatively high copper content have such extensive intergranular corrosion that they crumble between the fingers. When stress is applied to many ancient gold items, cracks can develop along the grain boundaries that lie roughly at right angles to the direction of stress. Over archaeological time spans, cracking can also occur in solder joins between substantial thicknesses of gold where there are inherent cooling stresses and possibly localized higher copper levels due to the solder alloys used. Modern gold alloys of over about 80% purity are considered immune to stress corrosion[12] but it seems, as with cracking caused by distortion, that higher purity alloys are affected over long time spans.

Occasionally, under the microscope, it can be seen that entire grains seem to have been raised up from the gold surface. Presumably this is some type of strain reduction or even recrystallization; it does not seem to be related to modern distortion.

Cathodic Protection

Gold items with silver parts such as inlays or other decorations now often appear paler than other contemporary gold items, as the presence of almost pure silver has retarded the leaching of the silver component of the gold alloy.

Copper corrosion products often occur on ancient gold items buried in contact with copper alloys, either integral parts of the object, such as cores, or associated burial items. The gold surface under such copper corrosion products can retain fine tool marks seldom seen on unprotected gold surfaces and the color of the gold is often paler because the presence of copper has prevented the more noble silver from being leached out of the alloy (fig. 6).

Precipitation Coatings

In some burial environments, substances can be deposited onto the gold surface from the ground waters. Iron oxide or hydroxide is a very common example (as occurs on gold placer nuggets). This iron oxide layer can have a thick, almost resinous appearance, and probably represents limonite in the mineralogist's widest use of the term.

Small-scale botryoidal growths of calcium carbonate are another common coating, particularly on jewelry from parts of Turkey. As with corrosion films, these precipitated coatings tend to protect the gold surface. Under calcite coatings, ancient gold often retains a paler and often brilliantly shiny surface or shows the fine scratches that must represent original polishing or wear on the gold. Such markings are most unusual on ancient gold not so protected.

Solution and Reprecipitation of Gold

Any loss of gold due to chemical action in the ground water would be expected to show as a rounding off of the edges and points on the artifact—an effect commonly noted even on ancient goldwork without much other indication of wear. This is in contrast to the sharp edges and points so often noted on fakes.

The aggregation of isolated gold atoms into islands, perhaps with crystalline structure, has been predicted as a possible result of the aging of gold alloys.[13] Some Mytilene electrum coins showed definite gold-rich areas on the surface, possibly a chemical effect of this type.[14]

Calcite coatings on ancient gold sometimes contain minute gold particles dislodged by some natural process from the gold surface. The sandy soil attached to a South Arabian gold bead of about 200 B.C. contained small yellow, metallic crystals that were seemingly reprecipitated gold. Similar reprecipitated gold crystals have been noted in proximity to gold veins in mines.[15]

Attack by Organic Agents

There seems to be some visual evidence to suggest that gold jewelry corrodes in a special way if buried in contact with decaying animal matter such as its deceased owner. Detailed examination has not yet been made but it is known that some gold alloys can be corroded when in contact with human sweat and that some micro-organisms have the ability to slowly dissolve gold.[16] It is also theoretically possible for decaying vegetable matter to result in strongly acidic burial environments that could corrode gold.[17]

Ordering and Self Annealing

We might expect some formation of ordered compounds such as gold-copper to occur, particularly in the grain boundary regions. In one published case the lack of order has actually been taken to be indicative of the modern origin of a fake coin.[18] Because ordering is most likely to be found in copper-rich alloys, a recent unpublished study has concentrated on Pre-Columbian copper-rich gold items but no firm conclusions have been drawn. Ordering in modern ternary gold-silver-copper 18ct alloys is well documented. This area deserves more study.

Gold, when mechanically worked, hardens and becomes internally stressed. Equilibrium is restored by annealing. First the flexibility is regained and then, with increasing temperature, recrystallization takes place and eventually grain size increases. The first step—the return of flexibility—seemingly occurs at room temperature over long periods of time and accordingly, ancient gold, with low copper and other base metal levels, is usually soft and lacks springiness (whereas the contrary is often true of forgeries). Recrystallization has not so far been demonstrated to occur with time alone, though this might be expected and might even be evidenced in some of the phenomena briefly described above.

References

1. J. M. Ogden, "Platinum Group Metal Inclusions in Ancient Gold Artifacts," *Historical Metallurgy* 11, 2 (1977): 53-72.

2. For a recent study and full references see J.M. Ogden, *Jewelery of the Ancient World*. (London and New York: Trefoil Books, 1983.)

3. The presence of more than minute traces of cadmium in gold is usually taken as proof of recent manufacture despite some recent suggestions to the contrary, (some of which were based on analyses of items of dubious antiquity): see for example: G. Demortier, "New Approach of Ancient Gold Brazing", (paper to 2nd International Symposium, "Historische Technologie der Edelmetalle"), Meersburg, Germany, April 1983.

4. See note 2.

5. Dr. Gladstone in, *The Chemical News*, 11 January, (1901); A. Lucas, *Antiques, Their Restoration and Preservation* (London, 1932) pp. 108-110.

6. Even recent 14ct ($58.33°_0$) gold alloys can tarnish over a period of years and high-purity Roman gold rings uncleaned in collections for a generation can show tarnish very different from that due to burial.

7. D. Clark, T. Dickinson, and W. N. Mair, "The Interaction of Oxygen and Hot Air," *Trans. Faraday Society*, 55 (1959): 1937; A. F. Moodie, "A Contribution to the Analysis of the Effects Observed on Heating Gold in Air," *Acta. Cryst.* 9, (1959): 995.

8. For ancient gold see: E. T. Hall, "Surface Enrichment in Buried Metals," *Archaeometry*, 4 (1961), pp. 62-66. For the surface scientist's view see: R. Bouwman, L. H. Toneman, M. A. Boersman, and R. A. Van Santen, "Surface Enrichment in Ag-Au Alloys," *Surface Science*, 59 (1976), pp. 72-82; R. Bouwman, "Surface Enrichment in Alloys", *Gold Bulletin*, (1978), 11, 2.

9. G. A. Desborough, "Silver Depletions Indicated by Microanalysis of Gold from Placer Occurrences in the Western U.S.," *Econ. Geol.*, 65, (1970) pp. 304-311.

10. See for example: E. T. Hall, G. Roberts, "Analysis of the Moulsford Torc," *Archaeometry*, 5 (1962), pp. 28-37; C. Johns and T. Potter, *The Thetford Treasure* (London: British Museum Publications, 1983), p. 59.

11. J. F. Healy, "Greek White Gold Electrum Coin Series," *Metallurgy in Numismatics*, 1 (1980).

12. J. M. M. Dugmore, and C. D. Desforges, "Stress Corrosion in Gold Alloys," *Gold Bulletin*, 12 (1979), 4.

13. R. Bouwman, private communication, April 1979.

14. Healy, "Greek White Gold Electrum Coin Series."

15. J. M. Ogden, *Society of Jewellery Historians Newsletter*, 1 (1977).

16. G. E. Gardam, "Why Jewellery Sometimes Blackens the Skin or the Clothing," *Special Report*, 4 (1969), Technical Advisory Committee, the Worshipful Company of Goldsmiths; Rapson, W. S., "Effects of Biological Systems on Metallic Gold", *Gold Bulletin*, 15 (1982), 1:19-20.

17. W. G. Fetzer, "Humic Acids and True Organic Acids as Solvents of Minerals," *Econ. Geol.*, 41, (1946), pp. 47-56.

18. H. C. Bhardwaj, *Aspects of Ancient Indian Technology*, (Delhi, 1979) p. 124.

Atomic Emission Spectrographic and Scanning Electron Microscopic–Energy-Dispersive X-Ray Studies of European, Middle Eastern, and Oriental Metallic Threads

The identification and study of metallic threads has, in the past, been the province of the textile curator and conservator. These scholars attempt to accomplish their task through *in situ* visual observation of thread color in order to distinguish between threads of different metallic composition, and by microscopic observation of individual sampled specimens to determine thread construction. Using these techniques, four general types of metallic threads have been identified and reported:[1] (1) small-diameter solid metal wires; (2) narrow strips of gilded or ungilded metal, called lamellae, spun around a fine linen or silk thread, called the core (these strips were sometimes cut from thin sheets of metal, or produced by rolling or hammering gilded or ungilded wires); (3) thin sheets of metal applied onto animal membrane—such as leather—or paper, then cut into narrow strips and wound onto a core thread; and (4) gilt leather or paper strips, not wound on a core, but attached to the textile as flat strips.

Visual and microscopic determinations of thread and strip construction by *in situ* examination of textiles are difficult at best. The microscopic examination of sampled threads provides more information about thread type and strip construction, but neither of these methods provides a positive identification of the metallic composition of the thread (if it is a wire) or the strip. The best the textile expert can do is to speculate that two wires or strips are composed of different metals, or mixtures of metals, based on the respective colors of their surfaces. The situation is further complicated by the frequently subtle color differences. Thus, in order to be certain of metallic composition, small specimens must be submitted for material analyses.

To the best of our knowledge, only a few analytical studies of metallic threads have been reported.[2,3] The Indianapolis Museum of Art is fortunate in having an extensive collection of textiles, numbering some 7,000 pieces, with approximately one-third of them incorporating metallic threads. These pieces date from about 200 B.C. to the present, and represent many countries and cultures—Chinese and Japanese, Western European, Islamic, Indian, and pre-Columbian. It was therefore our purpose to investigate the construction and metallic composition used in the most interesting of these pieces with a combination of optical microscopic, atomic emission spectrographic and scanning electron microscopic–x-ray analytical methods. The types of information generated by these analyses, and the significance of the data in providing a more detailed picture of thread and strip composition and construction, will be discussed here. Fifty-four individual objects were sampled by the museum's textile curatorial and conservation staff, and as many different thread types as could be visually recognized were taken from each piece. Thus, specimens were chosen for study based on surface color differences (golden or silver), obvious differences in construction (wires, narrow strips wound on cores, and narrow strips without cores), and differences in use within the piece (woven into fabric, used in embroidery, or in the trim). In all, 124 specimens were submitted for analysis.

Analytical Studies: Microscopic Observation of Samples

Prior to the spectrographic analyses, 1-3mm samples of each specimen were studied under reflected tungsten light, at magnifications of either 40X or 100X, using a Nikon optical microscope. The specimens observed fell into one of the following general categories: (1) strips clearly exhibiting a thin golden-colored surface layer on top of a silver or black (corroded silver) substrate—strongly indicating a gilded silver strip; (2) golden-colored stips with surfaces so intact that no surface metal layer or metal underlayers were observed—suggesting either a very well preserved gilded silver strip, a gold metal strip, or a gold alloy strip; (3) silver-or black-colored (corroded silver) strips with no definite golden surface layer detected; (4) small-diameter apparently solid metal wires, either golden- or silver-colored; (5) strips with thin, golden-or silver-colored layers on much thicker fibrous backings (presumably of paper), and (6) strips consisting of thin golden- or silver-colored surface layers on top of a colored claylike material, which was itself on top of a fibrous (paper?) backing.

Although the metallic identification could not be determined by this method, these data were not only very helpful, but indeed often critical, to classifying threads in a later part of this study.

Atomic Emission Spectrographic Analyses

Immediately after microscopic analysis, each specimen was transferred to a graphite sample electrode, packed with pure graphite powder, carefully positioned in an Applied Research Laboratories Spectrographic Analyzer, and heated for two minutes at about 4000°F by means of a DC electric arc. The use of this instrument for qualitative and quantitative metal analyses has been discussed[4] and will not be reviewed here. In the present study, semiquantitative estimates of the relative amounts, i.e., major, minor, or trace, of the three most important metals present in these samples—gold, silver, and copper were obtained by comparison of spectral line intensities in the emission spectrum of each sample with the same lines resulting from the excitation of a series of standards containing equal weights of gold, silver, and copper. The most sensitive emission lines available for these elements were used.

From preliminary studies with gold-silver-copper standards it was found that the size of the vaporized specimen affected the absolute intensity of silver and copper spectral lines but not their intensity. For the gold lines, however, no significant increase in absolute intensity occurred with sample size even when this was increased by two orders of magnitude. Accordingly, to correct for this phenomenon estimates of gold content in thread samples were made by comparison with the standard of approximately equal size. Semi-quantitative determinations of the other elements, usually present in trace quantities, were estimated using the line sensitivity values given in the literature.[5]

Classification of Metallic Threads

The visual, microscopic, and compositional data generated by the methods described above were combined (see table) and analyzed in order to classify the strips. The first column gives the Indianapolis Museum of Art accession number of the textile sampled and the number and use description of the individual threads. The second column lists the provenance of each piece. The third column shows the color of each thread or strip as visually determined from a long, uncorroded length of thread during sample selection and removal. The microscopic description of each laboratory specimen submitted for analysis is given in the fourth column, followed by its elemental composition as determined by spectrographic analysis. The final column lists the class to which each thread was assigned, based on the foregoing data.

The various classes established were:
I. Gold metal layer on silver strip
II. Silver metal strip, containing other metals
III. Solid wire threads—of gold, silver or copper metal, with other metals present
IV. Strip made of other kinds of metals, or of metal alloys
V. a. Gold or gold alloy layer on fibrous backing strip
   b. Silver or silver alloy layer on fibrous backing
VI. a. Gold or gold alloy on claylike layer over fibrous backing strip
   b. .Silver or silver alloy on claylike layer over fibrous backing
   c. Other type of metal or metal alloy over claylike layer and fibrous backing
VII. Miscellaneous strips or threads

Specimens were considered to be of Type I if there was clear microscopic observation of a thin golden-colored layer over a silver, or a black corroded silver, strip. In some cases, observation of the back of these types of strips revealed a silver-colored surface that greatly helped to confirm the assignment. For the majority of cases, however, the back surface color of the strip was rendered very questionable by the presence of a thin layer of corrosion product that imparted a golden color to the surface under the lighting conditions employed. All Type I strips were spectrographically determined to contain a major amount of silver, a major or minor amount of gold, and a minor or trace amount of copper. These data demonstrated that the surface metal layer was indeed gold, and that the substrate was silver metal containing copper, perhaps as an impurity. The elements silicon, calcium, magnesium, and lead were also commonly found as impurities in such strips. There were twenty-one definite Type I strips identified. In many other cases, the strip surface was intact, and it could not be definitely determined that it was really a gold layer over a silver substrate. These kinds of specimens (twenty-five of them) were given the designation I or IV, since even though the metallic profile resembled definite Type I gilded silver strips, this composition could also indicate a strip made of an alloy of silver containing large enough amounts of gold and copper to give the material a golden to yellowish color. (The color of gold-silver-copper alloys of various compostions is discussed below).

The twenty-eight Type II specimens clearly exhibited either definite silver-colored or blackened surfaces, and contained major amounts of silver together with minor to trace amounts of gold and copper. With the exception of three Indian (38.63-#2, 33.735-#4, and 33.748) and one Chinese (69.69.2-#3) specimens, all silver strips contained appreciable amounts of one or both of these other metals. The presence of gold and copper could, of course, be attributed to accidental inclusion and lack of their removal during refining of the silver—but it is also possible that they were deliberately added, either to change the appearance of the strip from a bright to a softer yellowish silver hue, or to increase the hardness and durability of the strip (the reason that 10% copper is added to silver coinage). If

either or both of these modifications of the properties of these materials had been intended by its manufacturers, then these strips would have to be considered silver alloys. The presence of copper in the silver substrate underlying the gold layer in most of the Type I strips tends to argue against purposeful addition for color change, and for accidental inclusion, since the substrate was never meant to be seen. However, in both Type I and Type II strips, the decreased ductility imparted by the copper might be an important advantage in safely handling these thin, narrow, and thus very fragile strips when being wound on to core threads. Unfortunately, the data available do not allow us to distinguish between these possibilities.

Six examples of wire threads were found—two definitely gold-colored (76.165-#1, and 81.211-#3), one possibly gold-colored (82.60-#3) one a coppery gold (81.210-#1), and two definitely silver-colored (81.211-#8 and 80.378-#2). All of the golden-colored specimens contained major amounts of silver and minor amounts of copper. Two contained minor gold and one (81.211-#3), contained gold in a major amount—again making it uncertain whether yellowish to golden-colored silver alloys were in evidence, or alternatively, that gilded silver wires had been encountered. The two silver-colored wires contained major amounts of silver and minor of gold and copper. The coppery-gold colored wire is basically copper metal containing minor amounts of iron and zinc, and only a trace of silver. Now corroded, this wire may have originally resembled brass in color. The diameters of these wires were measured during microscopic analysis—the golden-and silver-colored specimens were approximately 0.06mm in diameter and the coppery (brass) wire was 0.33mm in diameter.

Thirteen golden to coppery-golden colored specimens are listed as belonging to the Type IV category of strips, based mostly on their metallic compositions. All but one of these strips, (33.735-#3) contained copper as the major metal. In this one case, silver was the major and copper the minor element present; gold was not detected. Five of these specimens (28.299-#1, 77.360-#1, 81.210-#2, 81.338, and 76.338-#3) were almost pure copper, with only trace amounts of other metals detected. The objects from which these samples came were, in all but one case, nineteenth- or twentieth-century textiles. The remaining specimen (81.210-#2) was taken from a Swiss piece dated to 1520. Two Indian strips (33.735-#5 and-#6) may also belong in this category. Both may very well be silver-copper-lead alloys, rather than silver strips with copper and lead impurities. Here again, the spectrographic data does not allow us to definitely choose between these two possibilities.

Type V strips were all clearly identified by microscopic observation. The thin layer on the strip surfaces was sometimes partially missing, and, where intact, was usually extensively cracked. The metallic layer appeared to have been applied directly on the backing, without the aid of an adhesive. In all, six such samples were encountered—all golden-colored and all from Japanese textiles. Gold was the major element in all but one specimen (33.504), and that one had a major silver concentration and a minor amount of gold and copper.

Type VI strips were also easily identifiable microscopically. In all but three specimens, a layer of what appeared to be a particulate, red-colored, claylike material was observed between the thin metallic surface layer and the fibrous backing. The other three specimens exhibited a yellow-colored, claylike material underneath the surface of the strips. The red underlayer was similar in appearance to red bole, a fine-ground clay containing small amounts of red Fe_2O_3 (hematite) and often used in the gilding of paintings, and of stone and wooden objects, to provide a smooth surface layer upon which the gold leaf is then applied. The detection of minor amounts of silicon in several of the larger specimens analyzed would tend to support this speculation. The apparent absence of iron is explained by the fact that most of the Chinese specimens submitted for analysis were extremely small—and therefore the iron, which would be present in rather small amounts in even the largest samples, escaped detection. Twelve of these strips were golden colored: the compositions of their metallic layers ranged from almost pure gold (76.541 and 62.204) and pure copper (80.43c-#1) to various mixtures of gold, silver, and copper. Only one Type VI silver-colored strip was provided for analysis (69.69.2-#4), and its metallic layer contained a major amount of silver, a minor amount of copper, and only a trace of gold.

Scanning Electron Microscopic X-Ray Analysis Studies

It has already been pointed out that the assignment of a particular strip to a given class on the basis of optical microscopic and spectrographic compositional data was not always definite. Most notably, some gold-colored strips could not be unequivocally determined to have gold metal layers on silver substrates and had to be identified as being either Type I or IV. Also, laboratory specimens of several strips that were silver-colored in the textile exhibited quite corroded surfaces, and were found to contain minor amounts of gold—making it difficult to decide whether these specimens were Type II strips containing some gold, or Type I strips with the underlying silver corrosion products covering the gold metal surface layer.

An attempt to resolve these problems, and to further our knowledge of the details of metal-strip construction, was made by submitting several representative specimens for Scanning Electron Microscopic (SEM) examination.

The instrument used in these studies was an Etec Autoscan Scanning Electron Microscope fitted with an energy-dispersive x-ray analyzer. The specimens to be examined were mounted in epoxy resin and sectioned perpendicular to the length of the thread with an Ultramicrotome employing a diamond knife. Successive thin slices (0.9 to 0.1 μm) were removed from the mounted specimen surface until, whenever possible, an uncorroded section was obtained. The sectioned specimen was then coated with a very thin layer of conducting material, placed in the instrument's vacuum chamber, and scanned with a high-energy beam of focused electrons, causing the generation of secondary electrons from the sample. These electrons reach a detector and the resulting signals produce a secondary electron (SE) image on an oscilloscope screen, yielding information about the surface of the sample from images of any desired magnification, up to about 1,000,000X.

In addition to secondary electrons being emitted, some electrons are also back-scattered from the incident beam by the sample. This allows element differentiation capabilities since elements having a higher atomic number back-scatter more electrons than lighter elements do. Thus, for example, a sectioned surface of a strip having a thin layer of gold on a silver substrate with the SEM operating in the "back-scatter" mode, would result in a back-scatter electron (BSE) image with the gold layer observed as a bright line next to a gray-col-

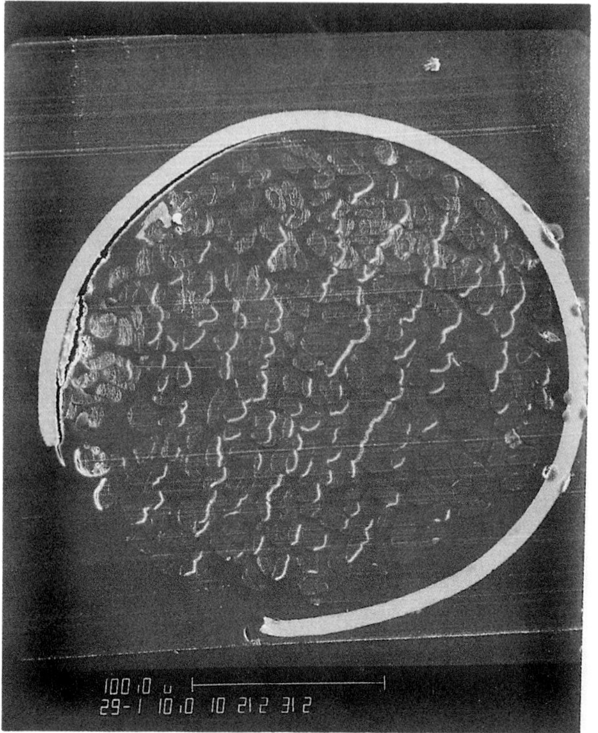

Fig. 1. 290X SE image of sectioned gold-colored thread 74.114-#1 (Italian, 14th-15th century)

ored layer, since gold has a much higher atomic number than silver (fig. 2). Furthermore, since the magnification of any SE or BSE image is known, the dimensions of features shown in the photograph of the image can be determined by simple measurement.

Another result of the interaction of the focused electron beam with the sample is the generation of x rays that are characteristic of the elements being excited, and the SEM fitted with an x-ray analyzer detects and separates emitted x rays on the basis of their various energies, to allow identification of the elements present. Also, the intensities of the x rays of different energies are used to determine the relative amount of each element present in the sample scanned.

The SE image (290X magnification) of a mounted and sectioned specimen of thread (74.114-#1) is shown in figure 1. The individual fibers of the core thread, as well as the cross-section of the strip, are clearly visible in this photograph. Measurements made from this photograph revealed that the diameter of the thread is of the order of 0.30mm. Figure 2 is a higher magnification (5000X) BSE image of the strip cross-section. The bright line on the left is the surface gold layer, as proved by x-ray analysis of the spots indicated on the photo. Analysis of area 1, the outermost part of the surface layer, and area 2, shows the presence of a large amount of gold, and only traces of silver and copper. Analysis of area 7 (just beyond the gold layer) reveals that only silver and a small amount of copper are present. Analysis of area 5 (back surface of strip) and area 6 (another part of gold layer) gave the same results as areas 7, 1, and 2, respectively. Area 4, the dark gray spot in the interior of the silver substrate, contains a much higher ratio of copper to silver than the lighter gray

Fig. 2. 5000X BSE image of 74.114-#1. Indicated area analyzed by energy dispersive x-ray analysis.

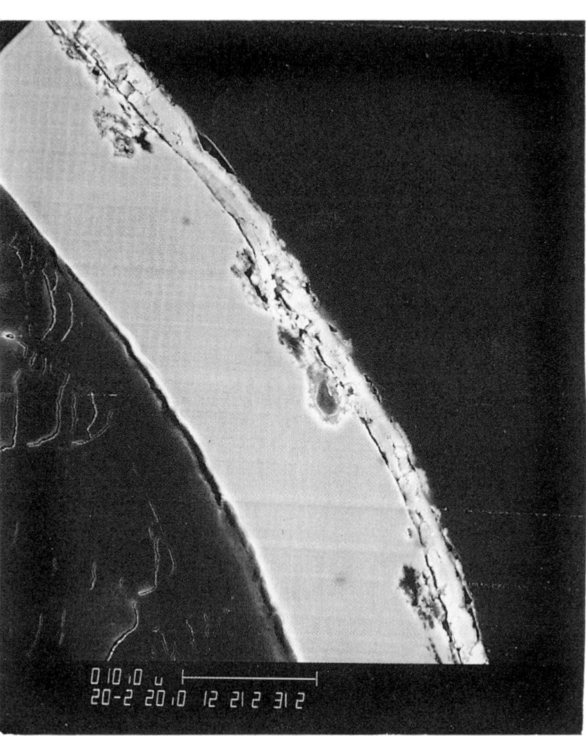

Fig. 3. 2000X SE image of highly corroded section of sample 74.543-#3 (French, ca. 1730).

areas (5 and 7) analyzed. The thickness of the gold layer was measured from the BSE image and found to be 0.7μm at the thickest point and 0.2μm at the thinnest point. The thickness of the entire strip is approximately 8.4μm. These results show that without a doubt, thread sample 74.114-#1 is a core thread wrapped by a strip having a very thin layer of gold on a much thicker silver substrate.

Several other threads were analyzed by SEM techniques. Figures 3 and 4 show the SE and BSE images (2000X and 5000X magnification, respectively) of sample 74.543-#3. The section of the strip scanned is highly corroded but the gold layer (bright line, right surface underneath corrosion layer in figure 4) is intact. The identification of this material as gold was again done by x-ray analysis. The corrosion layer above the gold, and the interior of the strip, contained silver. Copper was not detected by this technique, which cannot detect elements in concentrations of less than ~1%. A trace of copper was, however, detected spectrographically. A further result of this analysis was the detection of another gold layer on the inside of the strip (bright line, bottom left of figure 4). This second surface layer was proven by x-ray analysis also to be gold. This gilded silver strip must be one of those that was forged from a gilded silver wire. Such a strip would necessarily have a gold layer on four surfaces—the front and back and the two thinner sides. Furthermore, all four gold surfaces would be cut during sectioning and thus the two gold surfaces on the sides of the strip should be observable in the thread

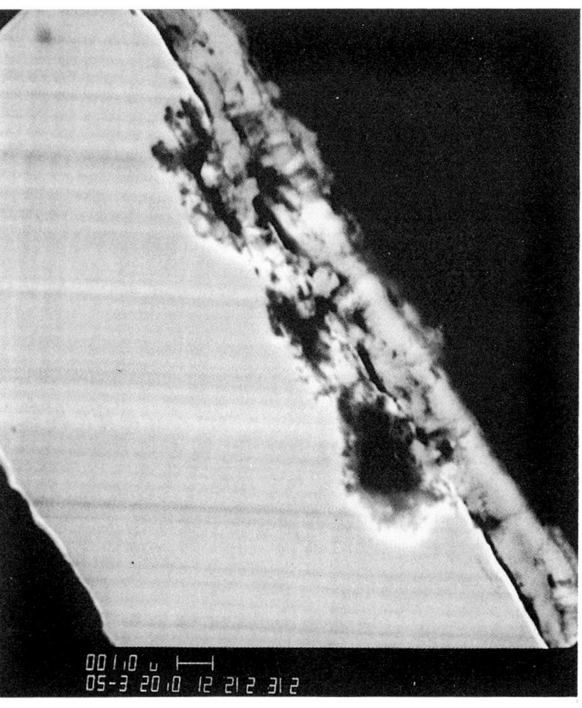

Fig. 4. 5000X BSE image of 74.543-#3 showing gold surface layers and corrosion.

cross-section. Figure 5, a 1000X magnification BSE image, shows one of these two sides (bright line on short edge, at bottom of image). The gold layer on the opposite edge was also observed, but is not shown here. The thread was 0.22mm in diameter; the strip (measured in an uncorroded area) was about 10μm thick and 0.20mm wide. The front and back gold layers, as measured from figure 4, were about equally thick—0.22μm. The side layers were also of this order of thickness.

Figure 6 shows the SE image of a golden-colored Japanese strip (33.498). The fibrous backing strip split into two pieces during mounting in the epoxy resin. The metallic layer is observed on the top surface of the upper fragment. A 10,000X magnification BSE image of this top surface is shown in figure 7. X-ray analysis of this surface layer revealed the presence of major gold and a minor amount of silver. Copper, undetected by x-ray analysis, was shown to be present in a trace amount by spectrographic analysis. The thickness of this layer is 0.1 to 0.2μm. The entire strip is about 0.62mm wide and approximately 0.11mm (110μm) thick.

Figures 8 and 9 show the 300X magnification SE and BSE images, respectively, of the mounted and sectioned golden-colored Chinese thread (76.541). It appears that the strip was under tension before mounting and, like a stretched-out watch spring, it relaxed and coiled up when placed in the epoxy resin. The surface metal layer is clearly seen (bright line) in the BSE image. The less bright layer underneath the top layer is the material previously suspected of being a red claylike layer. The darkest bottom layer is the fibrous backing. Figure 10 shows a 3000X magnification blow-up of these layers. The areas indicated were analyzed and found to contain only gold in the surface layer (area 1), a large amount of silicon, and smaller amounts of aluminum and iron in areas 2 and 4, and only very small amounts of silicon, aluminum, and iron in area 5, the center of the fibrous backing strip. One of the higher average atomic number particles (bright spot—area 3) was shown by x-ray analysis to contain a large concentration of iron. These results are consistent with the previously postulated construction of Chinese strips. Silicon, and to lesser extent, aluminum, are always present in clays. Red bole is colored red by the presence of iron oxide compounds—especially red colored hematite—Fe_2O_3. It can also be seen in these images that the gold layer lies quite evenly on the smooth surface of the claylike layer underneath, which fills in irregularities in the underlying fibrous backing. The gold layer on this strip is about 0.3μm average thickness; the claylike layer varies from about 3 to 10μm at its thinnest and thickest points. The thread diameter is 0.22mm and the strip 0.40mm (400μm) wide and 23μm thick.

Discussion and Conclusions

Both emission spectrographic and energy-dispersive x-ray analyses have revealed that most of the specimens studied are composed of varying amounts of gold, silver, and copper, with very few other metals present in more than trace quantities. A study of the colors of alloys containing different proportions of these three metals was published by Josef Leuser in 1949.[6] A ternary diagram of the color system for gold-silver-copper alloys was produced by examination of 1089 prepared alloys covering the full compositional range of the ternary system (fig. 11). It can be seen from this diagram that progres-

Fig. 5. 1000X BSE image of 74.543-#3 showing gold layer on strip edge, and front and back surfaces.

sive addition of silver metal (only) to gold changes the color of the resultant alloy from golden (red-yellow) to yellow, the point of demarcation being at about 96% Au and 4% Ag. Further addition of silver changes alloy color from yellow to green-yellow (at about 85% Au, 15% Ag), from green-yellow to a lighter green-yellow (about 70% Au, 30% Ag) and then to a whitish color (about 50% Au, 50% Ag). Addition of copper only to gold shifts alloy color toward coppery-red tones, with up to about 10% copper yielding a still golden-colored alloy, from ∼10 to 27% producing a reddish alloy, and more than ∼27% giving a progressively more coppery-red material.

It is instructive to determine the compositions of ternary alloys that contain the minimum amount of gold metal, but still exhibit a golden to yellowish color. An alloy containing 88% Au, 6% Ag, and 6% Cu would still be a golden color; a 55% Au, 30% Ag, and 15% Cu alloy would still be yellow in color, and a 24% Au, 48% Ag, 28% Cu alloy would be yellowish. Spectrographic estimates of the composition of the last two yellow alloys, if encountered as specimens in this study, would be reported as major gold and silver, minor copper, and major silver, minor gold and copper, respectively. The diagram also shows that alloys containing less silver but more gold and copper will be yellow as well. Thus, an alloy of composition 60% Au, 20% Ag, 20% Cu will still be yellow, and an alloy containing 32% Au, 28% Ag, and

Fig. 6. 150X SE image of 33.498 (Japanese, 16th-17th century) showing gold-silver layer on top of upper fragment.

Fig. 7. 10,000X BSE image of gold-silver layer on 33.498 (fibrous backing below layer).

40% Cu will still be yellowish colored. Here again, spectrographic analysis would give relative metallic content as major gold, major to minor silver and copper, and major gold, silver, and copper, respectively. It can therefore be seen why, in the absence of a clearly visible gold-colored layer on a silver- or black-colored substrate, many golden to yellow-colored strip specimens containing major amounts of silver, and appreciable amounts of gold and copper were listed in the table as belonging to either Type I *or* Type IV. For the same reason, non-silver colored wire threads shown to contain major silver and minor gold and copper could not be positively identified as gilded silver wires on the basis of spectrographic data alone.

Back-scatter electron mode scanning electron microscopic analyses of sectioned threads and strips conclusively identify the Type I gilded silver strips, even in highly corroded specimens where the silver (and copper) corrosion products completely cover the surface gold layer (figs. 3 and 4). By using energy-dispersive x-ray analysis, we are also able to determine the distribution of the major and minor metals throughout the sectioned specimen, which is not easily done by spectrographic techniques. Emission spectrographic analyses are however valuable in being capable of determining the presence of a wide range of elements in concentrations too low to be detected by the SEM-x ray technique.

It is obvious from the preliminary results presented above that much more information can be obtained through mounting, sectioning, and SEM-x ray studies of more of these thread and strip specimens. The positive identification of many other gilded silver strips, perhaps of gilded silver wires and also of other strips originially forged from gilded silver wires, are all to be expected. These studies are currently proceeding. It is however possible to make some general statements concerning the similarities and differences between metallic threads produced by different cultures during different time periods based on the information already available.

1. Sixteenth- to nineteenth-century European, French, Italian, Spanish, German, and Swiss (62 samples taken from 18 different textile pieces): The strips consisted of many gilded silver (Type I) and silver (Type II) specimens, both types containing minor amounts of copper; several also contained traces of lead. None of the silver strips were pure silver, but always contained definite amounts of copper and gold. Six gold-and silver-colored wires were detected among this textile group, but not in other textiles; five specimens had diameters of 0.06 mm—one Swiss specimen (ca. 1520) had a diameter of 0.325mm, and was an alloy of copper with minor amounts of zinc and iron.

Fig. 8. 300X SE image of sectioned gold thread 76.541 (Chinese, 18th-19th century).

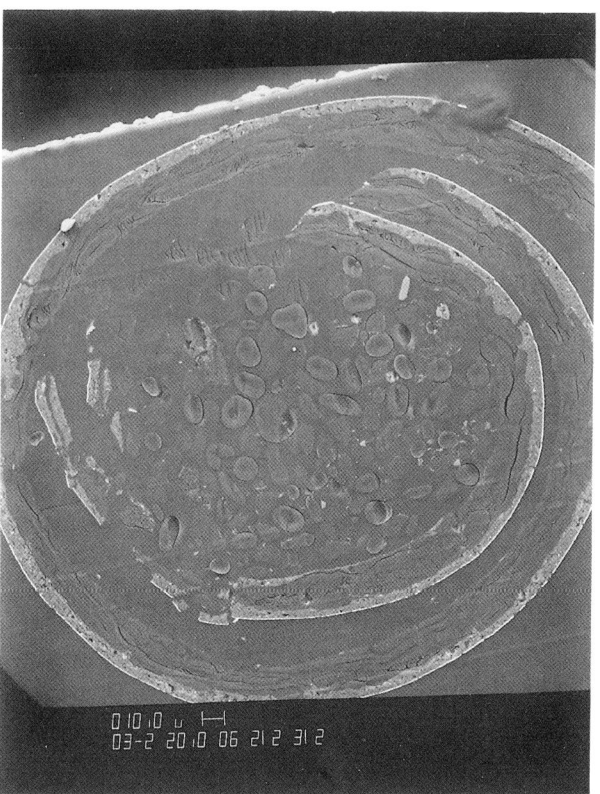

Fig. 9. 300X BSE image of 76.541 showing gold layer on top of claylike layer on fibrous backing.

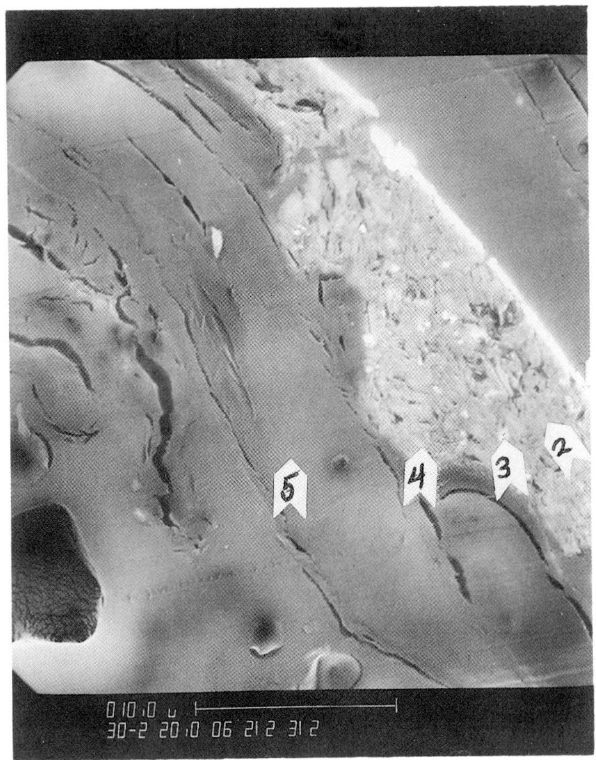

Fig. 10. 3000X BSE image of 76.541 showing areas analyzed by energy-dispersive x-ray analysis.

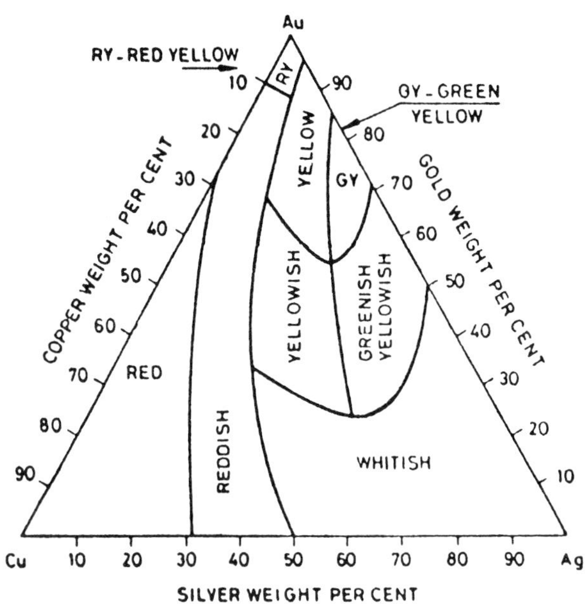

Fig. 11. Relationship between the color and composition of gold-silver-copper alloys (after Leuser[6])

2. Twentieth-century European, French and American (7 samples from 3 textiles): All strips were either pure copper or copper-based alloys; several contained minor to trace amounts of gold and silver. One strip appeared to be an alloy of copper and iron.

3. Seventeenth- to nineteenth-century Japanese (6 samples from 6 textiles): These strips were either pure gold or gold/silver alloy on fibrous (paper?) backings; minor to trace amounts of copper were also present.

4. Eighteenth- to early twentieth-century Chinese (13 samples from 7 textiles). The strips were pure gold, silver or copper metal, or various binary and ternary alloys of these metals almost always over a red or yellow claylike layer on a fibrous backing. Only two specimens, one silver-and one gold-colored, were metal strips with no claylike underlayer or fibrous backing.

5. Nineteenth-century Moroccan and Tunisian, sixteenth- to nineteenth-century Turkish, and nineteenth-century Iranian (18 samples from 11 textiles): Most of these strips resembled Type I and II early Western specimens; only one Type IV strip, containing a major amount of copper with minor gold and silver, was detected.

6. Nineteenth- to early twentieth-century Sumatran and Javanese (6 samples from 4 textiles): Among the different strips represented were pure silver and pure gold over a resin-like layer on a fibrous backing; pure copper and copper-silver alloy strip; one possible gilded silver strip; and one "Chinese" (Type VI) specimen having a gold-silver alloy layer over red underlayer on a fibrous backing.

7. Eighteenth- to early twentieth-century Indian (11 samples from 4 textiles): Some pure silver, and (possibly) gilded silver strips were detected. Also found were a few strips thought to be silver-gold, silver-copper, silver-gold-copper, and silver-copper-lead.

Table 1:
Data from Visual, Microscopic and Emission Spectrographic Analyses of Metallic Threads.

Sample Number	Curatorial Description	Strip Color (Visual)	Microscopic Description	Elemental Composition of Strip			Strip/Thread Type
				Major	Minor	Trace	
81.211	European, ca. 1730-40						
#1 (embroidery)		gold	gold-colored strip	Au, Ag	Cu	Si, Al, Pb Ca, Mg	I or IV
#2 (embroidery)		gold	gold-colored strip, black corrosion under	Au, Ag	Cu	Si, Pb, Mg	I
#3 (embroidery)		gold	0.06mm diameter golden-colored wire	Au, Ag	Cu	Si	III
#4 (embroidery)		gold	golden-colored strip, black corrosion under	Au, Ag	Cu	Si, Pb, Mg	I
#5 (embroidery)		gold	golden-colored strip, corrosion underneath	Au, Ag	Cu	Si, Pb, Ca, Mg	I
#6 (embroidery)		gold	golden-colored strip, black corrosion	Au, Ag	Cu	Si, Pb, Ca, Mg	I
#7 (embroidery)		silver	silver-colored strip, some black corrosion	Ag	Cu	Au, Si, Pb, Mg	II
#8 (embroidery)		silver	0.06mm diameter silver-colored wire	Ag	Au, Cu	Si	III
#9 (embroidery)		silver	narrow golden-colored strip	Au, Ag	Cu	Si, Pb	I or IV
#10 (embroidery)		silver	very corroded black strip	Ag	Au, Cu	Ca, Mg	II or IV
80.378	European, 1775						
#1 (weft)		gold	few golden patches on black strip	Ag	Au, Cu	Mg, Ca, Si	I
#2 (embroidery)		silver	0.06mm diameter silver-colored wire	Ag	Au, Cu	Mg	III
28.299	European, ca. 1910						
#1 (embroidery)		gold	dark golden surface, black corrosion	Cu	—	Ag, Zn	IV
#2 (embroidery)		gray-gold	strip, golden-colored top surface, silver-colored on reverse	Cu	Ag	Au, Si, Ca	IV
#3 (embroidery)		gold	bright golden-colored strip	Cu	Fe	Au, Ag, Ni, Cr, Si	IV
81.380	French, ca. 1707						
#1 (weft)		gold	golden-colored strip, black corrosion under	Ag	Au	Cu, Si, Ca, Mg	I
#2 (weft)		gold	very corroded strip; possible golden surface	Ag	Au	Cu, Si, Ca, Mg	I
#3 (weft)		silver	bright silver-colored strip with yellowish highlights	Ag	Au, Cu	Si, Ca, Mg	II
#4 (weft)		silver	black, corroded surface, only one possible golden patch	Ag	Au, Cu	Si, Fe, Ca, Mg	II
82.29	French, ca. 1710						
#1 (weft)		gold	golden-colored strip	Ag	Au, Cu	Si, Ca, Mg	I or IV

Sample Number	Curatorial Description	Strip Color (Visual)	Microscopic Description	Elemental Composition of Strip			Strip/Thread Type
				Major	Minor	Trace	
#2 (weft)		gold	golden-colored strip, some surface corrosion	Ag	Au, Cu	Ca, Mg	I or IV
74.543	French, ca. 1730						
#1 (weft)		silver	bright silver-colored strip, some corrosion	Ag	Au	Cu, Fe, Si, Ca, Mg	II
#2 (weft)		silver	narrow silver-colored strip	Ag	Au	Cu, Fe, Si, Ca, Mg	II
#3 (weft)		gold	bright golden-colored strip	Ag	Au	Cu, Fe, Si, Ca, Mg	I*
#4 (weft)		gold	golden-colored strip, corrosion underneath	Au, Ag	Cu	Si, Ca, Mg	I
#5 (trim)		gold	golden-colored strip	Ag	Au, Cu	Si, Ca, Mg	I or IV
#6 (trim)		gold	golden-colored strip	Ag	Au, Cu	Si, Ca, Mg	I or IV
#7 (trim)		gold	golden-colored strip, black corrosion under	Ag	Au, Cu	Si, Ca, Mg	I
#8 (trim)		gold	dark colored surface	Ag	Au, Cu	Si, Ca. Mg	I or IV
81.5	French, ca. 1735						
#1 (weft)		silver	silver-colored strip; some corrosion	Ag	Au, Cu	Si, Ca, Mg	II
#2 (weft)		silver	possible silver-colored strips; very corroded	Ag	Au, Cu	Si, Ca, Mg	II
#3 (trim)		silver	entire strip surface corroded (black)	Ag	Au, Cu	Si, Ca, Mg	II
#4 (trim)		silver	strip surface corroded (black)	Ag	Au, Cu	Si, Ca, Mg	II
70.114.2	French, ca. 1735		black corroded strip with gold patches	Ag	Au, Cu	Si, Cu, Mg	II
#1 (weft)		silver	black strip surface, no gold visible	Ag	Au, Cu	Si, Al, Fe, Ca, Mg	II
#2 (weft)		silver					
30.1096 (embroidery)	French, ca. 1780	gray-silver	very thin fiber loosely wrapped around core which appeared to have resin-like coating on bundle of fibers. Crystalline (Ag?) substance in spots on resin layer	Ag	—	Si, Mg	VII
77.360	French, ca. 1925		darkened strip (golden?), interior and exterior of strip same color; highly corroded	Cu	—	—	IV
#1 (embroidery)		gray-copper					
#2 (embroidery)		gold	black (corroded) surface, golden-colored in places; also some blue corrosion product	Cu	Ag	—	IV
#3 (embroidery)		gold	golden-colored strip; darkened in patches, with small spots of light blue or green corrosion product	Cu, Au	Ag	Si, Mg	IV

Sample Number	Curatorial Description	Strip Color (Visual)	Microscopic Description	Major	Minor	Trace	Strip Thread Type
82.7 (weft)	Italian, ca. 1475	—	golden-colored strip surface, black corrosion underneath	Ag	Au. Cu	Si. Pb. Ca. Mg	I*
74.114	Italian, velvet — ca. 1475 orphrey — 1500-1550						
#1 (weft)		gold	golden-colored surface layer-corrosion underneath	Au, Ag	Cu	Si, Pb, Mg	I*
#2 (embroidery)		gold	golden-colored strip	Au, Ag	Cu	Si, Ca, Mg	I or IV
#3 embroidery)		gold	corroded (black) strip surface with few golden patches	Au, Ag	Cu	Mg	I
#4 (trim)		gold	golden-colored strip surface, reverse side silver-colored	Au, Ag	Cu	Si, Mg	I
78.160 (weft)	Italian, ca. 1550	gold	golden-colored surface, strip largely intact	Ag	Au, Cu	Si, Pb, Sn, Ca, Mg	I or IV
37.131	Italian, ca. 1700						
#1 (embroidery)		gold	dark golden-colored strip	Ag	Au, Cu	Si, Ca, Mg	I or IV
#2 (embroidery)		silver	black-colored strip w/few golden patches (tarnish?)	Ag	Au, Cu	Si, Mg	II
#3 (embroidery)		silver	silver-colored strip w/some golden-colored patches (tarnish?)	Ag	Au, Cu	Si, Ca, Mg	II
#4 (embroidery)		silver	blackened strip surface w/ very few golden-colored patches (tarnish?)	Ag, Au	Cu	Si, Mg	II
82.60	Italian or French, ca. 1735						
#2 (trim)		gold	golden-colored strip	Au, Ag	Cu	Si, Pb, Al, Ca, Mg	I or IV
#3		gold	0.055mm diameter silver-colored wire, slightly tarnished	Ag	Au, Cu	Pb, Mg	III
#4 (embroidery)		gold	golden-colored strip, darkened in spots	Ag	Au, Cu	Pb, Ca, Mg	I or IV
82.59	Italian or French, ca. 1740						
#1 (trim)		gold	golden-colored surface layer over black corrosion	Au, Ag	Cu	Si, Ca, Mg	I
#2 (trim)		gold	golden-colored strip	Au, Ag	Cu	Si, Ca, Mg	I or IV
76.165	Spanish, ca. 1500						
#1 (weft)		gold	0.060mm diameter black-colored wire	Ag	Au, Cu	Si	III
#2 (weft)		gold	golden-colored surface w/ very few corrosion spots	Ag	Au, Cu	Si, Pb, Al, Ni Ca	I or IV
#3 (weft)		gold	golden-colored strip; surface intact	Ag	Au, Cu	Si, Pb	I or IV

Sample Number	Curatorial Description	Strip Color (Visual)	Microscopic Description	Elemental Composition of Strip			Strip Thread Type
				Major	Minor	Trace	
#4 (weft)		gold	bright golden-colored strip with corrosion underneath	Ag	Au. Cu	Si. Pb	I
#5 (weft)		silver	silver-colored strip w some black corrosion	Ag	Au. Cu	Si. Pb. Mg	II
#6 (trim)		silver	uncorroded silver-colored strip on both sides	Ag	Cu	Au. Si. Pb. Ca. Mg	II
30.916	Spanish. ca. 1700						
#1 (weft)		gold	golden-colored strip: surface intact but "lumpy"	Au. Ag	Cu	Si. Ca. Mg	I or IV
#2 (weft)		gold	intact golden-colored strip: lumpy surface	Au. Ag	Cu	Si. Al. Mg	I or IV
70.46	German(?). ca. 1760						
#1 (embroidery)		gold	golden-colored intact strip	Au. Ag	Cu	Si. Pb. Ca. Mg	I or IV
#2 (embroidery)		silver	golden surface layer with corrosion underneath	Ag	Au. Cu	Si. Ca. Mg	I or II
#3 (embroidery)		silver	black corroded strip surface	Ag	Cu	Au. Si. Pb. Ca. Mg	II
#4 (embroidery)		silver	silver-colored strip	Ag	Cu	Au. Si. Mg	II
81.210	Swiss. ca. 1520						
#1 (embroidery)		gray	0.325mm diameter copper-colored wire	Cu	Fe. Zn	Ag. Si. Pb. Mn. Sn. Ni. Ca. Mg	III
#2 (embroidery)		gold	golden-coppery strip: black corrosion in some spots	Cu	—	Ag. Pb. Mn. Sn. Fe. Ni. Mg	IV
81.338 (embroidery)	American. ca. 1925	copper	blue corrosion product over reddish-golden surface — strip badly deteriorated	Cu	—	Ag. Zn. Ca. Mg	IV
33.498 (weft)	Japanese. 17th century fabric — 16th century	gold	golden-colored surface layer over strip of fibrous backing — no red underlayer	Au	Ag	Cu. Si. Ca. Mg	V(a)*
33.504 (weft)	Japanese. 17th-18th century	gold	golden-colored surface layer (w/gray-green corrosion) on fragmentary fibrous backing	Ag	Au. Cu	Si. Ca. Mg	V(a) or V(b)
4893.83.1 (embroidery)	Japanese. 1675-1700	gold	golden-colored surface layer on fibrous backing	Au	—	Ag. Cu. Ca. Mg	V(a)*
33.488 (weft)	Japanese. ca. 1775	—	red-golden surface layer on red underlayer on fibrous backing	sample too small			VI(—)
33.514 (embroidery)	Japanese. 1775,1800	gold	shiny red-golden surface layer on fibrous backing	Au	Ag	Cu. Ca. Mg	V(a)
13.47 (embroidery)	Japanese. 1775-1850	gold	red-golden surface layer on fibrous backing	Au	Ag	—	V(a)
13.48 (embroidery)	Japanese. 1800-1850	gold	golden-colored surface layer on fibrous backing	Au	Ag	—	V(a)

Sample Number	Curatorial Description	Strip Color (Visual)	Microscopic Description	Elemental Composition of Strip			Strip/Thread Type
				Major	Minor	Trace	
76.541 (weft)	Chinese, 1775-1825	gold	golden-colored surface layer, cracked in spots exposing red-colored underlayer on top of fibrous (paper?) backing	Au	Si	Ag, Cu, Ca, Mg	VI(a)*
80.43c	Chinese, ca. 1820						
#1 (embroidery) core thread		copper	yellowish-colored surface layer over red underlayer on fibrous backing; green	Cu	—	Ag, Si	VI(c)
#2 (embroidery)		gold	golden surface layer over red underlayer — fibrous backing fragmentary	Ag	Cu, Si	—	VI(b)
62.203 (weft, seat)	Chinese, ca. 1820	gold	golden-colored surface over brick red underlayer over fibrous backing; red tinted core thread	Cu	Au, Ag	Si	VI(c)
62.204 (weft, back)	Chinese, ca. 1820	gold	golden-colored layer over red underlayer over fibrous backing	Au	—	Ag, Cu, Fe, Al, Si, Ca, Mg	VI(a)
TR9024	Chinese, ca. 1870						
#1 (embroidery)		copper	gold-red strip, corroded; underlayer present (yellow?) fibrous backing	Ag	Cu	—	VI(b)
#2 (embroidery)		gold	golden surface, red underlayer, fibrous backing	Ag	Cu	—	VI(b)
69.69.2	Chinese, ca. 1880						
#1 (embroidery)		gold	golden-colored layer over brick-red underlayer over fibrous backing — reddish core thread	Au	Ag, Cu	Si, Ca, Mg	VI(a)
#2 (embroidery)		gold	golden-colored strip, jagged around edges; no fibrous backing — some black corrosion	Ag	Au, Cu	Si, Ca, Mg	I or IV
#3 (embroidery)		silver	silvery-colored strip, black corrosion underneath; no fibrous backing	Ag	—	Si, Ca, Mg	II
#4 (embroidery)		gray-silver	black corrosion product over yellow underlayer over fibrous backing	Ag	Cu	Au, Si, Ca, Mg	VI(b)
33.365	Chinese, ca. 1900						
#1 (embroidery)		gold	bright golden-colored surface layer over red underlayer over fibrous backing	Cu	Au	Ag, Si, Mg	VI(c)
#2 (embroidery)		copper	reddish-golden-colored surface layer over red underlayer over fibrous backing and surface intact	Cu	Ag	—	VI(c)
1983.66	Moroccan, 19th century						
#1 & 2 (embroidery)		gold	golden-colored surface with black corrosion underneath	Au, Ag, Cu	—	Si, Pb, Ca, Mg	I
#3 & 4 (embroidery)		gold	golden-colored strip; black corrosion underneath	Au, Ag, Cu	—	Si, Pb, Ca, Mg	I

Sample Number	Curatorial Description	Strip Color (Visual)	Microscopic Description	Elemental Composition of Strip			Strip/Thread Type
				Major	Minor	Trace	
33.281 (embroidery)	Moroccan, 19th century	gold	golden-colored strip with some corrosion	Cu	Au, Ag	Zn, Fe	IV
33.243 (weft)	Moroccan, 19th century	gold	golden-colored surface with black corrosion underneath	Au, Ag	Cu	Si, Pb, Ca, Mg	I
33.313	Tunisian, 19th century						
#1 (embroidery)		gold	silver-colored strip	Ag	Au, Cu	Fe, Ni	I or IV
#2 (embroidery		silver	silver-colored strip, some corrosion	Ag	Cu	Si, Al, Ca, Mg	II
18.142 (weft)	Turkish, 16th century	dark gray	black (corroded) strip	Au, Ag	Cu, Mg	Si, Al, Pb, Mn Fe, Ca	I*
33.1457	Turkish, ca. 1600						
#1 (weft)		gold	golden-colored strip; uneven edges	Au, Ag	Cu	Si, Ca, Mg	I or IV
#2 (embroidery)		silver	silver-colored surface	Ag	Cu	Si, Al, Ca, Mg	II
#3 (embroidery)		silver	golden-colored strip	Au, Ag	Cu	Si, Ca, Mg	I or II or IV
55.57 (embroidery	Turkish, 19th century	gold	brass-colored strip	Au, Ag	Cu	Al, Mg	I or IV
48.29	Turkish, 19th century						
#1 (embroidery)		gray-silver	silver-colored strip	Au, Ag	Cu	Si, Mg	II
#2 (embroidery)		gold	golden-colored strip, black on reverse side	Au, Ag	Cu	Si, Pb, Mg	I
#3 (trim)		gold	light golden-colored strip	Au, Ag	Cu	Si, Ca, Mg	I or IV
33.834 (embroidery)	Iranian, ca. 1875	—	black strip, badly corroded	Ag	Cu	Si	II
33.821	Iranian, 19th century(?)						
#1 (embroidery)		gold	golden-colored surface layer on dark corrosion underneath	Ag, Au	Cu	Si, Ca, Mg	I
#2 (embroidery)		silver	tarnished silver-strip surface	Ag, Au	Cu	Si, Ca, Mg	II
18.139 (weft)	Iranian, 18th or 19th century	gold	golden-colored strip; surface intact	Ag	Cu	Au, Si, Ca, Mg	I or IV
82.149 (weft)	Sumatran, 19th century	silver	silver-colored layer over yellow resin-like layer on fibrous backing	Ag	Si	Ca, Mg	VII
76.338	Sumatran, ca. 1900						
#1 (weft)		gold	silver-colored strip; scoring of surface observed	Ag, Cu	Au	Si, Ca, Mg	I or IV
#2 (trim)		gold	black surface with some golden-colored patches	Cu	Ag	Au, Fe, Si, Ca, Mg	IV

Sample Number	Curatorial Description	Strip Color (Visual)	Microscopic Description	Elemental Composition of Strip			Strip Thread Type
				Major	Minor	Trace	
#3 (fringe)		gold	black surface on flat stri; some reddish-golden spots	Cu	—	Au. Ag. Zn. Si. Ca. Mg	IV
33.681 (weft)	Sumatran, ca. 1900	gold	golden-colored layer over brick red layer over fibrous backing, specimen badly deteriorated	Au. Ag	Cu. Si	Pb. Al. Fe. Ca Mg	VI(a) or VI(b)
33.686 (gold leaf)	Javanese, 19th century	gold	golden-layer over cracked bright yellow layer (crevasses present)	Au	—	Cu. Ag	VII
33.748 (embroidery)	Indian, ca. 1775	gray	three small-diameter threads wound together (not braided); each thread wrapped with dark metallic strip	Ag	—	Au. Cu. Fe, Si, Ca, Mg	II
82.12 (weft) 38.63	Indian, ca. 1900 Indian, ca. 1900	gold	silver-colored strip with slight reddish-gold hue	Au, Ag, Cu	Ca	Pb, Fe, Si, Mg	IV
#1 (embroidery)		gold	golden-colored strip, slightly damaged	Ag	Au	Cu, Pb	I or IV
#2 (embroidery) 33.735	Indian, ca. 1900	silver	silvery gold strip, slightly damaged	Ag	—	Au, Cu, Pb	II
#1 (embroidery)		gold	golden-colored strip; red-colored core thread	Ag	Au, Cu	Si, Ca, Mg	I or IV
#2 (embroidery)		silver	silver-colored strip — some surface corrosion; small tears on strip edges	Ag	Cu	Si, Ca, Mg	II
#3 (hem trim)		gold	golden-colored strip with black corrosion — torn edges in some places	Ag	Cu	Si, Mg	IV
#4 (hem trim)		silver	silver-colored strip with black corrosion	Ag	—	Fe, Si, Mg	II
#5 (seam trim)		gold	steel-blue colored strip with black corrosion products	Ag	Cu, Pb	Si, Cr, Ca, Mg	II or IV
#6 (seam trim)		silver	silvery-colored strip; tarnished	Ag	Cu, Pb	Si, Ca, Mg	II or IV
#7 (seam trim)		gold	golden-colored strip; corrosion on surface	Au, Ag	Cu	Si, Fe, Mg	I or IV

*Assignment also based on SEM analysis for these samples.

References

1. A. Geijer. *A History of Textile Art*. (New York: Sotheby Parke Bernet Publications 1982). pp. 11-13.

2. E. Hoke. and I. Petrascheck-Heim. "Microprobe Analysis of Gilded Silver Threads From Medieval Textiles." *Stud. in Conserv..* 22 (1977): 49-62.

3. E. Hoke. "Microanalysis of Metal Threads in Grave I of the Parish Church Traismauer." *Fundberichte aus Österreich.* 16 (1978): 255-259.

4. L. Stodulski. "The Use of the Emission Spectrograph in the Conservation Laboratory." *Bulletin of the American Institute for Conservation.* 15 (1975): 66-100.

5. Harvey. C.. *Semiquantitative Spectrochemistry*. (Glendale. California: Applied Research Laboratories. Inc., 1964).

6. J. Leuser. *Metall*, 3. 128 (1949): 105-110. Leuser's diagram was reproduced in: *Gold: Recovery. Properties and Applications*. edited by E. M. Wise. (New York: D. Van Nostrand Co. Inc., 1964) p. 262.

F. R. MATSON

Compositional Studies of the Glazed Brick from the Ishtar Gate at Babylon

Blue, turquoise, white, and yellow glazed brick fragments collected at the Ishtar Gate in 1954 have been analyzed by electron beam microprobe and x-ray fluorescence spectrometry. Petrographic thin sections of the brick were studied and neutron activation analyses were made of four clay samples from Babylon. Additional pieces of brick were carefully examined visually. Using the data obtained, the brick manufacturing processes used in the city around 570 B.C. were discussed.

A. KACZMARCZYK and C. LAHANIER

Ancient Egyptian Frits and Colored Faience Bodies: Problems of Classification

The discussion presented below is based on the results of an extensive study of Egyptian faience from several museums. Unless indicated otherwise, the accession numbers cited in this paper refer to faience from the Ashmolean Museum (Oxford) collection. The analytical data were computed from x-ray fluorescence and atomic absorption spectra recorded at the Research Laboratory for Archaeology at Oxford and the Louvre Research Laboratory in Paris.[1,2]

Types of Body Material

The basic ingredients of an ordinary faience body are: silica (from crushed quartz or sand), lime (from limestone or calcareous sands) and alkali (from natron or plant ashes). Occasionally, underneath the glassy skin (glaze) one finds a colored body of what Lucas labelled faience variants B, C, and D.[3] The letter A was assigned to faience in which a white layer was interposed between the glaze and the bulk body material. Variant E was characterized by a more vitrified body without a distinct surface layer.

We wonder if Lucas realized, when he devised his classification scheme, how difficult it is at times to distinguish "ordinary faience" from the variants. If the object is undamaged, none of the variants can be readily recognized, but even damaged faience is not always easy to classify. If either cobalt or manganese is responsible for the color, deliberate pigmentation may be presumed, since the normal ingredients of a core seldom contain more than trace amounts of these elements.[4,5] Copper is as uncommon in Egyptian sands as cobalt, but its presence in a blue or green body does not imply that a colored core was desired, because had the object been made by the so-called "self-glazing" process,[6-8] incomplete efflorescence to the surface would have left highly variable residual amounts of CuO. The question arises how much copper should be considered more than accidental, particularly if a significant body-surface concentration gradient is observed. Lucas offers little guidance in this matter, for after defining Variant D as blue or green faience with a like-colored hard and compact body, he lists in his Appendix four examples, three of which hardly fit the definition.[5] Thus, his No. 1 was originally described as brown-bodied (2.4% MnO, 0.8% CuO) with a violet blue glaze, while Nos. 3 and 4 (0.4% and 0.5% CuO, respectively) had bodies described as "white, sandy, coarse-grained and soft" and "very fine-grained and soft," respectively.

Iron presents other problems. Ferruginous sands abound in Egypt and were undoubtedly used in the manufacture of faience. Equally common are various ochres, and they too must have been used occasionally to create black (variant B) and red (variant C) bodies. Yet how many of the gray-bodied and red-bodied faience are the result of deliberate pigmentation?

Variant B

Lucas defines variant B as faience consisting of a black glaze over a dark gray or dark brown body colored by iron. In deciding to call such faience a distinct variant the author stated that in most of the examples examined by him, "It is most probable that the oxide of iron was added intentionally."[3] Considering that some Egyptian sands contain as much as 5.6% Fe_2O_3, how can one tell whether the iron oxide was added to a quartz powder (or purer sands) or whether ferruginous sands were intentionally or unintentionally used in the preparation of black faience?

Table 1
Egyptian Chronology 4000-30 B.C.

Historic Period	Dates, B.C.
PREDYNASTIC PERIODS	
Armatian (Naqada I)	4000-3500
Gerzean (Naqada II)	3500-2900
ARCHAIC (PROTODYNASTIC) PERIOD	
Dyn. I-II	2920-2649
OLD KINGDOM	
Dyn. III	2649-2575
Dyn. IV	2575-2465
Dyn. V	2465-2323
Dyn. VI	2323-2150
FIRST INTERMEDIATE PERIOD	
Dyn. VII-X	2150-2040
Dyn. XI	2134-2040
MIDDLE KINGDOM	
Dyn. XI	2040-1991
Dyn. XII	1991-1783
SECOND INTERMEDIATE PERIOD	
Dyn. XIII-XIV	1783-1640
Dyn. XV-XVI (Hyksos)	1640-1532
Dyn. XVII	1640-1550
NEW KINGDOM	
Dyn. XVIII (includes the Amarna Period)	1550-1307
Dyn. XIX-XX (Ramesside Period)	1307-1070
THIRD INTERMEDIATE PERIOD	
Dyn. XXI	1070-945
Dyn. XXII	945-712
Dyn. XXIII-XXIV	822-712
Dyn. XXV	750-657
LATE PERIOD	
Dyn. XXVI (Saite period)	664-525
Dyn. XXVII	525-404
Dyn. XXVIII-XXIX	404-380
Dyn. XXX	380-343
Dyn. XXXI	343-332
PTOLEMAIC PERIOD	332-30

We have analyzed almost 150 specimens of black faience of different periods and found that the majority had a white or creamy core. Our oldest example with a black glaze on a dark gray core was a barrel bead (E.E.456A) from an Old Kingdom tomb at Mahasna. Lucas too failed to find any examples of this variant antedating the Third Dynasty.[3] During the next millennium the popularity of such faience seems to have declined steadily. For example, while such faience accounted for about 50% of the examined Old Kingdom black beads, it represented only about 10% of the New Kingdom and later beads.

For reasons elaborated above it is unlikely that all of the gray and black cores examined by us had been colored intentionally, particularly when the glaze owed its blackness to manganese and the body was colored by iron. Deliberate pigmentation is less open to question when the same black pigment is dominant in both the glaze and the core. For example, a Twelfth Dynasty tomb at Abydos yielded a bead consisting of a blue black glaze (3.3% MnO, 0.6% Fe_2O_3, and 4.1%

CuO), over a gray black body with similar manganese and iron contents but a lesser amount of copper (E.E.633D in table 2). From the same tomb came a vase fragment (E.3306) consisting of a brown glaze containing 1.8% MnO, 0.3% Fe_2O_3, and 0.2% CuO over a gray core in which the concentrations of the three oxides were 1.2%, 0.3% and below 0.1%, respectively. These compositions make sense if one assumes that the efflorescence process was attempted with a pigment containing manganese among other ingredients.

Our data suggest that if the basic criterion distinguishing variant B from ordinary faience is the presence of a black glaze over a deliberately colored dark gray or brown core, then it is not iron but manganese that is most frequently found as pigment, except during the Ptolemaic Period. The Ramesside Period also saw the production of a peculiar gray-bodied gray faience colored by a mixture of transition elements similar to that found in the cobalt-colored faience discussed below.

Variant C

Variant C is defined by Lucas as faience consisting of a red body deliberately colored by Fe_2O_3 and covered by a red or colorless glaze. Lucas cites examples from as early as the Third and as late as the Thirtieth Dynasty.[3] The earliest examples of red-bodied red faience analyzed by us were the red beads from the foundation deposit of the Eighteenth Dynasty temple of Tuthmosis III at Coptos. Their rough-textured bodies contained as much red pigment (3-6% Fe_2O_3) as the glazes (Appendix C in ref. 10). A typical body composition is illustrated by bead E.E.238M in table 2.

Red faience of similar composition was used for inlaying miscellaneous objects of the Amarna and Ramesside periods (table 2). This variant must have gone out of vogue after the Twentieth Dynasty, because none of the more than fifty specimens of red faience of later date seen by us had a red body. Among the eighty-five specimens of red faience examined we failed to find a single example pigmented by copper in lieu of iron, even during times when all of the red glass seems to contain copper as colorant. For the composition of the glazes of the faience discussed above and below the reader is referred to a recent book by Kaczmarczyk and Hedges.[10]

Variant D

The name "Variant D" was assigned by Lucas to a hard and compact blue-bodied or green-bodied faience coated with a glaze of similar color. For a long time such material was attributed to the Saite Period, until Lucas pointed out that the Cairo Museum contains examples dated to the Third Dynasty.[3] We have analyzed blue and green vessel fragments from the First Dynasty royal tombs at Abydos, which have all the characteristics of this variant. The bodies are characterized by abnormally high concentrations of soda, magnesia, alumina, and lime, and the presence of small but significant amounts of cobalt (table 2). The glazes are more vitreous and darker than the bodies, contain slightly more CuO and SnO_2 and almost three times more K_2O (Appendix C in ref. 10). The concentrations of other elements have comparable values in the glaze and in the core.

The overall composition of blue-bodied faience of the next 1500 years is not significantly different from the specimens discussed above, except that tin is seldom seen and a number of Old Kingdom specimens exhibited very high levels of copper (up to 10%). For example, a blue bead (1935.171a) from a Fourth Dynasty grave at Armant contained: 0.8% K_2O, 13% CaO, 6% CuO, 0.13% As_2O_3 and less than 0.01% SnO_2. During the New Kingdom and subsequent time periods the levels of both calcium and copper declined significantly, while those of tin rose.

From the time of Tuthmosis III in the Eighteenth Dynasty the production of the traditional copper-colored faience is supplemented by the creation of two sub-variants. The first type has a green body covered by a green glaze, both colored by a mixture of copper oxide and lead antimonate (kohl tube 1889.148, table 2). Under high magnification yellow crystals, identified as $Pb_2Sb_2O_7$ by x-ray diffraction, are clearly visible in this type of faience. The second has cobalt oxide as the principal pigment at comparable concentrations in the body and the glaze (fig. 2).

The colors of the cobalt-colored bodies range from distinctly blue to virtually black. In table 3 the average compositions of "ordinary faience" bodies recovered from Tell el-Amarna are compared with the average compositions of cobalt-colored bodies. It is noteworthy that the latter have much higher levels of alkali and of magnesium, aluminum, sulfur, manganese, iron, nickel and zinc, elements whose association with New Kingdom cobalt in faience glazes,[10] glass[11,12] and pottery pigments[13-15] is well documented. Particularly significant are the levels of alumina, as they suggest that the cobalt might have originated in an alunite deposit, in the form of asbolan. For example, while the highest concentration encountered in cobalt-free cores of Amarna faience was only 0.37% Al_2O_3, the lowest value among cobalt-containing bodies was 1.25%.

The Kharga and Dakhla Oases in the Western Desert contain extensive alum deposits that were exploited by the ancient Egyptians.[16] That some of those alums are rich in iron and others contain cobalt and manganese has been known for almost a century.[17,18] However, only recently has it been established that the cobalt is also accompanied by magnesium, iron, nickel, and zinc.[19] Consequently, the elemental correlations noted above point at the two oases as the most probable source of New Kingdom cobalt. It is possible that the alum was originally introduced to render the body harder, and some of the high-fired high-alumina cores of the Amarna faience are indeed exceptionally hard (7 on the Mohs scale).

The peculiar composition of cobalt-colored faience suggests that the efflorescence technique was used with a mixed copper-cobalt pigment. The high alkalinity, and the firing temperatures indicated by the texture and hardness of such faience, would result in virtually complete migration of copper from the interior to the surface, to produce the type of concentration gradients observed. Comparable surface enrichment would not be expected with the other transition metals and none is observed. In figure 2 one can see that while the MnO and CoO remain evenly distributed between the body and the surface, very little copper remains inside. The crosses (+) in figure 2 suggest that a different process was used for the copper-colored variant.

Cobalt-colored body materials ceased to be made for almost a millennium after the Twentieth Dynasty. During the Ptolemaic Period a different, purer kind of cobalt shows up occasionally in some of the high-iron black bodies (e.g., vase 1913.804, table 2). The green bodies colored by copper and lead antimonate also disappear after the

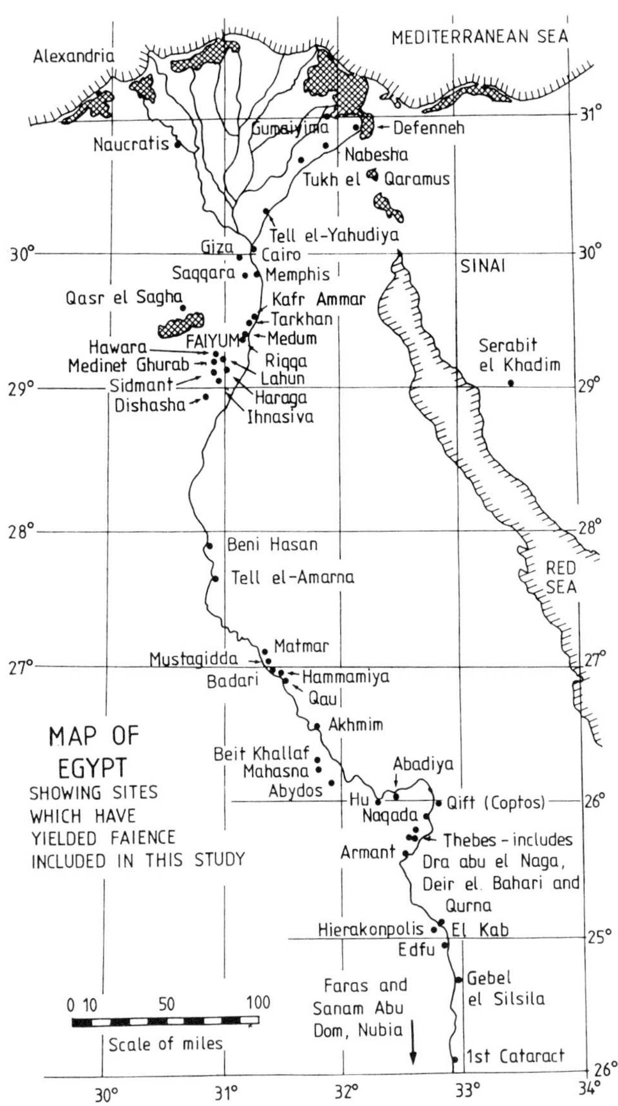

Fig. 1. Map of Egypt with sites that have yielded faience included in this study.

Fig. 2. The correlation between the surface and body concentrations of CuO, MnO, and CoO in New Kingdom faience with green and blue cores. The dots (•) and crosses (+) represent bodies colored by cobalt and copper, respectively. The solid lines represent 1:1 concentration ratios.

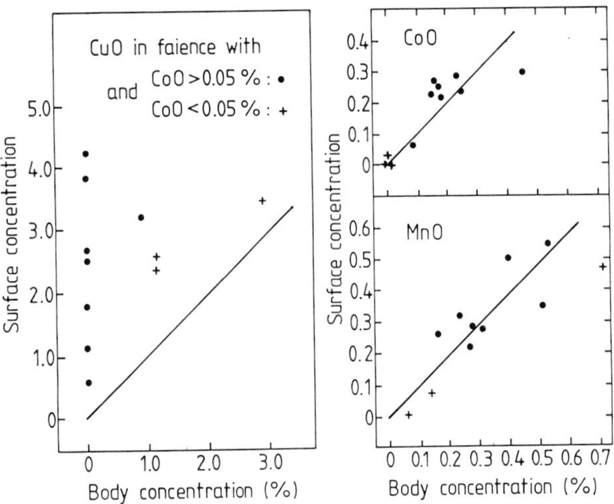

Table 3

Comparison of Ordinary and Cobalt-Colored Body Materials of Amarna Period Faience

Average % Composition of

	Na_2O	MgO	Al_2O_3	SiO_2	SO_3	K_2O	CaO
Ordinary:	0.50	0.32	0.28	93.3	0.05	0.36	1.21
Colored:	2.64	0.83	2.76	90.3	0.11	0.61	1.36

	TiO_2	MnO	Fe_2O_3	CoO	NiO	CuO	ZnO
Ordinary:	0.09	0.11	0.33	0.00	0.01	0.18	0.02
Colored:	0.11	0.28	0.73	0.18	0.10	0.15	0.18

Note: Excluded from the table are the concentrations of: Cr_2O_3, As_2O_3, SrO, SnO_2, Sb_2O_5, BaO and PbO, all below 0.02%.

Twentieth Dynasty, but they reappear and become very common in the Late and Ptolemaic periods (table 2).

Variant E

The so-called "glassy faience" is a misnomer, for though it may be glassy, it is not faience, and an alternative name—"imperfect glass" —has been suggested.[3] As long as there is no distinct outer layer different in texture or color from the interior, the term "faience" is inappropriate. The ambiguous nature of the material is more clearly expressed by the French name of "pâte de verre."

Lucas lists only one analysis of this variant.[5] The composition, intermediate between that of faience glazes and glass, is compatible with the physical attributes of this type of material.[3,20] Yet this blue-green fragment of undetermined provenance and date might have been misclassified by Lucas, who never saw it and relied on a rather vague description given by LeChatelier.

Few objects seen by us had the outward appearance of "glassy faience", as described by Lucas[3] and Cooney.[20] Some were green, e.g. a Twenty-Fifth Dynasty ushabti (1965.176) of unknown provenance and a Late Period sistrum (1909.1062) from Memphis. Others, such as the sistrum and vase from the Louvre collection listed in table 2, were blue. The few specimens analyzed by us had alkali and lime concentrations intermediate in value between the corresponding averages for contemporary faience glazes and glass.[10,22,23]

We were as unsuccessful as Lucas and Cooney in discovering examples of this variant dated to before the Twenty-First Dynasty. However, considering the problems of visual recognition, the exist-

Table 2

Chronological List of Selected Analyzed Colored Faience Bodies

Composition Expressed as % of Oxides

	Na	Mg	Al	Si	K	Ca	Mn	Fe	Co	Ni	Cu	Zn	Pb	Sn	Sb
E.1578	2.4	1.8	1.1	85	0.4	9.6	0.0	0.8	0.0	0.0	1.4	0.0	0.0	0.2	0.0
E.1577	2.9	2.0	0.8	84	0.3	10.0	0.0	0.7	0.0	0.0	1.3	0.0	0.0	0.2	0.0
E.3182	3.4	1.5	0.8	85	0.2	9.4	0.0	0.7	0.0	0.0	1.2	0.0	0.0	0.2	0.0
E.E.633D	—	—	—	83	0.4	0.9	3.3	0.8	0.0	0.0	0.4	0.0	0.0	0.0	0.0
1890.815	—	—	—	88	0.7	2.3	0.1	0.7	0.0	0.0	2.9	0.0	0.0.	0.0	0.0
E.E.238M	—	—	—	88	2.4	1.8	0.0	5.8	0.0	0.0	0.1	0.0	1.1	0.0	0.0
U.Coll.587	3.8	1.0	2.8	92	0.8	1.5	0.3	0.7	0.3	0.2	0.0	0.3	0.0	0.0	0.0
1889.148	—	—	—	80	2.0	3.9	0.1	2.4	0.0	0.0	1.1	0.4	7.5	0.1	0.9
1893.1-41 (471)	2.7	0.9	2.0	90	0.7	1.4	0.5	0.9	0.1	0.1	0.9	0.2	0.0	0.0	0.0
1893.1-41 (472)	1.1	0.6	1.2	94	0.3	0.9	0.2	0.4	0.2	0.1	0.0	0.1	0.0	0.0	0.0
1893.4-41 (945)	3.0	0.9	2.7	90	0.7	2.1	0.3	0.6	0.5	0.3	0.0	0.2	0.0	0.0	0.0
1924.114	3.9	1.2	3.5	87	0.6	0.9	0.5	1.7	0.2	0.1	0.0	0.3	0.0	0.0	0.0
1931.511	2.3	0.7	2.0	93	0.3	0.6	0.2	0.4	0.2	0.1	0.0	0.1	0.0	0.0	0.0
1893.1-41 (481)	3.8	1.0	2.5	94	1.0	1.2	0.4	0.6	0.2	0.1	0.0	0.2	0.0	0.0	0.0
1893.1-41 (439)	3.4	1.3	4.0	91	0.4	0.9	0.3	1.3	0.2	0.1	0.0	0.3	0.0	0.0	0.0
1935.590a	—	—	—	90	0.8	1.7	0.0	5.4	0.1	0.0	0.1	0.0	0.0	0.0	0.0
1936.635d	0.9	0.3	2.3	85	0.9	1.2	0.0	6.0	0.0	0.0	0.0	0.0	0.0	0.0	0.0
1890.906	—	—	—	86	0.7	1.3	1.0	1.8	0.2	0.2	0.0	0.2	0.1	0.0	0.0
1871.34A	0.9	0.6	1.2	92	0.4	1.9	0.7	1.3	0.0	0.1	1.1	0.0	0.0	0.1	0.0
1965.176	—	—	—	90	0.5	3.2	0.1	0.5	0.0	0.0	1.3	0.0	0.2	0.1	0.0
Louvre N.2264	<3	<0.5	<0.3	78	<0.2	4.3	0.0	0.6	<0.1	—	1.3	—	3.1	0.0	1.2
1916.2	0.4	0.3	0.4	91	0.0	2.7	0.1	0.5	0.0	0.0	1.4	0.0	0.0	0.0	0.0
Louvre E.3668	<3	<0.5	<0.3	88	<0.2	2.3	0.0	0.4	<0.1	—	1.3	—	0.0	0.0	0.0
E.4548	—	—	—	79	0.7	2.0	0.0	1.5	0.0	0.0	1.2	0.0	0.2	0.1	0.1
1909.1062	1.5	1.3	0.7	84	0.4	4.5	0.0	0.5	0.0	0.0	2.1	0.0	1.6	0.0	0.5
Louvre E.22355	<3	<0.5	<0.3	83	<0.2	5.3	0.0	0.5	<0.1	—	0.7	—	0.8	0.0	0.3
Louvre (no number)	<3	<0.5	<0.3	89	<0.2	1.0	0.0	0.6	<0.1	—	1.4 ·	—	0.1	0.3	0.0
Louvre N1673	<3	<0.5	<0.3	79.	<0.2	6.8	0.0	0.9	<0.1	—	1.8	—	2.5	0.1	0.8
1913.804	1.9	1.1	2.5	64.	0.6	4.1	0.0	9.8	0.7	0.0	3.0	0.1	8.3	0.1	0.1
1910. 551(2)	2.5	0.1	0.3	90.	0.2	1.5	0.0	0.5	0.0	0.0	0.0	0.0	1.8	0.0	0.7

E.1578: Pale blue core of blue vase Dyn. I (Djer), Abydos, tomb 0

E.1577: Pale blue core of blue vase Dyn. I (Djer), Abydos, tomb 0

E.3182: Blue green core of blue green vase Dyn. I (Udimu), Abydos, tomb T

E.E. 633D: Grayish brown core of black bead Dyn. XII, Abydos, tomb 416

1890.815: Green core of green vase Dyn. XVIII, Lahun, tomb of Maket

E.E.238M: Red core of red bead Dyn. XVIII, Coptos, temple of Tuthmosis III

U. Coll.587: Gray core of kohl tube Dyn. XVIII, temple of Amenhotep II, Thebes

1889.148: Green core of green kohl tube Dyn. XVIII, Tell el-Amarna

1893.1-41(471): Blue core of blue vase Dyn. XVIII, Tell el-Amarna

1893.1-41(472): Blue core of blue green bowl Dyn. XVIII, Tell el-Amarna

1893.4-41 (945): Blue core of green sistrum Dyn. XVIII, Tell el-Amarna

1924.114: Bluish core of vase Dyn. XVIII, Tell el-Amarna, house M50.33

1931.511: Blue core of blue crown Dyn. XVIII, Tell el-Amarna, house T.34.1

1893.1-41(481): Gray core of violet jar lid Dyn. XVIII, Tell el-Amarna

1893.1-41(439): Blue core of violet statue Dyn. XVIII, Tell el-Amarna

1935.590a: Red core of red inlay plaque Dyn. XVIII, Tell el-Amarna

1936.635d: Red core of red inlay Dyn. XVIII, Tell el-Amarna

1890.906: Gray core of blue kohl tube Dyn. XIX, Medinet Ghurab

1871.34A: Blue core, uraeus Dyn. XX, Tell el-Yahudiya, palace of Ramesses III

1965.176: Green surface of green ushabti Dyn. XXV (glassy faience?)

Louvre N.2264: Green core of green sistrum, Dyn. XXVI (Amasis cartouche)

1916.2: Greenish gray core of green ushabti Dyn. XXVI (Psammetichus II), Giza

Louvre E.3668: Blue interior of blue sistrum, Late period (glassy faience?)

E.4548: Green core of green model coffin Late Period, Naucratis

1909.1062: Green interior of green sistrum Late Period, Memphis, Apries palace

Louvre E.22355: Green core of green sistrum, Dyn. XXX (Nectanebo I cartouche)

Louvre (no number): Pale blue core of blue vessel, Ptolemaic Period

Louvre N1673: Blue interior of blue vase, Ptolemaic Period (glassy faience?)

1913.804: Gray streak, core of marbled vase Ptolemaic Period (glassy faience?)

1910.551(2): Yellow core of polychrome vase Ptolemaic Period, Memphis

Table 4
The Concentration Ranges (%) and Median Values of Selected Components of Egyptian Blue and Green Frits for Various Time Periods

	Na_2O	SiO_2	K_2O	CaO	CuO	SnO_2	PbO
Old Kingdom:	—	57-64	0.0-0.8	15-20	11-17	0.0-0.1	0.0-0.1
		(64)	($<$0.1)	(18)	(13)		
Middle Kingdom:	—	61-77	0.0-1.2	7-15	6-11	0.0-0.1	0.0-0.1
		(68)	($<$0.1)	(12)	(10)		
New Kingdom:	0.5-6.0	60-88	0.0-0.8	3-16	3-16	0.0-1.5	0.0-0.2
	(5)	(70)	(0.5)	(10)	(10)	(1.0)	($<$0.1)
Late Period:	—	70-88	0.0-0.4	2-13	2-10	0.1-1.0	0.0-1.5
		(75)	($<$0.1)	(7)	(6)	(0.4)	(0.1)
Ptolemaic Period:	—	70-85	0.0-0.6	6-8	2-13	0.1-1.0	0.0-1.7
		(80)	(0.4)	(7)	(7)	(0.3)	(1.3)
SOME OTHER ANALYSES OF NEW KINDGOM BLUE FRITS							
Ref. 24	1.62	60.6	trace	16.1	12.3	1.79	—
Ref. 28	0.8-7.6	57-89	0.0-2.0	8-14	2-21	—	—
		(70)		(9)	(16)		

Note: The median values are given in parentheses. In computing the median tin oxide concentration we excluded objects in which no tin was detected.

ence of older examples in some museum collection cannot be ruled out. Anyone who has studied Egyptian vitreous materials must be well aware how difficult it is to distinguish "glassy faience" from aged glass or polished frit.

Frits

The term "frit" refers to a material produced from silica, lime, and alkali with or without a pigment and heated high enough to fuse but not high enough to be capable of plastic flow. While numerous molded objects have been recovered, showing that frit was used as a substitute for glass and faience, the archaeological contexts within which the many lumps and cakes of raw material were discovered suggest that these were destined for the painter's palette.[24,25]

It has also been proposed that some frits may have served as intermediates in the manufacture of glass[4,26] and faience.[3,7,27] Lucas mentions that a number of imperfectly fired faience objects showed clearly discernible frit particles on the surface.[3] He lists six analyses of blue frits of various time periods,[28] but does not discuss the relationship between the elemental compositions of the frits and those of faience. He notes, however, that the composition of New Kingdom glass is incompatible with that of contemporary frits, and suggests that at least the frits recovered from Tell el-Amarna could not have served as precursors of glass.[26]

Kühne has suggested that powdered glass or some sintered material such as frit might have served as the binding agent for faience.[27] He computed the hypothetical composition of such a binder and its proportion in relation to silica and concluded that an 80:20 silica:binder mixture, where the binder contained about 16% CuO, 2.9% alkaline earths (MgO, CaO), 24% alkali (Na_2O, K_2O), 55% SiO_2 and 0.7% SnO_2 would yield a glaze of the composition he found in two specimens of Amarna faience. Our figures for the same elements in Amarna blue and green faience[10] suggest a binding frit resembling the binder proposed by Kühne, and a 70:30 mixing ratio.

Unfortunately hardly any of the Egyptian frits analyzed to date (more than sixty by us) have elements in the proportions indicated above. The overwhelming majority of blue and green frits have inadequate amounts of alkali and far too much lime (table 4) to have been used as suggested by Kühne.

That most faience glazes contain concentrations of alkali far in excess of what could have been derived solely from the frits can be explained by allowing for alkali migration to the surface (efflorescence), or by postulating that additional alkali were added to the silica-frit mixture. However, there is no way to explain how the very high CaO levels in most frits could have produced glazes with so little lime. Once the frit is made, how does one remove the lime selectively? If a higher proportion of silica is postulated to accommodate the low experimental levels of lime (a 90:10 mixing ratio is needed to yield the average concentration of CaO in Amarna glazes), the computed levels of CuO and most other ingredients fall far below their corresponding experimental values.[10]

Our analyses reveal that the composition of beads, amulets, vessels, and other objects made of unglazed frit, in general, resembles that of lumped or powdered frits that were definitely designed to serve as painting pigments.[24,29] A different recipe must have been employed to produce frits from which glass or faience were to be made.

Among our specimens, only three Ptolemaic frits from the Kom Qalana kilns at Memphis might have been destined for the manufacture of glass or faience. The three, two of which still adhered to the fritting pans, could have served as a binder in the manner suggested by Kühne, since they were not overburdened with the excessive lime levels seen in earlier frits (table 4).

In many respects the chronological changes in the composition of frits (table 4) parallel the changes in faience and glass,[10] with one puzzling exception. While the occurrence of arsenic, lead, and tin in the frits coincides in time with the occurrence of these same elements in faience and glass, one finds very little manganese or cobalt in the frits analyzed to date. Even if we allow for the fact that the specimens selected by us and others are unlikely to represent random samples, the elusiveness of such frits during times when manganese and cobalt were relatively common in glass and faience bespeaks their scarcity.

Alloyed copper scrap, "bronze scale," was used, apparently, in the same indiscriminate manner in the production of frits as in the manufacture of faience.[10] The fraction of pre-New Kingdom frits containing As_2O_3 and the absolute levels of arsenic are both much higher than in contemporary glazes, indicating that the loss of arsenic through volatilization was not as serious as in faience. These data accord well with the conclusions of Lucas[26] and Turner[4] that the fritting was done at or below 750°C, well below the temperatures needed to form alkaline glazes.

Concluding Remarks

It must be apparent by now that the present classification of faience into variants A through E is not satisfactory. We would like to suggest a modification of Lucas's original scheme, and the rationale for the proposed changes is presented below.

The term "ordinary faience" might be retained for faience whose body material shows no evidence of having been deliberately pigmented. Thus, the core of such faience can be white, creamy, gray, brown, pink, or any of the other colors encountered in Egyptian sands. If the body material is greenish or bluish and the copper concentration in the interior is far below that found in the glaze, the faience is probably of the ordinary variety, unless it can be demonstrated that the efflorescence process could not have been used on the object in question. The term "Variant A" might be re-

tained, but extended to faience in which an intermediate layer of any color intervenes between the body and the glaze.

When Lucas classified Egyptian faience into variants B, C, and D, the principal feature distinguishing the three types was the physical appearance of the objects — black-bodied black faience, red-bodied red faience, and blue- or green-bodied blue or green faience. Iron was held responsible for the colors of variants B and C, and copper for variant D. From our studies, however, the picture is more complex.

Now, if the colors of the body and the glaze are to be the chief criteria for classification, then many more letters of the alphabet will be needed to distinguish objects with cores of colors other than black, red, green, or blue, and cases in which the body and the glaze differ in color. For example, what letter would one assign to a bead (E.1183) from the Abydos tomb of Anedjib (First Dynasty) having a green glaze (0.8% MnO, 1.0% Fe_2O_3 and 7.2% CuO) over a purple core pigmented by: 1.3% MnO, 1.0% Fe_2O_3 and 0.5% CuO? Maybe a black body was intended, but unanticipated reduction of the manganese produced the purple color observed now. And what about the Ptolemaic vases whose bodies contained enough lead antimonate to indicate that the distinctive bright yellow core was created deliberately to provide a contrasting ground to the superimposed polychrome designs (vessel 1910.551 in table 2)?

Hence we propose that the use of the letters B, C, and D be abandoned and that such faience simply be described as being black-bodied, red-bodied, blue-bodied, etc. This has the additional value of by-passing the question whether a given core was colored intentionally or not.

We are no more happy with the term "glassy-faience" than Lucas was,[3] but it is preferable to the term "Variant E." The latter name implies that such material is just another form of faience, which it distinctly is not. In its properties it resembles glass much more than it does faience, as the alternate name "imperfect glass" clearly states. Were it not that the term "glassy-faience" is well enshrined in several important publications we would suggest it be discarded too.

Acknowledgments

We are most grateful to Drs. Helen Whitehouse, Barbara Adams and Jean-Louis de Cenival for allowing us to analyze objects from the Ashmolean, the Petrie and the Louvre Museum, respectively. Thanks are also due to Miss Helen Hatcher (Oxford) and Mr. Alain Leclair (the Louvre) for some of the analytical work cited in this paper.

References

1. E. T. Hall. F. Schweizer and P. T. Toller. "X-Ray Fluorescence Analysis of Museum Objects." *Archaeometry*. 15 (1973): 53-78.

2. Ch. Lahanier. "Compositions Optimales de la Perle au Borax de Silicates." *PACT*. 1 (1977): 41-46.

3. A. Lucas and J. R. Harris. *Ancient Egyptian Materials and Industries*. 4th ed. (London: Edward Arnold. 1962) pp. 157-164.

4. W. E. S. Turner. "Studies in Ancient Glasses and Glassmaking Processes. Part V.: Raw Materials and Melting Processes." *Journal of the Society of Glass Technology*. 40 (1956): 277-300

5. A. Lucas and J. R. Harris, op. cit., Appendix. pp. 474-475. 481.

6. J. V. Noble. "The Technique of Egyptian Faience." *American Journal of Archaeology*, 73 (1969): 435-39.

7. C. Kiefer and A. Allibert, "Pharaonic Blue Ceramics: The Process of Self-Glazing," *Archaeology*. 24 (1971): 107-117.

8. M. S. Tite, I. C. Freestone and M. V. Bimson. "Egyptian Faience: an Investigation of the Methods of Production," *Archaeometery*. 25 (1983): 17-27.

9. H. LeChatelier, "Sur les Poteries Egyptiennes." *Comptes Rendus de l'Academie des Sciences*, 129 (1899): 477-480.

10. A. Kaczmarczyk and R. E. M. Hedges, *Ancient Egyptian Faience* (Warminster, England: Aris and Phillips, 1983).

11. E. V. Sayre and R. W. Smith, "Analytical Studies of Ancient Egyptian Glass," in A. Bishay, ed., *Recent Advances in Science and Technology of Materials*, Vol. 3 (New York: Plenum Press, 1974) pp. 47-70.

12. E. V. Sayre, "The Intentional Use of Antimony and Manganese in Ancient Glasses," in F. A. Matson and G. E. Rindone, ed., *Advances in Glass Technology*, Part 2 (New York: Plenum Press, 1963) pp. 263-282.

13. W. Noll and K. Hangst, "Contributions to the Knowledge of Ancient Egyptian Blue Pigments," *Neue Jahrbuch für Mineralogie*. (Monatshafte, 1975): 209-214.

14. W. Noll, "Mineralogy and Technology of the Painted Ceramics of Ancient Egypt," in M. J. Hughes, ed., *Scientific Studies in Ancient Ceramics,* (British Museum Occasional Paper No. 19) (London: The British Museum, 1982) pp. 143-152.

15. H. G. Bachmann, H. Everts and C. A Hope, "Cobalt-Blue Pigment on 18th Dynasty Egyptian Pottery," *Mitteilungen des Deutschen Archaeologischen Instituts, Abteilung Kairo*, 36 (1980): 33-37.

16. A. Lucas and J. R. Harris, op. cit., pp. 257-259

17. H. J. L. Beadnell, "Dakhla Oasis: Its Topography and Geology," in *Geological Survey Report*, 1899, Part IV (Cairo: National Printing Department, 1901) p. 130.

18. H. Droop Richmond and H. Off, "Indications of a Possible New Element in an Egyptian Mineral," *Journal of The Chemical Society*, 61 (1892): 491-495.

19. A. Kaczmarczyk, unpublished data.

20. J. D. Cooney, "Glass Sculpture in Ancient Egypt," *Journal of Glass Studies* 2 (1960): 32-39.

21. H. LeChatelier, "Sur la Porcelaine Egyptienne," *Comptes Rendus de l''Academie des Sciences*, 129 (1899): 377-378.

22. A. Lucas and J. R. Harris, op. cit., pp. 476-480.

23. E. R. Caley, *Analyses of Ancient Glasses*, 1790-1957 (Corning, New York: The Corning Museum of Glass, 1962).

24. S. A. Saleh, Z. Iskander, A. A. El Masry and F. M. Helmi, "Some Ancient Egyptian Pigments," in A. Bishay, ed., *Recent Advances in Science and Technology of Materials*, Vol. 3 (New York: Plenum Press, 1974) pp. 141-155.

25. A. Lucas and J. R. Harris, op. cit. pp. 340-345, 362-363.

26. Ibid, pp. 186-191.

27. K. Kühne. "Aegyptische Fayencen." Part I. *Grabungen der Deutschen Orient. Gesellschaft* (1969): 5-27.

28. A. Lucas and J. R. Harris. op. cit.. p. 495.

29. Ch. Lahanier. Research Laboratory. Louvre Museum. Paris. unpublished data.

W. D. KINGERY and P.B. VANDIVER

The Technology of Tiffany Art Glass

The technology involved in the production of lustered and iridized art glass by Louis C. Tiffany has been investigated by analysis of fragments and replications, and found to be far more complex than had been believed. Glass compositions, surface reduction processes, and thin-film coatings are discussed within an historical perspective of the late nineteenth-century fashion for Art Nouveau glass and the contemporary understanding of glass technology. One vase is described in detail with regard to its composition, microstructure, and manufacture, as an example of technology focused to produce objects with striking visual impact.

Early in his career Tiffany realized the necessity of being at the forefront of glass technology to bring to fruition his artistic intent of investigating the interaction of light and glass. A complex materials technology was required to achieve the visual effects desired by Tiffany.

The optical effects achieved in the glass of Louis C. Tiffany are remarkable for their visual power and well known for their diversity. In contrast to general belief, we have found that Tiffany and his chemists did develop new compositions that can only enhance his established reputation as an artist-designer. His craftsmen developed new effects based on an experimental approach to processing, using variations in surface texture, color, and the multiple layering of materials in such a way as to emphasize the interaction of the glass with different lighting conditions. The best of their works seem to have a life of their own, changing color with the quality of light during a daily or seasonal cycle. In examining seventy fragments of Tiffany glass, we found many processes, often individually well-known to glass makers of the time, were combined, using an empirical and synthetic approach, in different sequences for different objects in order to produce a particular optical effect.

These processes include forming and decorating with different colored glasses, often juxtaposing opaque, transparent and opalescent glasses; varying not only composition but also heat treatment to affect opalescence, reflectance, and surface texture; as well as applying a reflective silver luster or an iridescent surface coating, or both. Rather than being tightly bound to the use of soda-lime-silica compositions for his window glass and lead silicate for his blown wares, Tiffany used a variety of compositions and processes at the forefront of glass technology to achieve striking optical effects.

Louis Comfort Tiffany (1848-1933), scion of a rich jewelry merchant of New York, began his career as a painter.[1] He had a good sense of form and a special aptitude for the use of light and color, and his talent was recognized while he was still young. He toured Europe, Africa, and the Near East, being much impressed with the brilliant colors and the quality of light in the leaded glass of Gothic cathedrals and Byzantine mosaics, the linear quality and textures of Islamic art, and the interplay of light and color in the iridescence on Roman glass. He returned to the United States at the age of twenty-two to continue painting, but was increasingly interested in making glass windows. The philosophy of the English Arts and Crafts Movement impressed Tiffany, and in 1879 he founded an interior decorating company that employed his window glass and mosaic tiles.

Tiffany was unsatisfied with the colors in glass produced by enamelling and sought to obtain the colors and textures he wanted, first by special order from glasshouses, and later from his own shops. In 1875 he was already experimenting with "drapery" glass, producing a layered effect by folding sheets of glass while hot. That Tiffany understood the optical effects he wanted is indicated by three patents from 1881 and 1882 that describe windows or mosaics composed of multiple layers with an opaque or opalescent backing to control reflectivity.[2]

In 1892 Samuel Bing persuaded Tiffany to design and build a chapel for the Columbian Exposition of 1893.[3] Bing commissioned other works, establishing Tiffany's reputation as the leading American contributor to the Art Nouveau movement. In the same year Tiffany met Arthur J. Nash, manager of the English firm Thomas Webb and Sons where iridized "bronze" glass had been manufactured in the 1880s. Tiffany established his own glasshouse in Corona, Long Island, where production began in 1893. Nash was hired to manage production, and Dr. Parker McIlhenny, a Scottish chemist, was hired to assist with research. Thus began a collaboration between Tiffany the artist, McIlhenny the chemist, and Nash the craftsman that was enormously productive.

Tiffany operated a decorating company for thirty-nine years (1885-1924), and a glassworks for thirty-one years (1893-1924). At its height, there were five glass shops producing windows and art ware. There were divisions of Tiffany's enterprise for design, glass production, pottery, enamels, metal work, and window fabrication. When Tiffany dissolved the company, some 600,000 pieces were sold at auction.

Although Tiffany glass is renowned for a wide range of optical effects, we know of no analyses of the compositions or processes employed, nor of any documentary evidence other than the limited information in the interview by R. Koch with gaffer Jimmy Stewart.[1] The window and lamp glasses are described as "traditional lime glasses, about 70% silica, 15% soda and 10% lime" and the vases as "lead glass, about 50% silica, 30% red lead (a lead oxide) and 15% pearl ash (a form of potash)."[4] In an 1881 patent Tiffany described the production of iridescent surfaces as "processes being well known to glass-manufacturers".[5] From this study, we hope to gain insights into the nature and extent of experimentation and into the relationship between artist, chemist, and craftsman. We also hope to learn about the level of glass technology at the turn of the century in an art glass factory operating at the forefront of the field.

Methods of Study

Glass as an art medium has enormous possibilities for variations of shape, color, reflectance, transmission, and surface texture that can be combined in innumerable ways to achieve the artist's purpose. Understanding how and why a glass object looks the way it does requires careful macroscopic and microscopic examination, determination of both composition and microstructure, and replication of processing. If a glass container containing 5% phosphate is rapidly cooled, it may be clear and transparent; if the same glass is treated to precipitate calcium phosphate particles, it can become slightly opalescent; longer heating may form an opaque milk glass. Microstructure, composition, and heat treatment are equally important in determining surface reflectivity.

Examination of an object or fragment began with observation of the surface characteristics and internal structure using a hand lens and optical microscope, enhanced by viewing cross-sections in various lighting conditions at high magnification. Scanning electron microscopy has been employed for magnifications up to 20,000X of a surface, or fractured or etched cross-section. Scanning electron microscopy has the advantage that simultaneous use of energy-dispersive x-ray analysis allows qualitative determination of local chemical composition. For higher magnifications and details of such structural characteristics as surface coatings and precipitates, transmission electron microscopy has been used after thinning the sample to electron transparency with an argon ion-milling technique.

Chemical compositions were first qualitatively determined using energy-dispersive x-ray analysis. Trace elements were determined by emission spectroscopy. Atomic absorption spectroscopy was used for boron. For other major and minor elements, electron beam microprobe analysis was employed, using wavelength dispersive techniques, geological standards, and Bence-Albee correction factors for differential matrix effects.

Seventy samples of Tiffany glass have been given or loaned to us for study by the Metropolitan Museum of Art, New York; Morse Gallery of Art, Winter Park, Florida; Lillian Nassau Gallery, Ltd., New York, and from the collections of Robert Koch and James Lundberg.

Optical Effects in Art Glass and How They Are Produced

Color

Glass can range from waterlike transparency to black opacity. The greatest part of the palette of colors results from the addition of a small amount (0.25-4%) of colorants, for example, oxides of copper, cobalt, iron, manganese and other transition metals. Iron is almost always present as a minor impurity that gives yellow green tints; to counteract this, a small amount of manganese, which in larger concentrations gives a purple color, may be added to the batch, giving the glass a neutral tint but diminishing the overall transmission. To obtain a transparent red, colloidal copper or gold particles are precipitated or formed by first cooling the glass to develop tiny nuclei and then reheating to "strike" or develop the color by growing the nuclei to a large particle size, on the order of the wavelength of light. A range of effects from translucence to opalescence to milk white opacity results from separating the glass during cooling, into either two liquids, or a crystalline and a liquid phase. The most common additives used to form opal glasses are fluorides and phosphates, although in certain instances two glasses without these additives may be immiscible (like oil and water) and have sufficiently different indices of refraction to produce an opal glass. All of these ways of producing color are found in Tiffany and other Art Nouveau glass, and many of the different glass colors were made sufficiently compatible in cooling behavior to be combined in the same object.

Surface Reflectivity

The specular reflectance (or highlighting) from the surface of a glass depends on its index of refraction and smoothness (fig. 1). The amount of light reflected from a smooth surface is about 4% of the incident light for a soda-lime-silica glass and about 6% for a lead-silicate. Increased specular reflectance gives glass a gemlike brilliance, the reason high lead and high barium are used in "crystal" glass compositions. This quality in Tiffany glass is subservient to qualities derived from diffuse reflectivity. A diffuse reflectivity is obtained from surface roughness or from subsurface opalescence. The incident

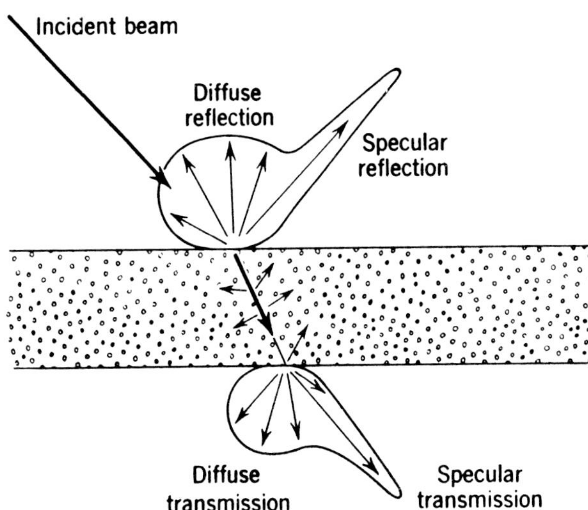

Fig. 1. Illustration of the influence of surface roughness and internal opalescence which give rise to a decrease in specular reflection and transmission and an increase in diffuse reflection and transmission in a plate of translucent glass (from W. D. Kingery, H. K. Bowen and D. R. Uhlmann, *Introduction to Ceramics*, ref. 6).

light beam is reflected almost equally in all directions or scattered so that surface gloss and reflected images are diminished (fig. 1), and surface textures may be described as having the appearance of satin or velvet. Opalescence and surface roughness also give rise to diffuse transmission.

Metallic Luster

Art glass commonly appears metallic, usually golden, due to fine particles of silver precipitated at and adjacent to the surface by reaction, for instance, of silver oxide in the glass being placed in a reducing flame and reacting to form the metal. In low concentration, these fine colloidal metal particles reflect and scatter the light, giving a soft luminosity to the partly translucent surface. Higher concentrations produce an almost opaque film, such as that found in Islamic lusterware. A continuous film, such as that found on Christmas tree ornaments, gives mirrorlike specular reflection but is opaque in transmission, whereas metallic lusters on Tiffany glass are not continuous and give a yellow to orange transmission color.

Surface Iridescence

When visible light (wavelength 0.35-0.75 microns) is reflected from a thin oil film on the surface of water, some parts of the light spectrum are reinforced and some tend to cancel as the light passes through the film and is reflected at the interface. A similar phenomenon occurs in the iridescent surface coatings on art glass. When viewed at differing angles, the wavelengths that are reinforced change. The particular rainbow effect depends on the curvature of the surface, film thickness, and the angle of observation. Iridescence is achieved primarily by forming a thin surface layer by deposition from a vapor phase. For instance, near the end of hot forming a glass object, the surface is reacted with an atmosphere containing tin and/or iron chloride vapors to form a thin layer of iron oxide and/or tin oxide. Pure tin oxide gives a silvery iridescence, whereas iron oxide gives a golden color. One Tiffany patent states that both vapor deposition and direct application were possible means of applying the iridescent coatings.

The Influence of Shaping and Decorating

An optical effect is determined as much by the techniques and sequences of shaping and decorating as by the compositions of the constituent parts. Sheet glass may be formed by pouring the molten glass on an iron plate and flattening it with a roller, by passing the glass between rollers, or by blowing. Textures on rolled glass can be imposed from a design on the roller. Mixtures of different colored glasses can be made by ladling different compositions onto the plate or into the roller reservoir, by putting fragments of glass on a table and pouring molten glass over them, or by trailing or dripping one glass over another before rolling in a variety of ways. Sheet glass is blown by one of two processes. A bubble can be blown, transferred to a solid pontil rod, and then spun out to open the glass into a disc or crown; or a cylinder can be blown. After the ends are cut off, the cylinder is cut along the axis and reheated until it opens and flattens.

For blown ware, a gather of glass on the end of a blowpipe is first marvered (rolled along an iron plate to cool the surface and make it tractable to blowing), then a bubble is blown into the marvered bullet shape at the end of the blowpipe. More gathers of glass or decorative designs may then be added to the inflated glass bubble or *paraison*. Designs are formed by attaching or trailing molten or preheated colored glass in a pattern, by wrapping a molten thread of glass around the paraison, or by adding small bits of molten glass called *prunts*. Trails or prunts can be manipulated on the surface with a simple hooked tool. Feather designs are made by wrapping parallel threads and then drawing (or combing) the hook alternately toward and away from the blowpipe (fig. 2). Lily pads or heart shapes are made by applying a prunt and using the hook to draw one edge inward and the opposite one outward. The viscosity of the glass and the placement of the hook determine the shape. These methods, combined with blowing the shape into optic molds (open, fluted dip molds), which require further blowing, as well as blowing into piece molds (closed, multiple part molds that produce final shapes), were used in the production of Tiffany and other Art Nouveau glass.

A Tiffany Vase

The Tiffany blown glass vase illustrated in figures 3 and 4 has a clear low lead-soda-borosilicate composition with green trailed and combed decoration of nearly the same composition, colored with iron and copper (table 1). There are eight heart or leaf shape prunts with a golden colored metallic silver luster. The entire surface has a satiny, iridized layer of iron and tin oxides. The vase is signed "3986 P L.C. Tiffany Favrile" and was acquired through the Lillian Nassau Gallery Ltd. The height is 12.5cm, maximum diameter is 10.5cm; the wall thickness varies from 0.3-1.0cm.

The probable sequence of operations consisted of gathering the clear base glass, marvering, and blowing to form a thick-walled paraison. Green glass was trailed and combed. The bubble was reheated and blown; the bottom was hotter than the top and the green bands near the bottom spread more than those at the top. Next the

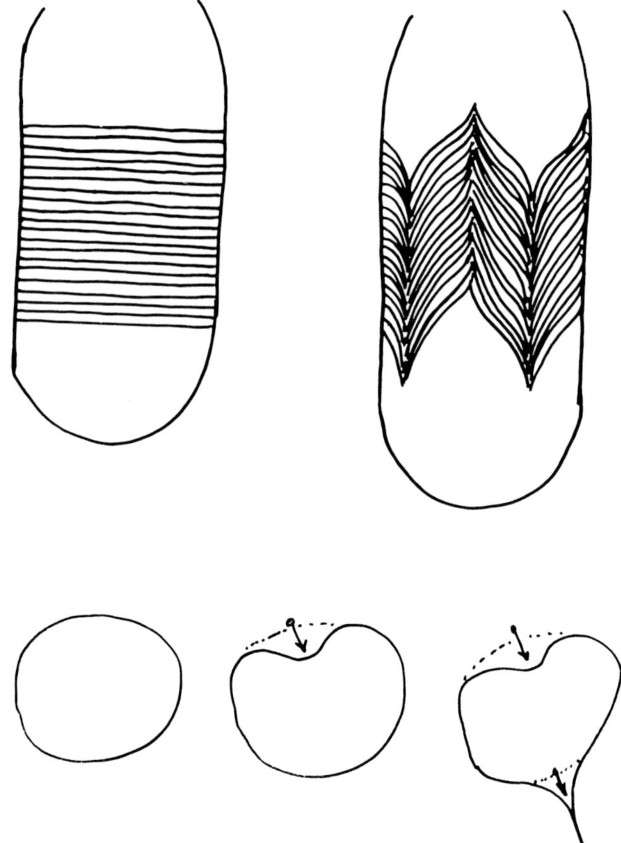

Fig. 2. (a) A paraison with threads applied is transformed into a feather design by drawing the hooked tool down through the threads and up through the threads alternately. (b) An applied prunt may be transformed into a lily pad by drawing the hook inward near the edge of the prunt and into a heart shape by using the hook starting at the very edge.

heart-shaped prunts, which show no such dissimilarity in size, were added in a silver-containing glass and combed to shape (fig. 4a). A cross-section (fig. 4b) shows how the glass was distorted by the pulling. The paraison was blown and rotated in a piece mold, as indicated by the base ring at the foot, radial chill marks on the bottom, and thickening of the wall on the interior surface beneath the heart-shaped prunts. The completed vase, however, is assymetrical. The first reason for this is that the initial paraison was blown off center and the wall thickness on one side is about twice that on the other, as shown at the lip in figure 4c; secondly, reheating caused the thinner wall to heat faster and to be drawn inward more, due to surface tension. After blowing in the mold, the vase was reheated in a reducing flame to develop the metallic silver luster on the heart-shaped prunts. These fine silver particles, which give a golden color in reflected light and a yellow color in transmitted light, are seen in figure 5 to be 100-200 Å in diameter. A bit of glass, sometimes called a false punty, was attached to reinforce the base, then the real punty, and next the mouth was opened and flattened. While on the punty, the vase was thrust hot into a fume chamber where iron and tin chloride vapors reacted with the surface to form an iridescent layer of tin and iron oxides, as shown in figure 6. This process was repeated two or three times with intermittent heating to develop a satiny texture, after which the vase was detached from the punty and placed in an annealing oven. When cold, the base was lightly ground in three places to provide a flat bottom surface.

The composition of this glass (table 1) is a low-lead soda-borosilicate, quite different from the compositions described by McKean[4] and used by turn-of-the-century glass houses. The low lead content leads to a soft copper green color with a bluish overtone that could not have been achieved in a high-lead composition. The low-lime content facilitates the formation of the silver luster, and the borosilicate composition assures durability and hardness without an increase in melting temperature. Although the green was added on top of the clear glass, the green bands are depressed compared with the clear glass. An explanation is found in the five-fold increase in alumina content, which made the green glass more viscous than the clear.

Tiffany Fragments and Replications

In addition to the vase, seventy Tiffany fragments and several replications have been examined. These include the one pressed tile and seven window glass samples illustrated in figure 7a and the six blown fragments in figure 7b. These fourteen selected fragments are described in table 2 outlining the details of processing, source of color, nature of opalescence, and such characteristics as metallic silver luster and iridescent coating. Chemical compositions were determined by electron microprobe analysis together with atomic absorption spectroscopy for boron (tables 3 and 4). Replications of Tiffany processes were generously given to us by James Lundberg of Lundberg Studios, Dominick Labino, and Steve Metcalfe of Uroboros. Other pieces were replicated in the M.I.T. Glass Lab.

The range and diversity of compositions, mechanisms for opalescence, and variation in the sequence and details of processing are summarized below. For instance sample I-A appears green in transmitted light, yet was made in a unusual way by casing a blue glass with a yellow glass, using a cylinder process. Other window glasses were made from multiple ladles of different colors and opacities of glasses poured and rolled together, such as the striated opals I-E and I-F, which have up to ten layers of different colors in a 3mm cross-section. Sample I-B has opaque gold paint on the reverse surface to increase the internal reflection and luminosity of the glass. The pressed tile (fig. 2(a)) was made by pouring blue glass over thin clear broken fragments mixed with a silver compound in a press mold. Such processes are unusual and would not be in use in any commercial glasshouse without an experimental focus on unusual optical effects.[7]

It is an oversimplification to characterize Tiffany glasses as either soda-lime-silica or lead-silicate. Most compositions are unusual and expensive for their time period, and are an indication of an empirically based experimental melting program being conducted at the Tiffany glassworks. For instance, the vase shown in figure 3 was made from a composition intermediate between a lead glass and a

Table 1

Chemical Composition of the Clear Base Glass, Green Trailed Bands and Heart-Shaped Prunts Used to Make the Vase Shown in Figure 3 (Mean of Six Analysis for Each Sample)

	clear glass*	green glass*	prunt
SiO_2	53.8 (53.5-54.3)	49.4 (48.9-49.5)	52.8
B_2O_3	10.5 (est.)	10.5 (est.)	10.5 (est.)
Al_2O_3	0.22 (0.21-0.22)	1.15 (1.13-1.16)	0.22
PbO	19.6 (19.3-19.9)	17.6 (17.2-17.9)	19.1
CaO	3.12 (3.07-3.16)	2.75 (2.74-2.79)	3.1
MgO	2.26 (2.25-2.32)	1.94 (1.91-1.95)	2.25
Na_2O	8.60 (8.39-8.69)	7.8 (7.68-8.00)	8.7
K_2O	2.83 (2.80-2.85)	2.7 (2.61-2.75)	2.8
FeO	0.07 (0.06-0.09)	2.52 (2.40-2.62)	0.08
CuO	—	2.4 (2.08-2.13)	—
TiO_2	—	1.1 (1.07-1.21)	—
BaO	—	0.17 (0.14-0.19)	—

*For the clear and green glass, the mean, minimum, and maximum experimental values are shown to indicate ranges of values typically obtained by electron microprobe analyses. Boron values were determined by atomic absorption.

soda-lime-silicate, together with a substantial addition of boron. It can be characterized as a lead-soda-lime borosilicate. Boron is described as a strong and useful flux by glassmakers during the late nineteenth century, but one that was not widely used except in experimental optical glasses because of its expense.[8] Other functions of borax are described as enhancement of color, durability, and hardness. About two-thirds of the Tiffany samples examined here are borosilicates with a substantial concentration of B_2O_3, and that finding confirms our understanding that Tiffany production was not concerned with economy.

Alumina concentrations are often high in Tiffany glasses. For instance, sample I-A is a soda-lime borosilicate with a relatively high alumina concentration and only moderate boron concentration. Microscopic examination indicates that Tiffany glasses are usually heterogeneous, containing much cord and striae, as shown in figure 4b. In some cases, the high alumina is due to the use of cryolite, Na_3AlF_6, as a source of fluorine for forming opals, but also results from the fact that fluid lead borosilicates, as well as fluorine opals rapidly attack crucibles. The higher than usual silica and alumina contents in samples I-A and I-E suggest that these glasses may have been left for some time in the pot.

The window and pressed glass compositions are unusual for their variation. Sample I-E is a soda-lime-silica glass with fluorine added as an opacifier. Samples I-B, I-F, and I-G are soda-lime-baria-silica glasses to which a fluorine opacifier has been added. Barium oxide has been used to obtain a brighter glass with a higher index of refraction, as described at the turn of the century.[9] I-F and I-G have a phosphate addition as well as fluorine, making them distinct compositions. Window and pressed glasses were also made from a lead borosilicate composition (I-C, I-D, and II-A). I-D also contains a small amount of P_2O_5, which helps to nucleate phase separation and opacification. Lead borosilicates are subject to emulsion formation in the heated state, and thus form the nearly transparent opals Tiffany found particularly attactive.

The blown glass compositions consist mainly of lead borosilicates

Fig. 3. Tiffany lead-soda borosilicate blown glass vase marked "3986 P L.C. Tiffany Favrile". The vase is 12.5cm high and has trailed bands of green overlaid with eight asymmetrically balanced silver luster hearts. The entire exterior surface has a golden iridescent coating of tin and iron oxides. The color and iridescence change with the nature and incidence of lighting, making it difficult to illustrate in a single photograph.

Fig. 4. (a) Enlarged view of one of the heart-shaped prunts illustrates how the prunt was pulled into shape. (b) 4mm cross-section from under one of the heart-shaped luster prunts in figure 3 viewed with transmitted light shows different layers resulting from trailing, combing and pulling (10X). (c) View of the top of the vase illustrates the variation in iridized metal luster appearance with angle of incident light and also the variation in the thickness. (d) The bottom of the vase shows the "false punty" added to thicken the base, along with the smaller diameter ring where the actual punty was attached.

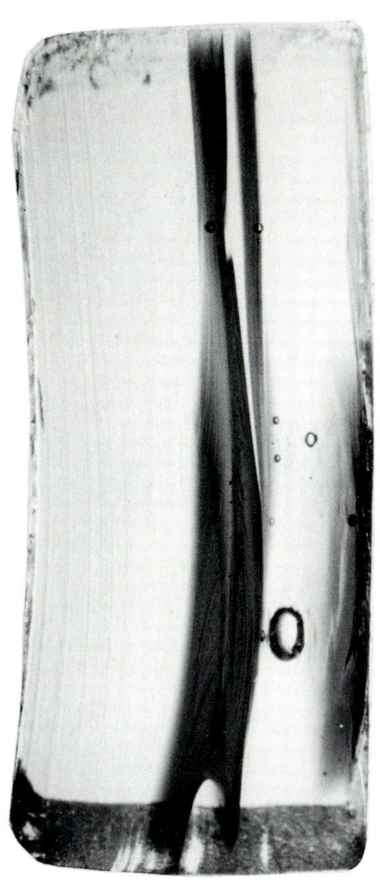

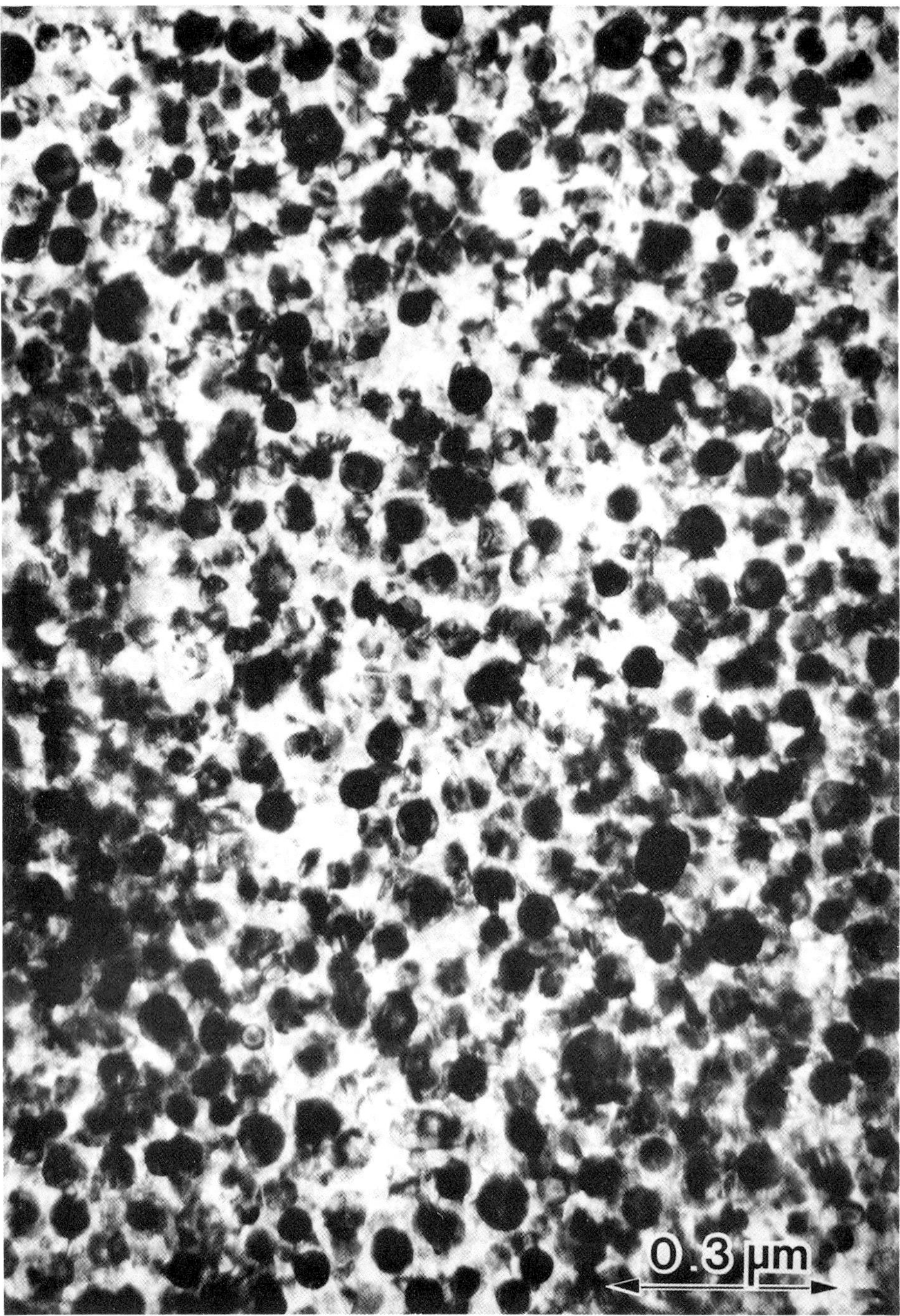

Fig. 5. The metallic luster of the heart-shaped prunts of the vase
consists of individual particles of metallic silver having a particle size about
100Å in diameter. In transmitted light the silver layer has a yellowish tinge, but
it remains partially transparent, which would not be the case if there were a
continuous metal film. (110,000X)

Fig. 6. (a) At high magnification (22.400X). the iridescent coating on the vase is seen to consist of a layer of submicron-particle sized crystals of iron oxide and tin oxide. (b) A qualitative energy-dispersive x-ray analysis indicates that both iron and tin are present in the coating. as does x-ray diffraction. This is a section over one of the heart-shaped prunts and the silver content of the glass is also indicated. The lead silicate base glass is seen through the thin coating. (A gold coating was added during sample preparation.) (c) At lower magnification. the iridescent coating is seen to be buckled. giving rise to a diffusely reflecting surface.

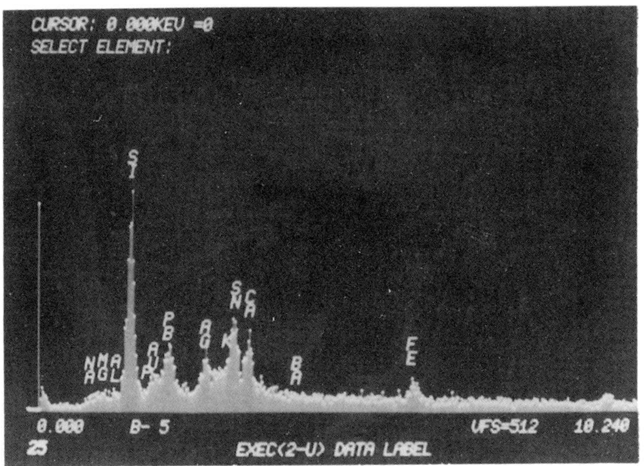

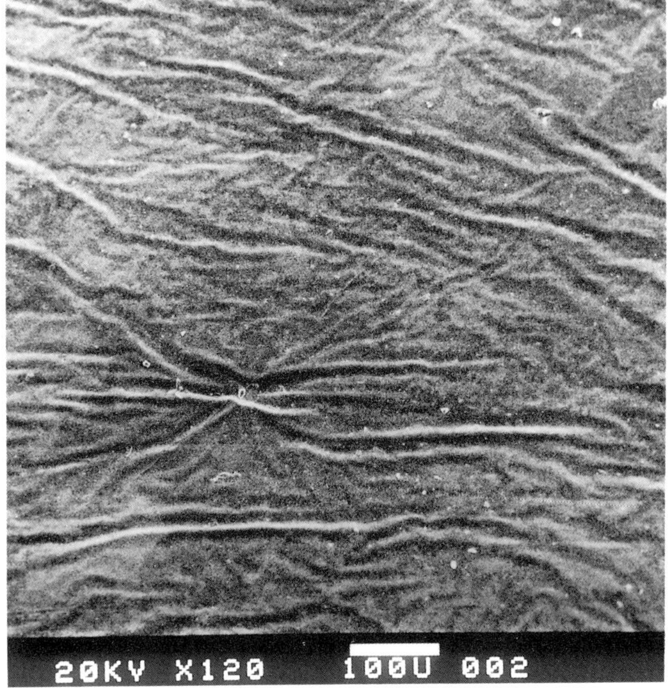

Fig. 7. (a) Fragments of Tiffany window and pressed glass analyzed by microprobe technique, described in table 2. (b) Fragments of analyzed Tiffany blown
glass, described in table 2.

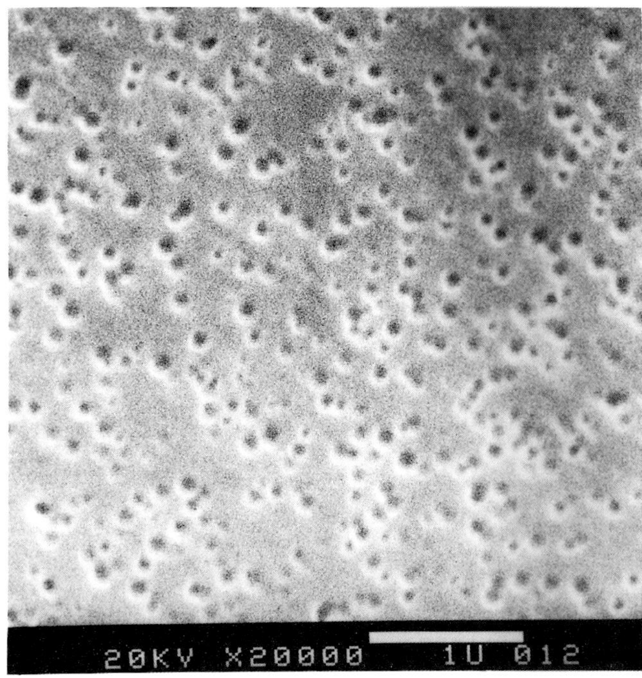

Fig. 8. Phase separation accounts for the opalescent effect of Tiffany glasses. Tiffany used a number of different compositions to achieve his desired results. (a) Sample I-B is a soda-lime-barium silicate containing about 6% fluorine (20,000X). (b) Sample I-D, white opal formed from phase separation in a lead borosilicate containing only 0.26% phosphate and no fluorine (20,000X). (c) Sample III-C, a phase separated lead-potassium silicate which has been re-heated a number of times during iridizing the surface to give a wide variation in droplet size. The two phases have similar indices of refraction so that a light opalescence results (20,000X).

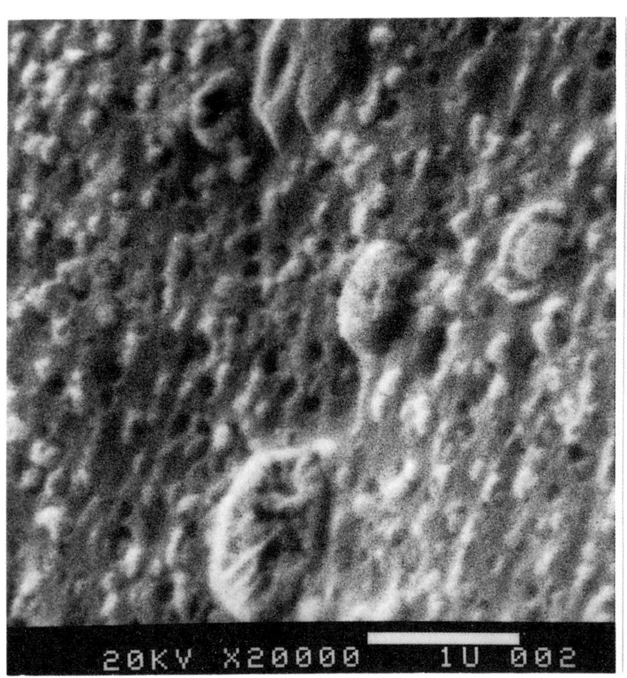

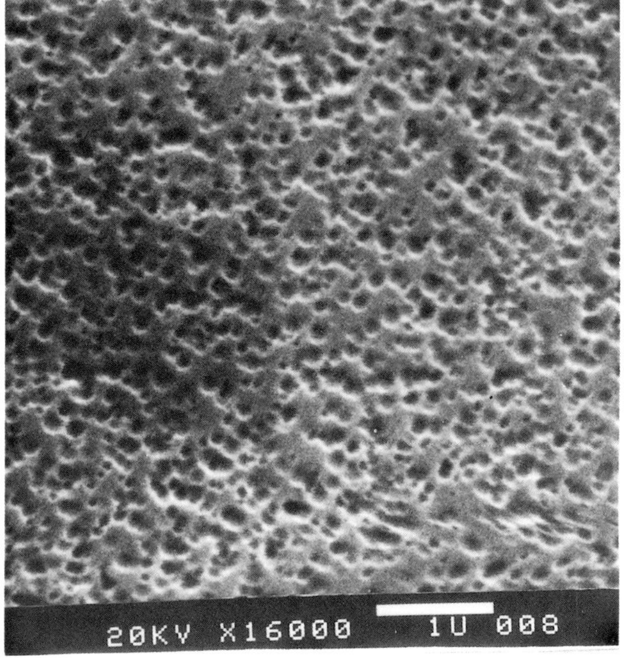

(III-B, III-D, III-E, III-F), but also include one sample of a lead silicate without boron (III-C) and a sample of soda-baria borosilicate (III-A). This sample has 2.6% phosphate, but has not had the necessary heat treatment to develop opalescence. The lead borosilicates with low alkali content phase-separate to form opals without the addition of phosphate or fluorine, as does the lead silicate with a high potassium content (III-C).

Nine of the fifteen samples are at least partly opalescent, varying from translucent to opaque white. Opalescence, a quality much admired by Tiffany, was produced by the variety of compositional deviations described in tables 3 and 4. To create opalescence, a two-phase mixture must be formed in the glass, consisting either of crystals in the glass or of a liquid-liquid emulsion. Development of phase separation requires time at certain temperatures, and for some compositions the glass must first be cooled for nucleation, followed by striking or growing of the second phase at higher temperatures. The most common opalescing agent was fluorine, well known at the time, in which a fluorine-rich phase separates as small droplets of about 0.1 microns, such as illustrated in figure 8a for sample I-B.[10] The lead borosilicate (I-D) phase separated by heat treating with the addition of a small amount of phosphate is shown in figure 8b. Opalescence from lead borosilicate phase separation was not commonly used at the time.[10] Liquid-liquid phase separation is also found in the lead-potassium-silicate, illustrated in figure 8c (III-C). Longer heat treatment has led to larger particle size droplets (1-2 microns). Since the opalescence is slight, the two phase-separated compositions must have similar refractive indices.

Tiffany glasses frequently included a silver metallic luster at the surface, usually under a superficial iridized coating. Four of the six blown glass fragments described in table 2, one of the seven window samples, and the pressed tile have such a coating. All of these were formed by introducing silver into the batch or onto the molten glass surface and then subjecting the surface to a reducing flame to crystallize metallic silver particles at and near the surface, as shown in figure 5.

In summary, it seems clear that Tiffany's chemist Parker McIlhenny and craftsman-manager Arthur Nash kept at the forefront of technology in devising and refining glass compositions and heat treatments for the effects desired by Tiffany, and that over the years they carried out an extensive experimental program, developing complex compositions of what are now thought of as different mechanisms for producing an optical effect. A variety of compositions and heat treatments were used in various ways to obtain clear, translucent, and lightly opalescent glasses. They are colored with iron, cobalt, copper, chromium, and silver together with phosphate and fluoride added as opalizing agents—sometimes in the same piece of glass. Spectrochemical analysis indicates that Tiffany compositions included arsenic as a fining agent and manganese as a decolorizer; similar results were found at the Corning Research Laboratory.[11] Thus, the chemical technology represents an empirical and eclectic panoply of approaches to producing rich, subtle visual effects difficult to control during processing.

Fig. 9. Replications of Tiffany iridized coatings by spraying tin chloride in an alcohol solution on hot glass produced (a) fine particle sized coating in a single application which (b) shows simple ridges and furrows by surface buckling. (c) Three subsequent coatings and heatings give a thicker layer of larger particle size, and (d) a more complex pattern of ridges and furrows in multiple directions.

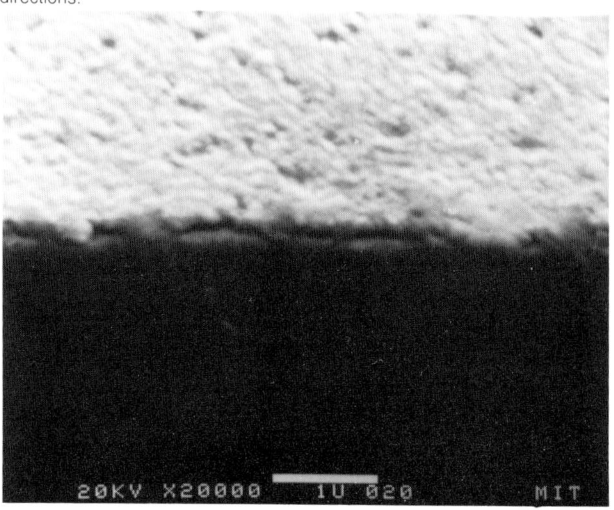

(a)

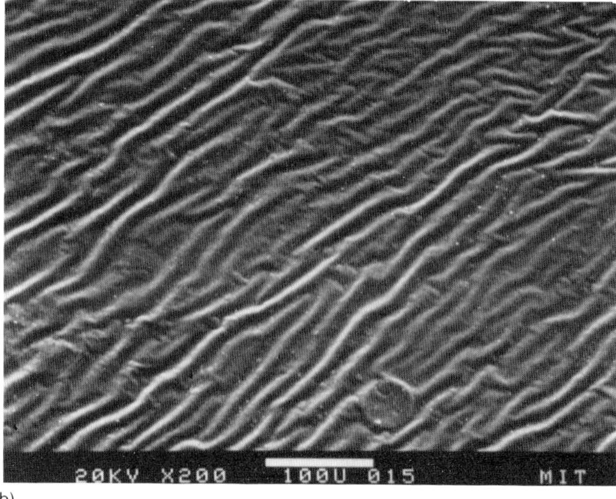

(b)

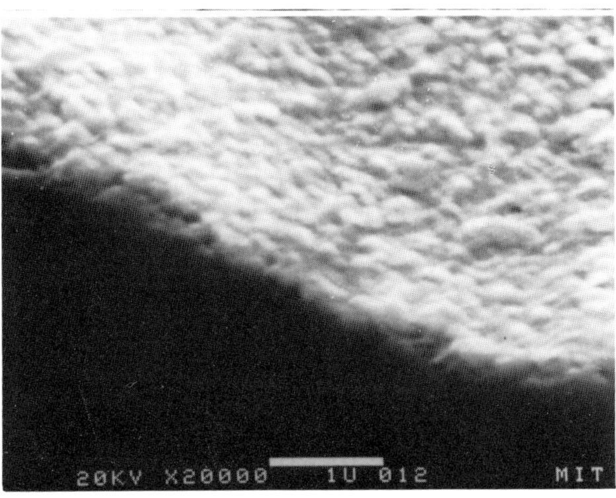

(c)

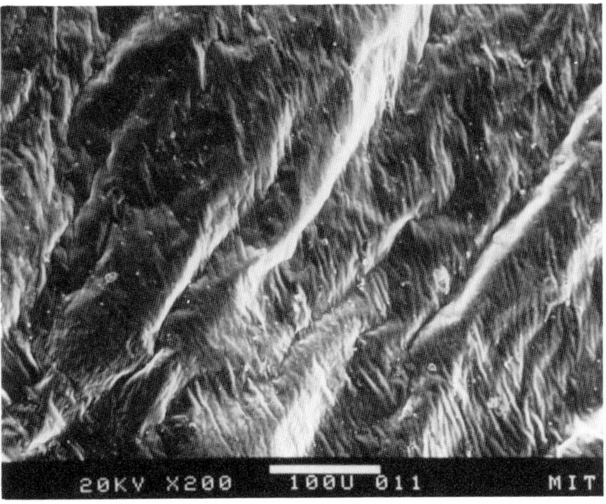

(d)

Iridescent Coatings and Rumpled Surface Textures

Tiffany was fond of the iridescent effects of weathered Roman glass, which he collected. Three of the window glass samples and six of the seven blown glass samples have an iridescent coating on the surface. Figure 6 shows a thin layer of oxide deposited from the vapor phase on the surface of the glass to give a rainbowlike effect. Pure tin oxide is silvery, a mixture of tin and iron oxides is golden, and pure iron oxide is more reddish. Tiffany's patent stipulates the oxide be applied by heating chloride crystals rather than by spraying with an alcoholic solution as is now more common and probably was done by F. Carder at Steuben.[12] The rough irregular surface and variation of color suggests that sample III-A was thrust directly into a crystal source. Analyses show that Tiffany used pure tin oxide or various mixtures of tin and iron oxides. These were frequently applied over a metallic silver luster to give a combination of high reflectivity and iridescence, as in the heart decoration of the vase in figs. 3 and 4.

While the microstructure seen at high magnifications is similar in different coatings, at lower magnifications there are substantial differences resulting from the buckling and rumpling of the iridized layer. Replications using tin oxide, shown in figure 9, indicate that a single sprayed application gives rise to a very fine particle size coating with a dull silver color, while subsequent coatings and reheatings produce a thicker layer of larger average particle size and more brilliant color. After a single application, the coating buckles in simple patterns and the surface texture looks silky; after three applications, a more complex set of ridges and cross furrows creates a surface texture that changes from that of silk through satin to that of velvet.

Substantial compressive stresses are introduced in the iridescent layer during cooling as the glass shrinks more than the coating. The thermal expansion coefficient of tin oxide is about $4 \times 10^{-6}\ °C^{-1}$, while that of glasses is about twice as high. In a standard ceramic glaze or glass enamel this incompatibility would lead to shivering (fracture and delamination of the applied layer), but the fine particle size of the tin and iron oxide coatings allows it to deform plastically in a buckling mode in what must be one of the first technological applications of what is now referred to as "superplasticity" of normally brittle materials, which develops when they have an extremely fine particle size. Limited working of the blown sample (III-F) shown in figure 10a produces simple ridges in a satiny, golden colored coating consisting of a mixture of iron and tin oxides. Sample III-A shown in figure 10b has been heated and iridized several times; buckling has occurred in multiple directions during several reheatings. There is variation in color and composition, but the texture remains velvety. If a glass is worked by rolling or blowing after the iridescent coating has been applied, the crystalline coating cannot deform as rapidly as the glass, and fractures to give a coarse surface texture, such as that illustrated in the spun lip of sample III-C or in the rolled sample I-C (fig. 10c). This investigation and control of the details of heat treatemnt, and their relation to the visual effect of the final product, were carried out by Nash and other glass craftsmen.

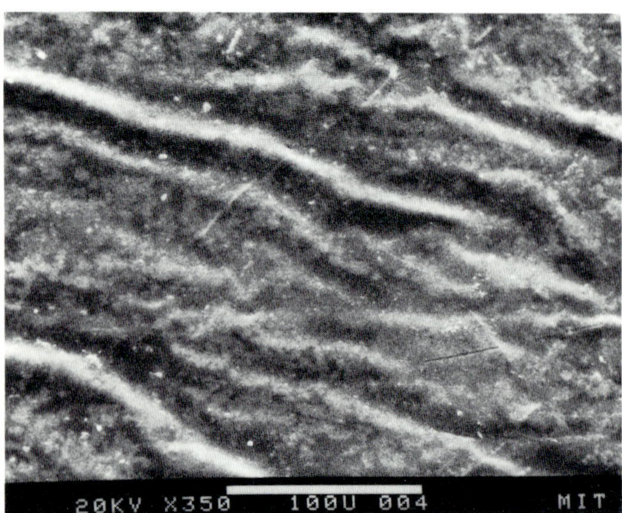

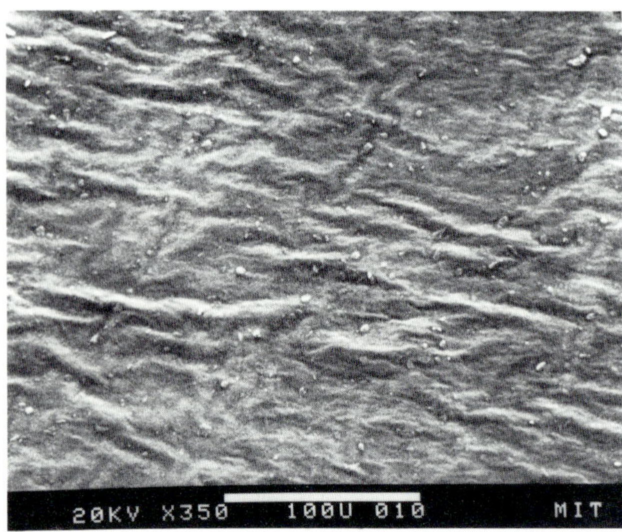

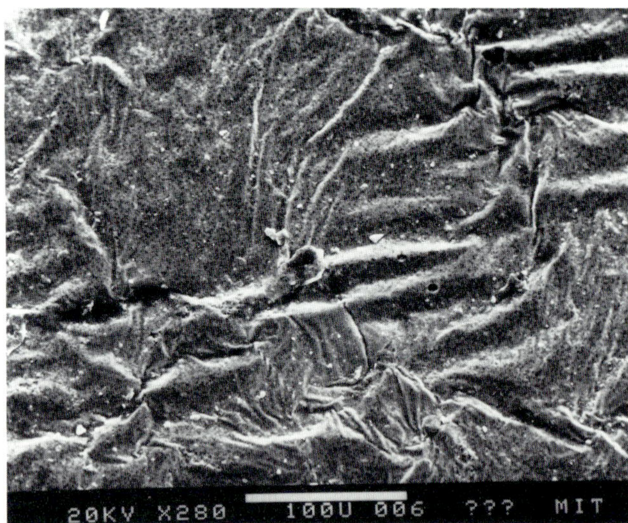

Fig. 10. Scanning electron microscopic observation of the surface of three Tiffany glasses showing different degrees of surface roughness and buckling. (a) Sample III-F shows limited working and simple ridges and valleys. (b) Sample III-A has been heated and fumed several times to create a quite rough surface which is velvety in texture. (c) In Sample I-C the iridized tin oxide layer has been fractured during flattening or rolling of the glass sheet to give a coarse, velvety texture.

Table 2
Description of Tiffany Glass Fragments

I. Window Glass (shown in figure 8a)

I-A. Soda-lime-silica window glass, green in transmitted light. Composed of two layers — a thin cobalt-containing blue layer has been cased with a thick silver-containing yellow layer. Surface has a dull silvery blue iridized coating over the yellow glass. No metallic silver luster is present, and the surface is quite smooth. A cylinder process of manufacture is probable as there is slight curvature; texture from rolling or chilling is absent; both surfaces are firepolished and there is an indented ridge along the axis on the interior caused by setting on an annealing oven support. (Metropolitan Museum of Art)

I-B. Yellowish green translucent glass having a mottled opalescent surface, phase-separated near surface layer, beneath which as many as eight layers of translucent brownish-yellow and transparent green glass are interspersed. Textured roller marks on one side and chill marks on the other side indicate it was formed by ladling the two glasses together and rolling. A layer of gold paint has been applied cold to the back surface, making the glass opaque to transmitted light, but having a luminous translucent quality of great depth in reflected light. (Metropolitan Museum of Art).

I-C. Cobalt blue window glass rolled with an uneven greenish blue iridized coating of tin oxide, heavily rumpled surface and transparent purplish blue color in cross section. No silver luster is present. In transmitted light the glass appears dark green. (Metropolitan Museum of Art)

I-D. White opal cased with a thin layer of silver-containing yellow glass. The surface has been reduced to precipitate metallic silver particles forming a mirrorlike metallic silver luster and then iridized with a combination of iron and tin oxides to give a golden surface color in reflected light. In transmitted light the glass is a yellow orange color. Probably made by cylinder process. (Metropolitan Museum of Art)

I-E. Striated opal of clear green, red and opalescent white glasses, opaque to transmitted light. Sheet has been rolled from multiple ladles of the three different glasses. The roller was smooth, and there are chill marks and some folds on the reverse side. (Metropolitan Museum of Art)

I-F. Opalescent blue, green, and yellow window glass almost opaque to transmitted light; rolled with a smooth roller from ladles of the two colored glasses and having up to nine layers of glass in a cross section. (Metropolitan Museum of Art)

I-G. Opalescent yellow and clear green glass rolled to form a translucent sheet glass. (Lillian Nassau Gallery, Ltd.)

II. Pressed Glass (shown in figure 8a)

II-A. Pressed blue tile having a surface of inlaid clear glass fragments with an uneven golden reduced metallic silver luster. We believe that a small amount of silver halide powder was set in the tile mold followed by some fragments of thin blown clear glass, then molten blue glass was added to produce this multiple layered tile. Due to the thickness of the tile, the transparent blue glass is almost opaque. (Robert C. Koch Collection)

III. Blown Glass (shown in figure 8b)

III-A. Vase fragment of transparent yellow glass with splotchy iridized surface, mostly golden with areas of red and blue. There is a metallic silver luster at the surface, and surface iridescence with a heavily rumpled texture. The iridized surface consists of tin and iron oxides. From the multiple colors and splotchy uneven coating it is probable that the vase was in direct contact with iron and tin chloride crystals. In cross section, there are two layers showing at least two gathers were made. (Morse Gallery of Art #1)

III-B. Opal lead borosilicate with no fluorine. A lamp or vase that has vertical ribs indicating it had been blown into a ribbed optic mold and then further blown. Similar to Carder's aurene glass, this glass is a white opal cased with a thin layer of yellow glass containing silver. The golden iridescence is dull with a low-relief surface texture, and varies in color with the cord in the glass. There are two white layers: the inner one is thin, opaque and well struck, and the next one is a translucent, evenly struck opal. The glass has a metallic silver luster and iridized coating of tin and iron oxides. (Morse Gallery of Art #2)

III-C. Lip fragment of an opalescent green with a thin clear copper-green casing a barely struck opal of lead silicate composition. This lead-potassium silicate phase-separated opal composition is a novel development of Tiffany. The thin green layer is colored with copper. The thickness of the layer varies, causing variation in the green color from light to dark. There is a silvery iridescent tin oxide coating on the surface. No metallic silver luster is present. (Morse Gallery of Art #3)

III-D. Stem fragment from a long-stem flower vase made of a clear silver-containing glass that has been struck at the surface to an opaque yellow and has a trailed and combed green decoration. In three vertical stripes clear glass has been combed over the green and opaque yellow areas, and in three alternate stripes the combed decoration has been pulled a second time through so that a very fine pattern of stripes is formed. (Robert C. Koch Collection)

III-E. Mold blown, metallic silver lustered and iridized Tiffany lily lamp. The glass is clear in cross section, brownish yellow in transmitted light because of the silver luster and iridized coating. (James Lundberg Collection)

III-F. Fragment of small lamp shade, the lower edge of which is scalloped. Evenly colored golden iridized surface with silver luster on the outer surface. Two layers of glass are apparent in cross section showing that two gathers of glass were made. Blown into a mold. (Lillian Nassau Gallery, Ltd.)

Table 3
Window and Pressed Glass Compositions* (weight %)
(Average of seven analyses)

Sample	I-A. Blue Layer of Green Glass	I-B. Yellow-Green Opal	I-C. Transparent Blue	I-D. Translucent White Opal	I-E. Greenish White Opal	I-F. Blueish Grey Opal	I-G. Greenish Yellow Opal	II-A Blue
SiO_2	63.1	48.7	45.0	43.3	69.9	44.5	44.18	45.1
B_2O_3	(6)	-	10.4	11.5	-	-	-	10.6
Al_2O_3	4.1	2.7	0.9	2.0	2.75	4.8	2.9	0.6
PbO	-	0.5	38.9	38.93	-	0.5	0.4	38.1
BaO	-	9.31	-	-	-	9.5	8.9	-
CaO	7.9	12.4	0.05	0.35	9.5	13.8	15.6	0.2
MgO	3.4	0.75	0.06	0.03	0.04	0.9	1.0	0.04
K_2O	1.5	1.0	2.8	2.7	0.9	1.3	1.2	2.9
Na_2O	13.3	15.2	0.2	0.2	10.6	12.9	14.3	0.13
FeO	0.15	0.26	0.16	0.09	0.04	0.28	0.30	0.12
CoO	0.11	-	0.28	-	-	-	-	0.22
F	-	6.3	-	-	6.10	7.1	8.4	-
P_2O_5	-	-	-	0.26	0.3	2.4	2.7	0.11
CuO	-	0.24	-	-	-	-	-	0.07
TiO_2	-	-	0.1	0.1	-	0.09	0.1	0.04
	99.56	99.2	98.8	98.9	100.2	98.1	100.1	98.1

*All others <0.1%

Table 4
Blown Glass Compositions*
(Average of six analyses)

Sample	III-A. Transparent Yellow	III-B. White Opal	III-C. Translucent Opal	III-D. Colorless Transparent	III-E. Colorless Transparent	III-F. Colorless Transparent
SiO_2	43.3	44.8	45.8	44.0	40.5	46.4
B_2O_3	11.0	11.4	0	11.4	11.5	11.2
Al_2O_3	2.8	0.4	0.43	0.66	0.32	0.6
PbO	0.5	38.3	37.6	38.7	36.6	37.5
BaO	9.0	-	-	-	-	-
CaO	15.6	-	0.8	0.3	-	0.3
MgO	1.0	-	-	-	-	-
K_2O	1.1	3.1	11.1	2.9	3.9	2.7
Na_2O	13.3	0.07	0.4	0.2	0.1	0.1
FeO	0.26	-	-	0.25	0.05	0.14
P_2O_5	2.6	-	-	0.3	-	0.3
Cr_2O_3	0.24	-	-	-	-	
CuO	-	-	2.2	-	-	
Total	99.7	98.3	98.4	98.0	93.0	99.2

*all others < 0.1%

Conclusions

A careful examination of Tiffany glass indicates that a wide range of compositions, forming methods, heat treatments, and surface treatments were used in the manufacture of Tiffany window glass and art glass in a creative and imaginative way to achieve a great range of visual effects. Tiffany's fascination with the interaction of colors with light led to a much more complex manipulation of glass as a material than found elsewhere. In addition, Tiffany promoted compositions that used emulsion formation in a way not well known at the time and not scientifically understood until about twenty years ago. Because emulsions were mixtures of glasses with similar refractive indices, Tiffany glass displays better control of the light opals and translucent glasses than did traditionally well-known compositions. Such compositions as the soda-lime-baria borosilicate used in Tiffany glass were really only matched by optical instrument glassmakers such as Abbe and Schott of Jena who carried out systematic compositional researches.

Although Tiffany was secretive about the range of compositions and methods used to achieve his result, examination of objects and fragments with modern laboratory techniques has given an insight into the sophistication and range of variation of Tiffany's art glass, to the workings of Tiffany's glass enterprise, and to the application and development of technology by Tiffany and other members of his staff in the service of artistic intent.

Acknowledgments

The authors gratefully acknowledge the advice and generosity of many persons for entrusting us with a wide variety of Tiffany fragments for analysis: Alice Freylingheusen, American Decorative Arts and Steven Weintraub, Conservation, Metropolitan Museum of Art; Robert C. Koch; David Donaldson, Morse Gallery of Art; Lillian Nassau Gallery, Ltd., New York; James Lundberg, Lundberg Studios, Davenport, California. Replications of optical effects and practical advice were contributed by James Lundberg, Dominick Labino, and Steve Metcalfe of Uroboros, Portland, Oregon. We acknowledge the kind assistance of Susan Rosevear with the historical review and manuscript preparation and of Dr. Yusuke Moriyoshi of the National Institute for Research on Inorganic Materials, Japan, with the TEM micrographs. The microprobe analyses were performed at the Department of Geological Sciences, Harvard University, with the assistance of David Lange. Ken Shull and Francis Liu carried out undergraduate research projects replicating iridized coatings, which contributed to our appreciation of Tiffany glass. This study has been funded in part by the M.I.T. Sloan Research Fund.

References

1. Robert C. Koch, *Louis C. Tiffany: Rebel in Glass* (New York: Crown, 1964). R.C. Koch. *Louis C. Tiffany's Glass, Bronzes, Lamps* (New York: Crown, 1971). A. Christian Revi, *American Art Nouveau Glass* (Nashville, Tenn.: T. Nelson, 1968) pp. 21-92.

2. U.S. Patent No. 237,417, Fe. 8, 1881: juxtaposing a colored and opalescent window with intervening air space to increase brilliance and iridescence; U.S. Patent No. 237,416, Feb. 8, 1881: placing a reflective, absorbent surface behind opalescent glass to increase brilliance and iridescence; U.S. Patent No. 255,910, March 21, 1882: using heat to laminate a glass sheet to perforated glass sheets to produce a pleasing effect from multiple layers.

3. Samuel Bing, *Artistic America, Tiffany Glass and Art Nouveau* (Cambridge: M.I.T. Press, 1970).

4. Hugh F. McKean, *The "Lost" Treasures of Louis C. Tiffany* (Garden City, N.Y.: Doubleday, 1980).

5. U.S. Patent No. 237,418, Feb. 6, 1881: applying a highly iridescent metallic luster on or in the glass by forming a film of metal, oxide, or metal compound by either of two methods: exposure of the glass to vapors or by direct application.

6. W.D. Kingery, D.R. Uhlmann, and H.K. Bowen, *Introduction to Ceramics*, second edition (New York: Wiley, 1976) p. 667.

7. Shaping. Thomas Gaffield, the first head of the chemistry department at M.I.T., described in his *Notes on Glass*, Vol. 4, 1875, p. 235, some unusual processing of window glass. In his *Glass Journal*, Vol. 3, 1881, pp. 278-279, a report of a detailed discussion of processing with J. Dobinson, a glass technologist who had worked with Tiffany, is given. On page 282, Gaffield reports a visit to the workshops of both Tiffany and Lafarge. These notes support the view that Tiffany glass was considered unusual in its processing and optical effects but that it was not considered entirely new and without precedent. Many of the processes in the Gaffield manuscripts are the same as the processing descriptions given in this paper.

8. Boric Oxide. Dionysius Lardner, "The Manufacture of Porcelain and Glass," Vol. 16, *The Cabinet Cyclopedia* (Philadelphia: Carey and Lea, 1832) p. 124, "Borax is used in preparing only the finest descriptions of glass ... It is too expensive to admit of its forming part in the composition of common descriptions, although its use in all cases would be desirable, as its efficacy in promoting the fusion of vitrifiable substances is unrivalled." Thomas Gaffield, *Notes on Glass*, Vol. 1, 1867-1869, p. 179, states prices paid for raw materials at the Boston Crystal Glass Works in South Boston were sand, 0.0032$/lb; red lead, 0.0122; antimony, 0.1600 and boron, 0.1854, the most costly raw material listed.

   In the batchbooks attributed to the Boston and Sandwich Glass Factory (Collection of the Sandwich Glass Museum, transcription by P. Vandiver of the batchbooks of James D. Lloyd, Mr. Reed, T. Kern and Pat Mahoney, containing the dates of 1866 and 1874), about 25% of the glass batches are reported with borax contents of 1/4 to 2%. Only a few rare optical glass batches reported in other texts contain borax above 2% (for instance, see Biser and Koch, *Elements of Glass and Glassmaking*[9], pp. 102 or 156). T. Gaffield, *Notes on Glass*, Vol. 4, 1863-1875, pp. 32-33, in a report on research in the use of boron in optical glass summarizes, stating, "These observations ... warrant the conclusion that boric acid must before long contribute to the perfection of glass for optical purposes."

9. "Barium Oxide," Benjamin Biser and J.A. Koch, *Elements of Glass and Glassmaking*, (Pittsburgh: Glass and Pottery Publishing Co., 1899) p. 42, "The claim for the use of barium carbonate in glass is as a substitute for lead, lime, or the alkalies, potash or soda, in that it imparts luster, aids fusion, adds hardness, increases density, tends to reduce the liability of devitrification, and produces a glass that is very slightly affected by the atmosphere."

10. "Opalescence," The Sandwich batchbooks, *Elements of Glass and Glass-making* and several other turn-of-the-century publications contain batches for opals made from cryolite or fluorspar additions to provide fluorine. However, the fluorine retention in the Tiffany glass (I-G), high by modern standards, is also high by comparison with contemporary recipes.

The function of P_2O_5 is twofold as found in the literature. "Boneash is used principally in the manufacture of opal glass, but the addition of a small quantity to a lime batch assists in dispersing the impurities, and is recommended as a remedy for 'cordy' glass" (*Elements of Glass and Glass Making*, op. cit., p. 47, other references on pp. 59, 75, 103). On p. 151, there are three examples of lead silicate batches (nos. 18-20) with no obvious opalizing agent, in which opacity is probably caused by phase separation. McIlhenny and Nash surely would have been familiar with this book giving practical information as well as analytical procedures to the glass technologist. It was advertised in and published by the same company as the "Commoner and Glassworker," a newspaper of the glass trade.

11. D.P. Hood, "Identification of Gold Aurene Type Glass, A Preliminary Report", *Corning Glass Works Research Report No. R,3166*, Jan. 21, 1966, 4 pages.

12. "Iridescence," Three entries describe Thomas Gaffield's knowledge of the processes for forming iridized coatings on glass. In *Notes on Glass*, Vol. 3, 1886, pp. 192-193 describe the possible mechanisms of iridescence on ancient glass. In the *Glass Journal*, Vol. 2, 1873, pp. 106-107, a description of iridescence and mechanisms for its production is given with regard to Lobmayer and Yahn glass in the 1873 Vienna exhibition. In the *Glass Journal*, Vol. 4, 1883, p. 24-25, a conversation with William Leighton details the production of the iridescence as follows: "The new iridescent glass is made by exposing the pieces hot from the furnace to the fumes of metallic oxides, which are placed in a red hot iron and send forth their coloring smoke or fumes to make a deposition on the articles exposed, while this makes a deposition [with] the effects of a thin plate, still its peculiar effect or iridescence is not entirely coming to the refraction, reflection or interference of light, because he noticed that a different coloration was made by strontium, by time and the metals ... "

RONALD L. BISHOP, GARMAN HARBOTTLE, DORIE J. REENTS, E. V. SAYRE, and LAMBERTUS VAN ZELST

Compositional Attribution of Non-Provenienced Maya Polychrome Vessels*

*A collaborative research project between the Research Laboratory of the Museum of Fine Arts, Boston, and the Chemistry Department of Brookhaven National Laboratory.

The Maya, who inhabited much of southern Mexico, Guatemala, Belize, and Honduras during their Classic Period (approximately A.D. 600-900), left a rich legacy of cultural achievement. They excelled in monumental architecture, calendrics, writing, and a varied array of sculptural and decorative art forms. Included among the latter are painted ceramic vases on which the Maya artisans depicted historical events and illustrated their view of man's relationship to natural and supernatural forces. It is no surprise, therefore, that the study of Maya pottery and its representational art is a focus of attention for the archaeologist and art historian as both attempt to understand the organizational principles and functioning of Maya society.

Although they were previously encountered mainly in natural history museums, the pictorial ceramic vessels created by the Maya are today increasingly found within the collections of fine arts museums. Their aesthetic appeal to the modern viewer as well as the appreciation of their iconographic messages has been greatly enhanced by the scholarly, well-illustrated books of Coe[1-3] and others.[4] Several scholars have suggested that the vessels may have been funerary in function, and were placed with the dead as part of the grave paraphernalia. This interpretaion is strengthened by the iconographic content: many of the major themes deal with death and apotheosis, or with gods and heroes; on the other hand, some do portray scenes or bear hieroglyphic texts concerning historical individuals and events. Maya pictorial pottery may be viewed both from a perspective of aesthetic appreciation or as an invaluable source of information about the natural or cosmological world of the Maya.

Unfortunately, few of the vessels have been found by scientific excavation. The resultant lack of spatial (as well as contextual) control dramatically reduces the potential information to be gleaned from a study of the pottery. For this reason, the Research Laboratory of the Museum of Fine Arts, Boston and the Department of Chemistry at Brookhaven National Laboratory have undertaken an intensive program of trace element analysis, seeking to characterize the ceramic paste and, hence, infer a delimited source for a vessel's origin. If a probable area of raw material procurement may be established, individuals, themes, or hieroglyphic texts found on the pottery may be placed into spatial perspective. In the absence of a probable source for the pottery, the paste compositional data still provides an empirical means of assessing the similarity between individual vessels, thereby yielding information concerning the relationships among different "styles" or "schools" of Maya ceramic art.

We have now completed five years of study and have established a paste compositional data bank for Maya pottery that contains more than 12,000 trace element determinations, including those of approximately 1,500 whole vessels. The massive accumulation of analytical data has required that increasingly more sophisticated and efficient methods be developed in order to present the data in an accessible format. Both the conceptual and practical aspects of data manipulation have commanded our attention. In this paper, we will illustrate how the Maya ceramic data bank can be used to assign a probable source of origin to non-provenienced, epigraphically and iconographically important vessels. While aspects of form and decoration will shortly be encoded in our data bank, our present comments deal primarily with the interpretation of the chemical data as it is seen to covary with selected styles of Maya painting. These covarying attributes, style and paste composition, will form the basis

for a hierarchy of vessel attribution that is decreasingly specific: site specific, sub-regional, and hypothetical. Taken together, these levels of attribution will illustrate how the trace element data can merge archaeology with art history and address questions concerning Maya art and society.

Chemical data for the project was derived by instrumental neutron activation analysis, following routine analytical procedures described in detail elsewhere.[5] Our method of sampling whole vessels was guided by an effort to minimize the size of the burred area (usually on the base) while achieving analytical reproduceability. Actual sample size varied from 50 to 200mg, depending on vessel thickness. Analysis carried out on the smaller samples frequently showed non-reproduceable manganese concentrations, and since that element is one of three that we routinely obtain from a short irradiation, we have relied on only those fifteen elements whose concentrations may be determined during a single gamma count eight to ten days after a long bombardment. A single count also expedited the production of a large number of analyses—a requisite for the compositional characterization of the vast geographical area of investigation, and the very large number of whole vessel samples. Since we have retained our previously analyzed samples, short irradiations could still be performed in the future if needed to resolve particular questions.

We should emphasize that we are not necessarily attempting to relate the pottery directly to the clays or other raw materials employed in its manufacture, although this approach has been successful in other investigations.[6] The multi-component nature of Maya ceramic fabrics, with their varied mineralogy, grain size distributions, and manufacturing histories, forces us to a higher order of abstraction; we are compositionally analyzing the choices and products of cultural activity rather than naturally occurring materials.[7] As an example, although the clay from a given source may be more or less homogeneous in elemental composition, the addition of components such as volcanic ash, crushed carbonate rock, sand, or other clays may significantly alter the compositional profile of the original clay. For this reason, greater reliance in the task of relating compositional characterization to locality of origin is placed on the "criterion of abundance."[8] Briefly stated, this rule holds that the type of pottery that is strongly represented or that has extended continuity through time at a site is more likely to be of local origin and thus can be used to establish a representative profile for locally produced pottery. Although there are weaknesses in this approach, it nontheless provides a point of departure in compositional investigations, and is usually acceptable to the archaeological community.

The most important factor in the successful characterization of a site's or region's locally produced pottery is the intensity of sampling; given the paucity of detailed excavations or even surface collections, our sampling is obviously incomplete, but it does represent a beginning. The principal Maya sites that have been the focus of more extensive ceramic sampling are shown in figure 1. From each of those sites indicated, 75-150 polychrome sherds produced between A.D. 300-900 were analyzed. The pottery from another fifty-some sites has also been sampled. In addition, some 2000 analyses have been carried out on pottery samples obtained from outside the heartland of Classic Maya polychrome production. Focused on the site of Palenque and outlying sites in northeastern Chiapas and adjacent Ta-

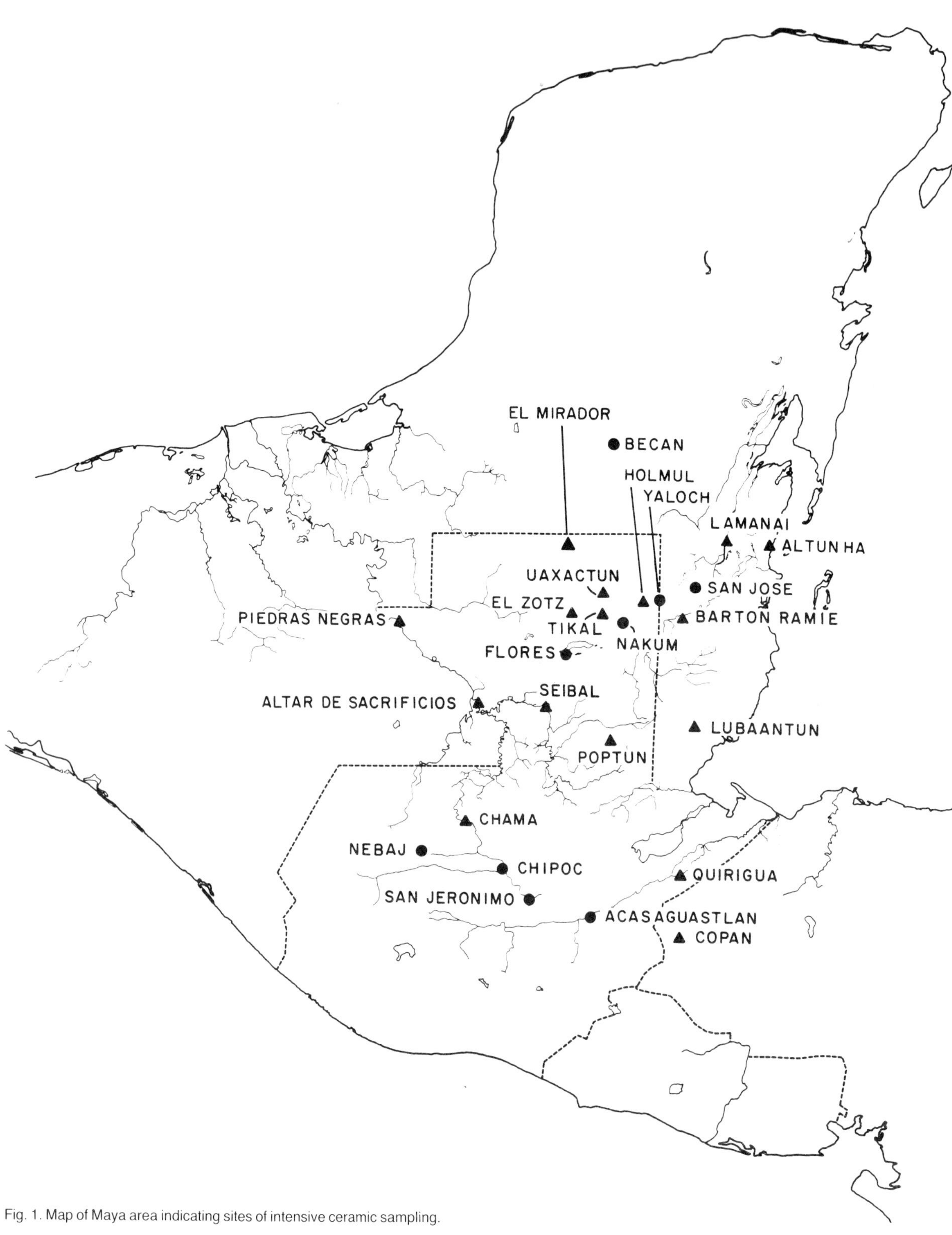

Fig. 1. Map of Maya area indicating sites of intensive ceramic sampling.

basco, these analyses broaden the compositional perspective. Our coverage is certainly uneven and yet distinct compositional trends across the Maya area are observed and "key" elemental concentrations are noted. For example, pottery produced from clays in the Usumacinta and Motagua river drainages are notable for their high chromium values: they exceed 1,000 parts per million—a factor of ten or more greater than the usual levels in Maya pottery.

Even with intensive sampling, the number of analyzed samples that may constitute a "homogeneous" group may not be large. It is not unusual that multiple reference groups or representative compositional profiles are formed in order to more realistically represent the local ceramic production at a site; clay beds may be exhausted or access to them restricted, different tempering materials may be utilized, and changes may occur in the production stages. The need for multiple reference units combined with the elimination of the "nonlocal" or intrusive samples reduces the number of analyses that will constitute a reference group. These smaller units, if archaeologically as well as chemically reasonable, may well be adequate for the purposes of attributing non-provenienced vessels. If they are too small to be considered "representative" they might be susceptible to merging with other geographically close groups to create a larger "synthetic" reference unit for sub-regional characterization.[9]

Stated differently, we seek to utilize the analytical and archaeological data in an attempt to subdivide the Maya area into the smallest divisable units that are maximally similar within themselves and demonstrably separable from other such units. The realities of incomplete coverage and selective sampling necessitate varying operational as well as conceptual levels of data synthesis. The successful manipulation of the trace element data is confirmed in part by the degree of *stylistic* similarity within our *compositionally* derived groups. These units, at the site-specific or sub-regional level provide the basis for the attribution of the non-provenienced vessels.

This concept of site-specific compositional characterization may be illustrated by the results on the pottery from the site of Tikal, Guatemala. Here, a single compositional unit has been obtained that is comprised of non-carbonate tempered pottery representing several different ceramic periods of that site but which leans heavily on the analytical composition of the Late Classic figure "dancer" plates (Fig. 2a).[10] As stated above, for this to be a useful reference unit, it must be relatively homogeneous and statistically distinct from the locally produced pottery at other sites in the vicinity. Although some elemental concentrations are similar to those of pottery from the site of El Sotz, approximately 35 kilometers distant, some concentration levels, particularly those of the rare earths and hafnium, serve to differentiate the compositional profiles of the two sites (table 1).

Given that one or more compositionally homogeneous and archaeologically meaningful groups can be formed to represent the pottery locally produced in or around a site, statistical procedures employed in the evaluation of the group can be extended to non-group samples such as non-provenienced vessels. (Many of the statistical approaches are discussed elsewhere).[5] One of the more powerful approaches is to calculate the variance-covariance properties and centroid for a group of specimens as well as the Mahalanobis distance of each sample from that centroid. From the distance, probabilities that individual samples could belong to the group are calculated. These calculations not only adjust for a varying number of

samples comprising the reference group, but also take into consideration the pattern of interelemental correlation as well as the sample's proximity to the group centroid.[11]

As noted above, a Tikal reference group was formed which, although stylistically biased toward the figure plates, still contained other sherd material also found at Tikal, and of presumed local origin. To this group the chemical profiles of the whole corpus of non-provenienced vessels were one by one multivariately compared, revealing several additional specimens that were observed to have high probabilities of belonging to the Tikal group. Included among the newcomers were several in the figure plate style (fig. 2b). These data strongly suggest that ceramic specialists at or around Tikal were engaged in the production of these distinctive plates. If our assumptions and statistical procedures are valid, then a probable source of manufacture for the unprovenienced museum specimens has been demonstrated. The bringing together using the compositional data of a number of these figure plates also provides a sound, empirically derived group of stylistically related vessels with which the art historian can now investigate aspects of stylistic variation within a restricted geographic area of production, with the relatively safe assumption that he or she is not dealing with widely separated production centers.

In the absence of a site-specific characterization, it is necessary to resort to a sub-regional perspective: an expanded compositional unit formed by a merger of groups or reflecting a particular pattern. Although some specificity of manufacturing location is lost, the compositional data provide an objective basis for the comparison of vessels. For example, members of a stylistically related group of vessels can serve, each in turn, as reference points from which to assess the extent of their similarity to every other sample in the data bank. Since a compositional group with its set of elemental variance-covariance relationships is not known, similarity can be based on simple Euclidean distances (or other measures of association). A computer program for chemical profile matching within the data bank has been developed, but not all "close" matches are of equal utility. For example, if a reference sample lies near the center of its actual (or natural) group, then samples close to it will most likely be other members of that group. However, if the reference point is tending toward the periphery of the group's distribution, non-group members may be as close or closer to the reference point than other members of the group to which the reference sample belongs. One application of this type of data searching has been to vessels that we refer to as the Area Group (figs. 2c,d).[12]

Conducted as a test of the data bank searching techniques, stylistically similar vessels were selected from the whole vessel corpus, using five-by-seven inch photographs. The group is defined by such shared attributes as a similar palette (red-orange, gray, and black slip painted on a cream background), a two-dimensional picture plane filled with iconographic forms and decorative details, and a red rim band with black painted glyphs. The painting technique employed quick brush strokes that produced loose black outlines and overlapping fill-in colors. The iconographic program is as consistent as the artistic style, for these vessels portray either anthropomorphic or zoomorphic versions of the *Underworld Paddlers,* or depict God N, one of the chief underworld gods.[12]

When each of these vessels was used in turn as a reference point

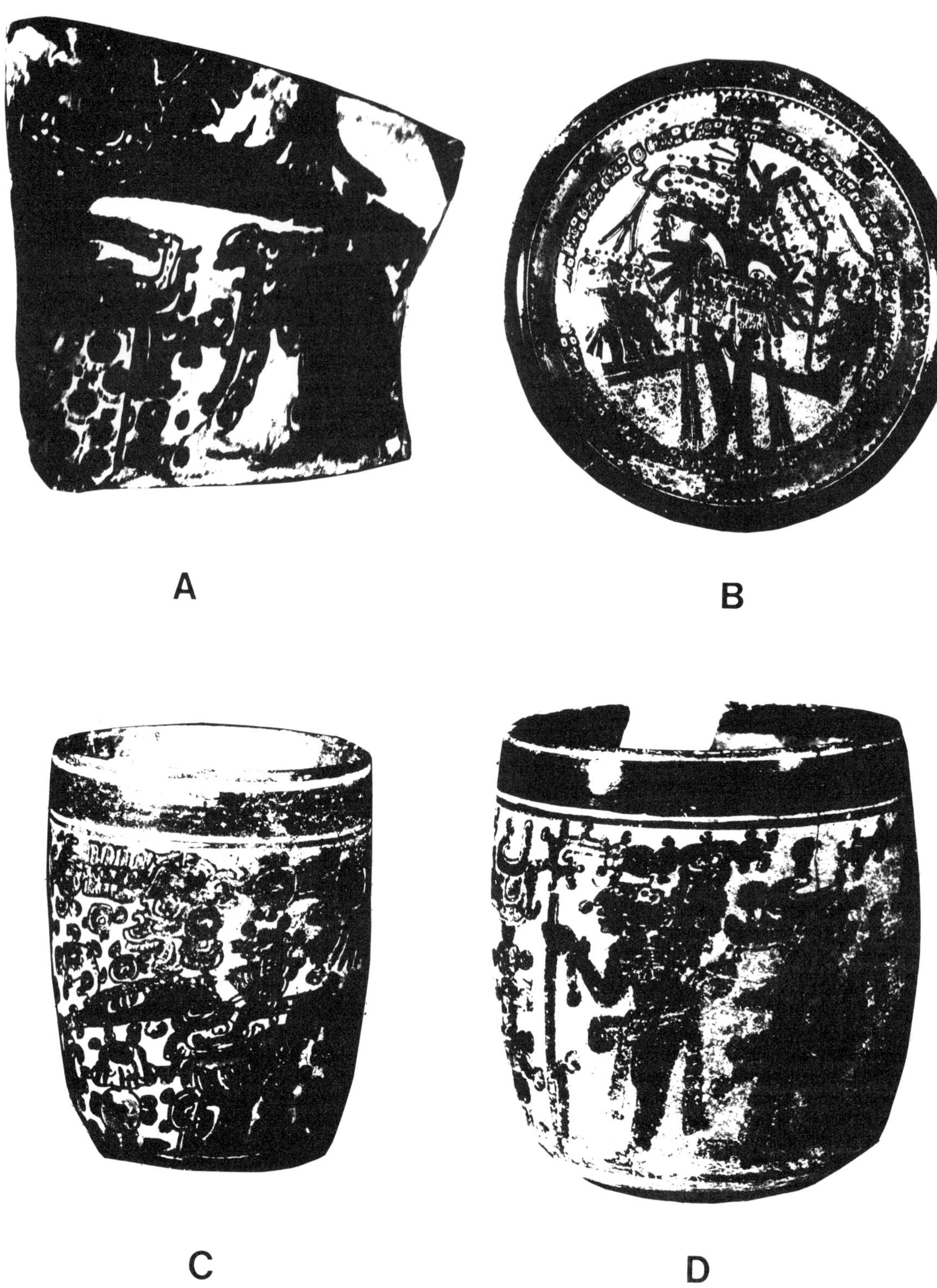

Fig. 2. (a) Figure painted polychrome plate excavated at Tikal, Guatemala; (b) Non-provenienced figure painted plate with compositionally based probability of having been made in Tikal area. (c,d) Examples of Area Group style of painted vessels.

some—but not all—of the stylistically similar vessels were found to fall within an "acceptable" level of chemical similarity. One vessel, which must have occupied a more central position in the sample distribution, revealed close similarity to most of the other vessels and to excavated sherds from two sites that showed iconographic elements resembling those of the Area Group specimens. Once a pattern of overlapping compositional similarity was observed, the related profiles could be brought together to form a group and thus evaluated by the same multivariate statistical procedures employed in the Tikal example. These procedures resulted in a group of twenty-seven specimens nine of which were of the Area Group style. Elemental mean concentrations and percent deviation values are given in table 1.

If the probability distribution about the twenty-seven samples is enlarged to include those specimens lying just outside of a 95% confidence interval, we begin to include many with a Tikal provenience. One can note the similarity between the mean elemental concentrations of the Area Group vessels and those of the Tikal reference unit in table 1. Although we cannot say that the specimens of the Area Group were made at Tikal, we can infer a manufacturing location somewhere in the general Tikal-Uaxactun sub-region as opposed to other characterized sites or sub-regions in the Maya Area.

Three vessels had been tentatively placed within the pictorially sorted Area Group, on the basis of their styles. However, each was divergent in at least one pictorial or iconographic attribute that brought into question its group membership. Indeed, each was afterward found to have a chemical profile that was divergent from the others in the Area Group, suggesting that they were formed from different clays and painted by artists other than those responsible for the Area Group specimens. Such data are of critical importance for investigations of Maya ceramic art and the nature of craft specialization. Additionally, these data attest to the strength of compositional and sytlistic covariation in Maya elite pottery.

Another group of ceramic vessels that illustrate chemical as well as stylistic homogeneity are those painted in a "codex style," characterized by the use of a fine black brown whiplash line on a cream slip with red bands at the rim and base (fig. 3). A little more than a decade ago few of these vessels were known; they now number more than two hundred and have been the subject of a recent book that hypothesized that the scenes on the vessels could be sequenced to illustrate an underworld tale or myth.[4] A previous compositional investigation of approximately thirty samples of codex style pottery suggested a single production area somewhere to the north of the Tikal-Uaxactun-El Sotz area.[13] The general location of the production area was inferred from an observed similarity of the codex pottery to non-codex pottery from El Sotz and Uaxactun and slightly less resemblance to Tikal ceramics. In other words, viewing the Maya area as a whole, it was possible to eliminate large geographical areas by reference to compositional trends. Subsequent analysis combined wih recent excavation has revealed chemically matching codex pottery from the sites of El Sotz and El Mirador in northern Guatemala—well within the sub-region previously suggested. As with the Area Group, the analytical data for the codex style vessels are consistent with a view of stylistic and compositional cohesiveness and sub-regional localization of production.

Not unexpectedly, we often encounter vessels apparently produced from raw materials for which a compositional reference group is currently lacking. In some instances the presence of a glyphic text may suggest a source for the vessel. The compositional and epigraphic analysis of Maya cylinder MSO651 illustrates what we refer to as the hypothetical level of vessel attribution. The text on MSO651 seen here in a rollout photograph by Justin Kerr (fig. 4) begins with the date 3 Ben 6 Kankin. This corresponds to the Maya Long Count position 9.12.11.9.12, or November 11, A.D. 683 (GMT correlation). The passage records a bloodletting ritual performed by the lord seated on the throne. His personal name is a glyph read as "Ah God K." This follows the nominal tradition of Petexbatun rulers, which includes such individuals as Flint Sky God K, Shield God K, and Scroll-head God K. Further, Ah God K's nominal phrase ends with the Petexbatun emblem glyph; this emblem gylph, also found on monuments from this region, is either the name of a particular royal lineage or denotes the site itself.

The Petexbatun region of the Guatemalan southern Peten Lowlands is archaeologically poorly known with no scientifically excavated ceramic information. However, the region has many carved stone monuments whose glyphic texts yield a rich trove of historical and political information. Although it is not possible to directly connect Ah God K with any of the individuals found on the monuments, the similarity in the pattern of the glyphs and the presence of the emblem glyph relate him to the ruling line.[14]

Table 1

Comparison of Mean Concentrations and Percent Deviations in Ceramic Reference Groups.

Element Analyzed	Area Stylistic Group N = 9	Tikal Reference Group N = 30	El Sotz Group* N = 28
Rb_2O	79.0 (33)	84.0 (25)	117.0 (22)
Cs_2O	2.5 (20)	3.4 (28)	4.4 (29)
BaO	.18 (28)	.12 (39)	.13 (50)
Sc_2O_3	14.0 (7)	20.3 (13)	16.0 (16)
La_2O_3	16.5 (6)	19.3 (18)	32.2 (24)
CeO_2	33.0 (13)	41.0 (18)	74.0 (25)
Eu_2O_3	.35 (9)	.65 (22)	.93 (34)
Lu_2O_3	.27 (13)	.37 (20)	.56 (16)
HfO_2	5.3 (10)	6.8 (12)	9.0 (30)
ThO_2	15.3 (6)	15.8 (11)	20.0 (20)
Cr_2O_3	46.0 (7)	73.0 (35)	91.0 (34)
Fe_2O_3**	3.3 (8)	4.7 (12)	3.1 (18)
CoO	6.3 (30)	6.7 (17)	8.9 (31)
Sm_2O_3	2.1 (11)	3.0 (27)	4.8 (24)
Yb_2O_3	1.4 (19)	1.9 (27)	3.2 (17)

*It is quite likely that the El Sotz group will undergo further refinement as work progresses. The Tikal reference group represents the more frequently encountered deviation ranges when working with Maya volcanic-ash tempered pottery. Parentheses contain standard deviation expressed as %.

**Fe_2O_3 concentration in %, all other in parts per million.

Fig. 3. Examples of codex-style painted ceramics.

Fig. 4. Rollout photograph of vessel MSO651; photograph copyrighted by Justin Kerr.

The glyphic text, then, suggests a hypothetical area of manufacture for vessel MSO651. Yet, the vessel could conceivably have been made in some other area and served only to commemorate an event that took place in the Petexbatun. When the chemical data for the vessel is used as a reference point for searching the data bank, only a very few non-provenienced vessels were found to lie within the Euclidean distance limits usually employed when working with pottery from other sections of the Maya area. This could imply either that the vessel was manufactured in a region from which we have not sampled the pottery or that the vessel was made from a different clay from the regions where we do have sampling coverage. When we increase the measure of Euclidean distance, lessening our criterion of similarity, we begin to extract from the data bank samples from the site of Seibal, Guatemala—the closest site to the Petexbatun from which we have analyzed pottery. While far from conclusive proof, the fact that MSO651's closest similarity with provenienced pottery occurs with that from Seibal supports the notion that MSO651 could have been made in the Petexbatun. As more vessels are analyzed, MSO651 can serve as a reference point representing pottery with a hypothetical origin in the region.

We have discussed the procedures and a few of the results of the Maya ceramic project from the perspective of non-provenienced vessel attribution ranging from site specific through a more inferential level to the rather hypothetical. The few examples presented serve to illustrate the manner in which we are attempting to view compositional and stylistic covariation in an investigation of Maya ceramic art. The large data base, including archaeologically recovered pottery as well as the stylistically and iconographically elaborate vessels, requires a continued refining of statistical analysis and a greater understanding of the sources of ceramic compositional variation in the Maya area. Above all, it emphasizes the mutually beneficial collaboration between science, art, and archaeology.

Acknowledgments

This research is part of the Maya Jade and Ceramics Project, supported by the Research Laboratory of the Museum of Fine Arts, and Mr. Landon T. Clay, both of Boston, in collaboration with the Department of Chemistry, Brookhaven National Laboratory. Work at the latter is conducted under the auspices of the U.S. Department of Energy. Reents's participation nas been made possible by an anonymous philanthropic donation to the project, with additional support generously provided by: the Prescott Fund, Department of Fine Arts (1982, 1983); Graduate Student Research Grants (1981, 1983); a University Fellowship (1982) administered by Dean William S. Livingston, Graduate School—all of the University of Texas, Austin; and Mrs. Marie B. Hanna of Austin, Texas. The rollout photograph of vessel MSO651 (fig. 4) is kindly provided and copyrighted by Mr. Justin Kerr. We also wish to express our appreciation to the Museo Popol Vuh, Guatemala, for permission to sample the vessels shown in figure 2c and 2d as well as to the Mint Museum in Charlotte, North Carolina, figure 3b.

References

1. M. D. Coe, *The Maya Scribe and His World,* (New York: The Grolier Club, 1973).

2. M. D. Coe, *Classic Maya Pottery at Dumbarton Oaks,* (Washington: Dumbarton Oaks, 1975).

3. M. D. Coe, *The Lords of the Underworld,* (Princeton: Princeton University Press, 1978).

4. F. Robicsek and D. F. Hales, *The Maya Book of the Dead: The Ceramic Codex,* (Charlottesville, University of Virginia Art Museum, 1981).

5. R. L. Bishop, G. Harbottle, and E. V. Sayre, "Chemical and Mathematical Procedures Employed in the Mayan Fine Paste Ceramics Project," in J. A. Sabloff, ed., *Analysis of Fine Paste Ceramics, Excavations at Seibal, Guatemala,* (Cambridge: Memoirs of the Peabody Museum of Archaeology and Ethnology, 15, no. 2 1982), pp. 272-282.

6. M. F. Kaplan, "The Origin and Distribution of Tell el Yehudiyeh Ware," in *Studies in Mediterranean Archaeology,* LXII (Göteborg: Paul Åstroms Förlag, 1980).

7. R. L. Bishop, "Aspects of Ceramic Compositional Modeling," in R. E. Fry, ed., *Models and Methods in Regional Exchange,* SAA Papers, 1, (Washington: Society for American Archaeology), pp. 47-66.

8. R. L. Bishop, R. L. Rands, and G. R. Holley, "Ceramic Compositional Modeling in Archaeological Perspective," M. B. Schiffer, ed., *Advances in Archaeological Method and Theory,* 5, (New York: Academic Press, 1982), pp. 275-330.

9. R. L. Bishop and R. L. Rands, "Maya Fine Paste Ceramics: A Compositional Perspective," in J. A. Sabloff, ed., *Analyses of Fine Paste Ceramics, Excavations at Seibal, Guatemala,* (Cambridge: Memoirs of the Peabody Museum of Archaeololgy and Ethnology, 15, no. 2 1982), pp. 283-314.

10. C. Coggins, *Painting and Drawing Styles at Tikal: An Historical and Iconographic Reconstruction* (Ann Arbor: University Microfilms, 1975), p. 266.

11. M. M. Tatsuoka, *Multivariate Analysis: Techniques for Educational and Psychological Research,* (New York: John Wiley and Sons, Inc. 1971), pp. 62-84.

12. R. L. Bishop, D. J. Reents, G. Harbottle, E. V. Sayre and L. van Zelst, "The Area Group: An Example of Style and Paste Compositional Covariation in Maya Pottery," (paper presented at Quinta Mesa Redonda de Palenque, 1983).

13. R. L. Bishop, G. Harbottle, E. V. Sayre and L. van Zelst, "A Paste Compositional Investigation of Classic Maya Polychrome Art," in E. P. Benson, ed., *Proceedings of the Quarta Mesa Redonda De Palenque* (Austin: University of Texas Press, 1984).

14. D. J. Reents and R. L. Bishop, "History and Ritual Events on a Petexbatun Classic Maya Polychrome Vessel" (paper presented at Quinta Mesa Redonda de Palenque, 1983).

GARY W. CARRIVEAU

The Examination and Analysis of Oriental Lacquer

The sap of the shrubby tree *Rhus vernicifera* has been referred to as
"the most ancient industrial plastic known to man." Objects covered
with a polymerized film of the sap have been used for at least 3700
years. Material having this extremely durable, often highly decorative coating ranges from utilitarian pieces such as boxes, trays, cups,
etc., to furniture, armor, masks, and sculpture. Technical examination produces information that can be used to determine the composition of materials and methods of manufacture, which sometimes
indicate where and when the lacquer object was made. Further, the
results may also provide guidance when considering conservation or
restoration treatment if the object is damaged. Analytical techniques
have included atomic absorption spectroscopy, infrared spectrophotometry, thin-layer chromatography, differential thermal analysis,
emission spectroscopy, x-ray radiography, optical and scanning
electron microscopy and mass spectroscopy. These methods and
results were reviewed. In addition, new methods were discussed, including the use of energy dispersive x-ray fluorescence, scanning
photoacoustical microscopy, laser microprobe, non-destructive infrared spectrophotometry, proton induced x-ray emission spectroscopy and infrared reflectography. Results from museum objects
illustrated the use of all experimental methods.

G. V. ROBINS, D. PENDLEBURY, R. A. FLETTON, and J. ELLIOTT

The Application of [13] Carbon-13 – Fourier Transform Nuclear Magnetic Resonance Spectroscopy to the Analysis of Art Objects and Archaeological Artifacts

The use of ^{13}C-FTNMR spectroscopy in the examination of archaeological artifacts and art objects in our laboratories arose from an interest in the organic components of historical non-metallic seals. The analytical problem posed was to identify the components of a suspected wax-plant resin mixture while taking into account the multicomponent nature of any of these natural products.[1]

A number of analytical strategies can be applied to this kind of problem. For example, gas chromatography (GC) and mass spectrometry (MS) provide separation of components and fingerprinting via identification of characteristic molecular fragments and components, whereas infrared spectroscopy can identify characteristic functionalities without prior separation. Alternatively, microcomponents such as sterols in waxes can be identified by GC or its combination with MS or another technique. It is also possible to fingerprint by analyzing for trace inorganic components in an organic material.

All of the approaches outlined above have their own particular advantages and disadvantages, and our own experience of analyzing organic materials from ancient artifacts encouraged us to pursue a holistic method that did not require prior separation or chemical modification for analysis, and that did not produce large quantities of data through a separation or fragmentation regime that could only be resolved comparatively. Initial studies with infrared spectroscopy and x-ray diffraction (to be described in a future publication) were moderately successful, but we found that ^{13}C-FTNMR spectroscopy was ideally suited to discriminate between target components without sample modification. In this paper we describe the extension of this preliminary study into a context of wider conservational interest.

Principles of ^{13}C-FTNMR Spectroscopy

NMR spectroscopy depends upon the possession, by the nuclei of certain kinds of atom, of a magnetic moment. In a strong magnetic field a nuclear magnet is permitted to adopt one of only a limited number of orientations (most commonly two), with respect to the field direction and these differ in energy. Thus, in an assembly of like magnetic nuclei, more nuclei will occupy the lower energy states than the higher ones. With an input of electromagnetic energy in the radiofrequency range the nuclei in the lower states may be promoted to higher states with a net absorption of energy. This process is nuclear magnetic resonance.

The utility of the NMR effect is a consequence of the fact that nuclei in all normal states of matter are surrounded by electrons that screen them slightly from the influence of the magnetic field. The magnitude of this screening depends upon the density and distribution of the surrounding electrons, and thus varies with the chemical environment. The spectrum of a sample containing the same type of nuclei in several distinct chemical environments thus consists of several absorption peaks. This dependence of the position of an NMR peak upon the chemical environment of the nucleus is known as the chemical shift and is the basis of the analytical power of NMR spectroscopy.

The most abundant magnetic nucleus is hydrogen (the proton). Over the last two decades proton NMR spectroscopy has had a profound influence on the development of NMR spectroscopy and has provided a unique insight into the subtleties of molecular structure. The chemical shift of the proton is small compared with larger atoms, but in organic molecules the carbon skeletons consist mostly of ^{12}C,

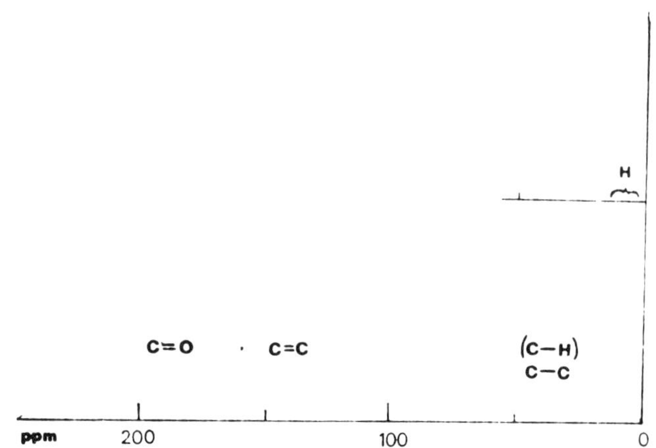

Fig. 1. The chemical shifts of major structural features in the ^{13}C-FTNMR spectrum. The proton (^{1}H) chemical shift is also shown for comparison.

which is NMR-inactive. The natural ^{13}C isotope is only present at the 1% abundance level and gives very weak signals. Recently, however, the development of pulsed Fourier transform methods of NMR spectroscopy has reduced the experimental problems associated with insensitive nuclei and observation of ^{13}C resonances has become a routine, if sophisticated, analytical tool in organic chemistry.[2,3] Production of a suitable spectrum requires some 10^4 acquisitions on approximately 30mg of sample dissolved in a suitable solvent (normally deuterocholoroform or perdeuterodimethylsulphoxide in this study) to make about 0.4ml of test solution.

Signals are found in characteristic regions of the ^{13}C spectrum as indicated in figure 1. Spectral regions are associated with alkane, alkene (and aromatic) and carbonyl resonances at increasingly greater shifts from a reference standard of tetramethylsilane (TMS), and the shifts are measured in parts per million (ppm). TMS is normally added to a test solution as an internal marker.

The ^{13}C-FTNMR spectra described in this paper are obtained with the simultaneous excitation of all the attached protons present with a strong radiofrequency field. This causes the rapid reorientation of the protons and averaging of the interactions with the ^{13}C nuclei to zero. This "spin-decoupling" avoids overcomplication of the spectra.

In the present study such spin-decoupled spectra were obtained with a Jeol FX100 NMR spectrometer operating at 25.05 MHz for ^{13}C and utilizing a 2.35 Tesla electromagnet. Other details of the experimental conditions and data acquisition regime have previously been reported.[4]

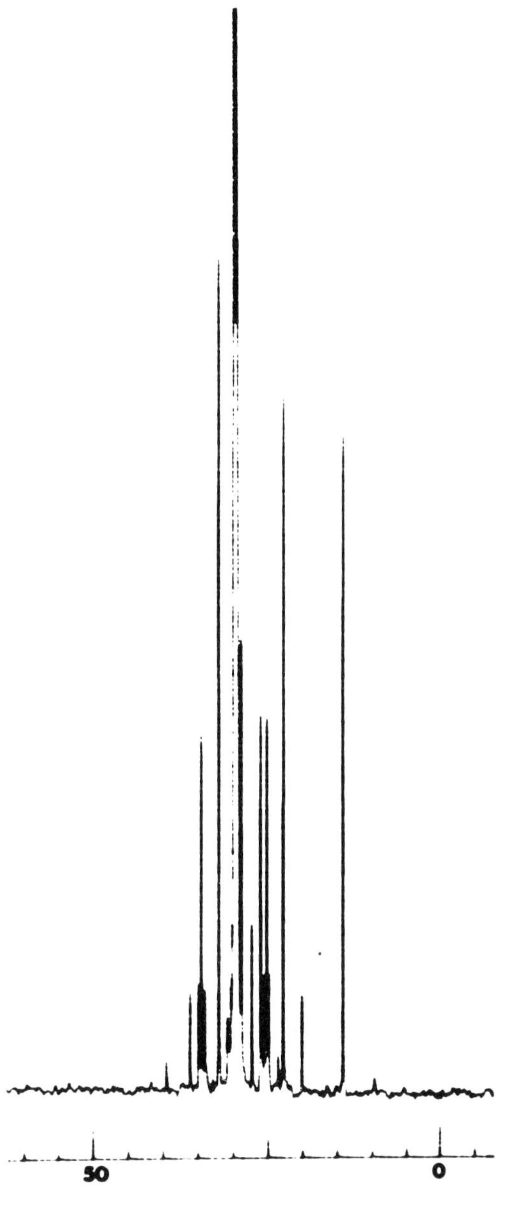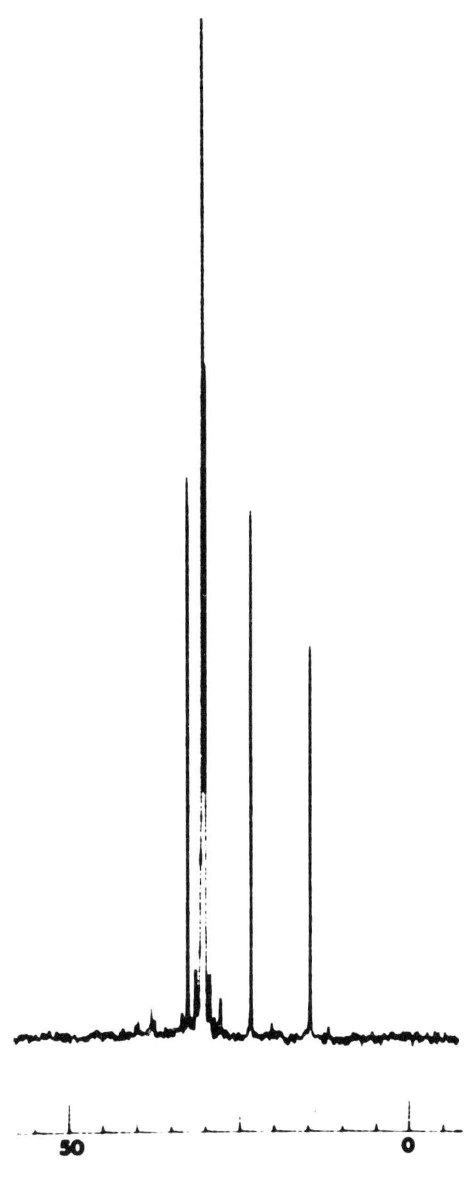

Fig. 2. Alkane region ^{13}C-FTNMR spectra for typical waxes (O-50ppm). a. modern beeswax. b. modern paraffin wax

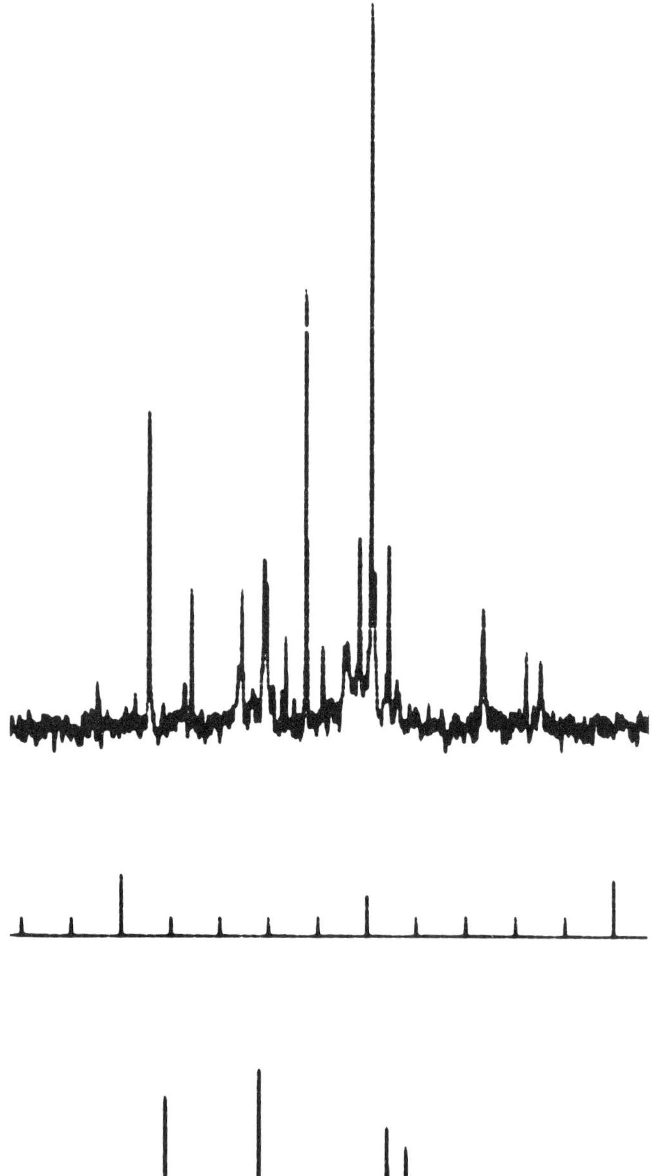

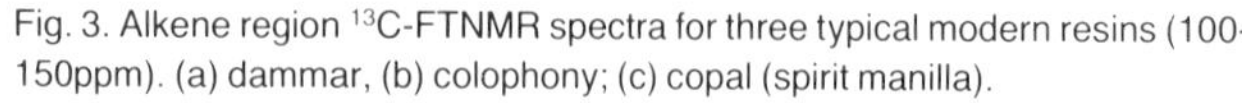

Fig. 3. Alkene region ^{13}C-FTNMR spectra for three typical modern resins (100-150ppm). (a) dammar, (b) colophony; (c) copal (spirit manilla).

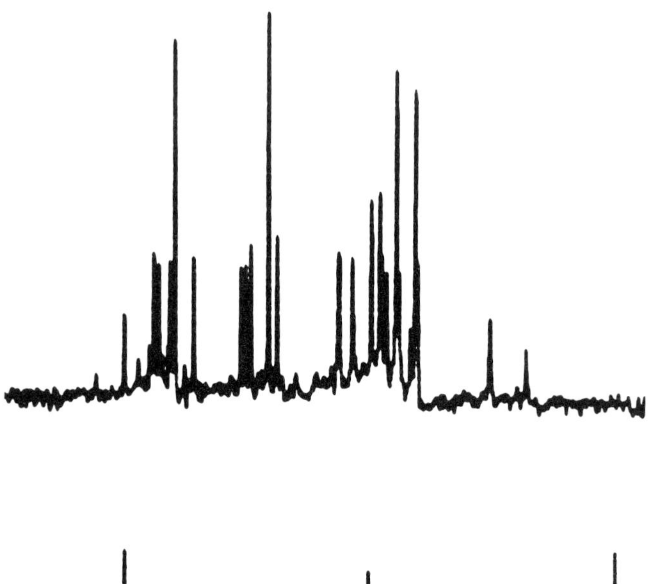

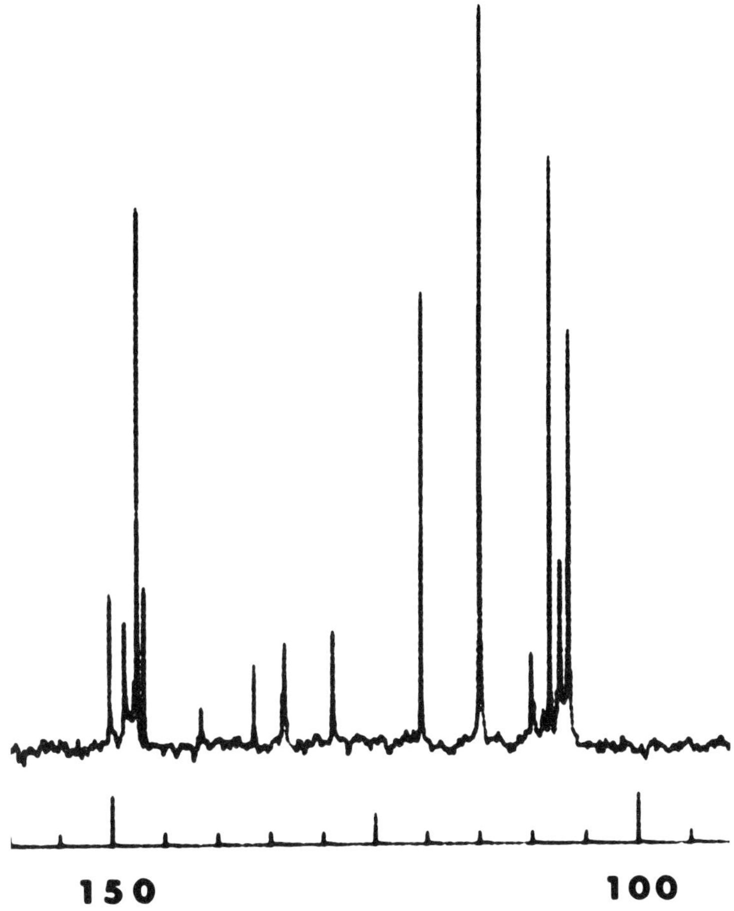

Chemical Shifts of Waxes and Resins

Most waxes consist of long chain hydrocarbons (C_{20}-C_{40}) with or without ester, carboxylic and carbonyl functionality, and there is always a small amount of unsaturation in the chain.[4] Waxes are, therefore, complex "near neighbor" mixtures that may rapidly be fingerprinted from their ^{13}C-FTNMR spectra. We have found the alkane region to be most informative here and figure 2 shows the spectra in the 0-50ppm range for beeswax and paraffin wax as examples. Despite the overall similarity there are distinct differences that are directly related to differences in the molecular profiles. Similarly tallow, shellac, and spermaceti wax all show distinct profiles in this region and the presence of these materials in ancient artifacts can be determined and readily differentiated. The most striking feature of our analyses so far is the stability and longevity of these materials—fingerprints are recognizable in artifacts up to 3000 years old.

Waxes have a vestigial alkene signal whereas most plant resins have a complex alkenic structure, both linear and cyclic.[5] These structures are associated with a varied alkanic complement as well, leading to considerable ^{13}C-FTNMR information on their structures. Our major interest in this context is in the alkenic spectrum since this can be readily identified in the presence of waxes or other lipids that have little absorption in this region. Identification can therefore proceed without separation of the components. While at present such identification depends upon visual matching we envisage further developments in the near future using computer sorting and matching that will be particularly applicable to mixtures of resins.

Figure 3 shows the alkenic spectrum of three typical resins: dammar, colophony, and copal. Each resin spectrum is quite distinct from the other. This specificity also extends into the carbonyl region, which falls outside the present discussion. Fingerprinting of resinous materials in this way shows considerable promise for rapid analysis of organic artifacts and our current program is aimed at producing a wide range of reference spectra to compare with target substances. Some of this work involving colophony is in press.[6]

Source-Typing of Copal

The spectra in figure 2 are of modern reference materials drawn from random supply sources, and the wide differences in the alkene absorption prompted us to examine different sources of the same resin. At present our work has centered upon copal since this is commonly encountered as a varnish material and a wide variety of sources were readily available to us.

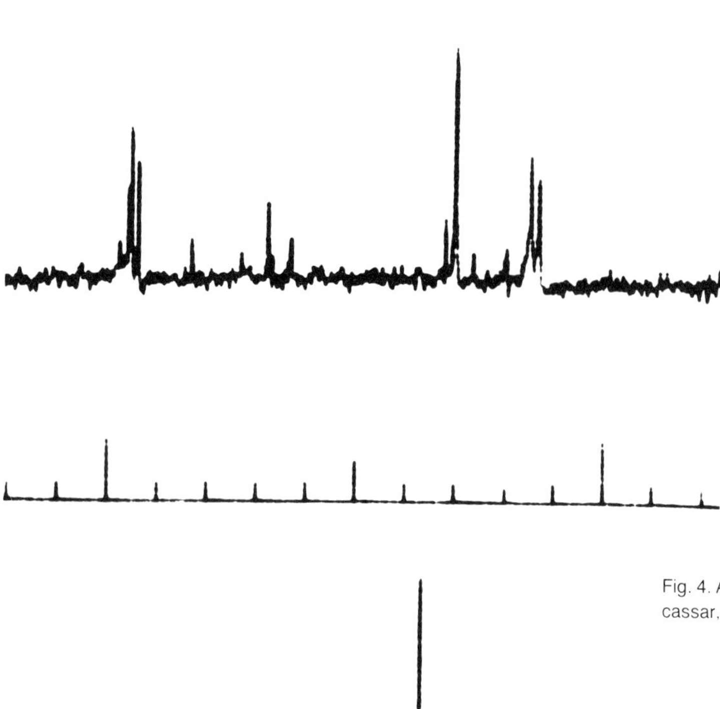

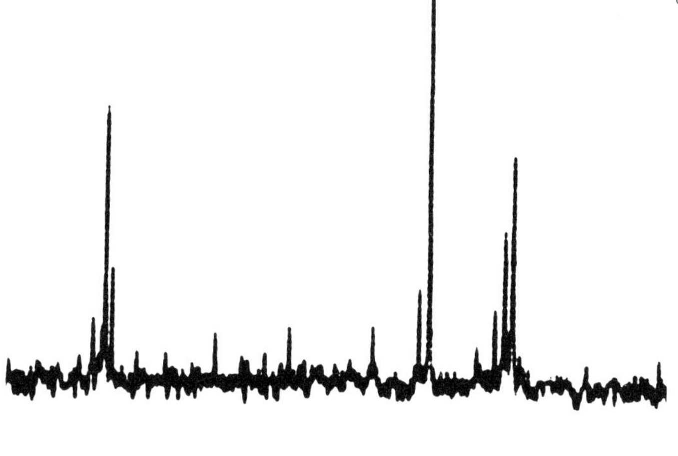

Fig. 4. Alkene region spectra for three sources of copal (a) Palawan, (b) Macassar, (c) spirit manilla.

Figure 4 shows three such sources of copal that are geographically distinct. Whilst we observe an overall similarity in the alkene region that extends to both alkane and carbonyl regions, there are distinct differences in the alkenic region that prompt us to suggest that this technique will allow fingerprinting to be extended to the source-typing of resinous materials used in art objects.

It may be pointed out that knowledge of the structure of these plant resins is normally based upon degradative and related procedures, and the application of NMR spectroscopy to the study of these materials may well contribute to the knowledge of their structure and nature.

The identification and potential source-typing of resinous substances encountered in varnished or lacquered materials opens up new possibilities in the authenticating of many art objects and will, at the very least, be complementary to studies made on these materials using degradive analytical methods.

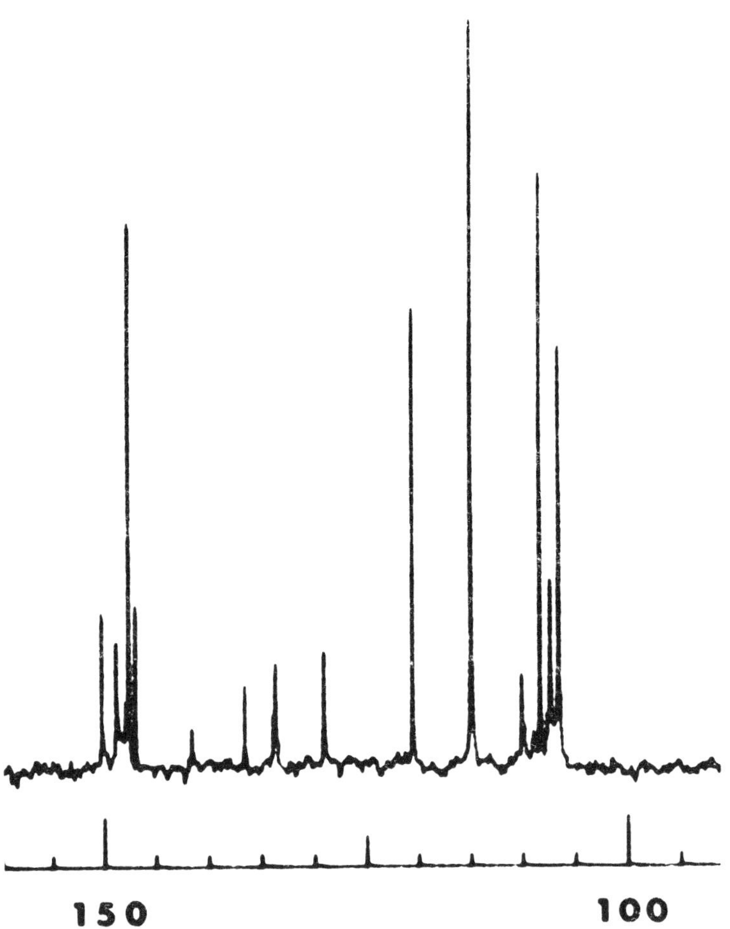

Acknowledgments

This program would not have been possible without the generous supplies of resins and waxes from J. Suter and Company of London, and their invaluable assistance is gratefully acknowledged. The use of ^{13}C-FTNMR spectrometers were made possible with the assistance of Glaxo Group Research and the University of London Intercollegiate Research Service. GVR also acknowledges assistance from SERC.

References

1. M. Cassar, G. V. Robins, R. A. Fletton and A. Alstin, Organic Components in Historical Non-Metallic Seals, *Nature*, 303, (1983): 238.

2. J. W. Akitt, *NMR and Chemistry*, 2nd Ed., (London: Chapman and Hall, 1983).

3. G. C. Levy, R. C. Lichter, and G. L. Nelson, *Carbon-13 NMR Spectroscopy*, 2nd Ed., (New York: John Wiley and Son, 1980).

4. P. E. Kolattukudy (Ed.), *Chemistry and Biochemistry of Natural Waxes* (Amsterdam: Elsevier, 1976).

5. E. A. Bell and B. V. Charlwood, *Secondary Plant Products*, (New York: Springer-Verlag, 1980).

6. G. V. Robins, The Assessment of the Deterioration of Resinous and Related Varnish Materials with ^{13}C-FTNMR Spectroscopy, *Proceedings of the First International Conference on the Non-Destructive Testing of Works of Art*. (Rome: Associazione Italiana Prove Non Distruttive, 1983, in press).

JEAN M. FRENCH, EDWARD V. SAYRE, and LAMBERTUS VAN ZELST

Nine Medieval French Limestone Reliefs: The Search for a Provenance*

*A collaborative research project between the Research Laboratory of the Museum of Fine Arts, Boston, and the Chemistry Department of Brookhaven National Laboratory.

The largest group of French Romanesque reliefs in the United States consists of nine figures of outstanding quality dispersed among four museums: two apostles in the Museum of Art of the Rhode Island School of Design; a St. Peter in the Smith College Museum, Mass.; an apostle and angel in the Memorial Art Gallery, University of Rochester, N.Y., and four apostles in the Duke University Museum of Art, N.C. (figs. 1-9).

The reliefs are approximately the same size; in their present state, the blocks range from 33 to 35 1/8 inches in height. Traces of a reddish-brown pigment are visible on several of the figures. The reliefs show, in varying degrees, the effects of time and the elements. Three of the heads have suffered severe breakage, the surfaces of the Rochester reliefs are considerably weathered, and there has been some infilling and minor reworking. Otherwise, they are remarkably well preserved and have lost none of the energy of the original conception.

St. Peter faces outward toward the viewer, his right hand raised, his left clutching a large key. Two of the extant apostles turn to the side, gesticulating eloquently. Several gaze upward, in the direction of the pointing apostles and angel, as though witnessing some central vision. The poses and gestures convey a sense of communication, excitement, and awe. Presumably an upper zone would have depicted the Christ of the Ascension flanked by two dynamic angels, similar to the one found at Rochester, dramatically linking the upper and lower zones.

The reliefs share certain formal and stylistic features: the rectangular block, the inclined ledge, the same degree of projection, the halos with pearled borders, the triple-ridged system of drapery patterns, as well as drilled pupils, large feet and hands, and expressively elongated fingers. Aspects of the styles of the Rouergue, Limousin, Quercy, and even western France have been cited by various writers,[1-5] but the stocky proportions of the figures, as well as their awkwardly endearing poses and gestures, have eluded classification within any particular school.

Until the late 1960s, the United States reliefs were not recognized as members of the same group; furthermore, little was known of their provenance other than that each had passed through the hands of Joseph Brummer, one of the foremost American dealers in medieval art. Examination of the Brummer files revealed that the reliefs had been purchased from the dealer Altounian in Mâcon (Burgundy) at the end of 1928 and they arrived in this country in February of 1929.

During the next few decades, the reliefs entered American collections; the Smith St. Peter was purchased in 1937, the Rhode Island apostles in 1941, the Rochester angel in 1943, the Rochester apostle in 1949 from the Brummer estate, and the remaining four apostles in 1966, when a significant part of the Ernest Brummer collection was acquired by Duke University. Both correspondence with Brummer in museum files and discussions with former museum personnel suggest that at the time of each sale no mention was made of existence of others of the group.

Thus, for many years the individual reliefs were viewed in isolation. It was only in 1969 that they were identified by Robert C. Moeller III as part of the same sculptural complex, a monumental Ascension, similar to those found today on the tympana of portals in south central France (Mauriac, Collonges, Cahors) or, in western

Fig. 1. Apostle, Duke University Museum of Art, Durham, N.C.; 1966.147.

Fig. 2. Apostle, Duke University Museum of Art, Durham, N.C.; 1966.148.

France, distributed over the surface of the facade (Angoulême, Ruffec). On the basis of stylistic comparisons with capitals *in situ*, Moeller attributed the United States figures to the church of Saint-Martin in Brive (Corrèze) or, alternatively, to a church in the immediate vicinity.[2-4]

While subsequent scholars agree that the individual figures represent participants in an Ascension, Moeller's identification of the United States reliefs as part of the same Ascension program has not been universally accepted. Certain stylistic discrepancies within the group, the range of variation in clarity and surface detail produced by weathering, as well as the lack of further corroborative evidence, have raised questions regarding the composition of the group. Moreover, Moeller's admittedly tentative attribution of the reliefs to the former western portal of the church of Saint-Martin in Brive has proved untenable.

Fragments of a monumental *Descent into Limbo*, discovered in 1878 during the demolition of the masonry of the western porch of Saint-Martin and now in the Musée Rupin in Brive, show little stylistic affinity with the United States group.[6] More importantly, the stone of the portal fragments, as well as that of the Brive capitals, and of the

Fig. 3. Apostle, Duke University Museum of Art, Durham, N.C.; 1966.149.

Fig. 4. Apostle, Duke University Museum of Art, Durham, N.C.; 1966.150.

Fig. 5. Apostle, Museum of Art, Rhode Island School of Design, Providence, R.I.; 41.045.

Fig. 6. Apostle, Museum of Art, Rhode Island School of Design, Providence, R.I.; 41.046.

Brive area in general, is distinctly different in composition from that of the United States apostles.

The objections outlined above raise two major questions that more traditional art historical methods seem unable to resolve. Do all nine reliefs in the United States collections indeed form a single homogeneous group, and what was the original location in France of this monumental complex? These questions form the basis of the present study.

The quality of the limestone of the United States figures is not so fine as to warrant transportation any great distance; therefore, a local origin can be presumed. Earlier work has shown that trace element characterization of limestone can be of use in provenance studies.[7] Consequently, it was decided that neutron activation analysis and petrographic study of samples from the reliefs, in conjunction with analysis of comparative samples from French quarries, might provide invaluable evidence in determining the character of the group and in attempting to localize the origin of the monument.

In the first stage of analysis, one of the Providence reliefs was sent to the Research Laboratory of the Museum of Fine Arts in Boston.

These and subsequent samples were analyzed by neutron activation analysis at the Department of Chemistry of Brookhaven National Laboratory. Powder samples of various sizes were taken from the back of the relief at three different locations to establish the relative homogeneity of elemental compositions within one block. It was found that samples of one gram were representative for the bulk composition with regard to most analyzed trace elements.

Subsequent analysis of samples from the lower back of all nine blocks demonstrated that the concentrations in all nine reliefs are remarkably similar (table 1). They are as closely related to each other as the members of almost any compositionally related group of other materials encountered in similar trace element characterization studies.

A more immediate measurement of the uniformity of the nine reliefs is given in table 2, which shows that the spread in results among the nine separate reliefs is not significantly different from the spread in results among the multiple analyses of the single relief RISD 41.045 initially tested.

These findings are complemented and corroborated by petrographic analysis of thin sections of all nine reliefs. The rock is a

Fig. 7. Apostle, Memorial Art Gallery of the University of Rochester, N.Y.; 49.6.

sandy biopelsparite. The physically reworked carbonate clastic grains are fairly well size-sorted: mostly they are broken and rounded fragments of bryozoa and indeterminate shelly fragments (possibly ostracods), but also foraminifera and coccoliths, in a crystalline calcite cement (as opposed to a fine-grained lime-mud). The limestone is very well recrystallized. Little clay is scattered around in the pore spaces. The carbonate is mainly calcite (confirmed by means of x-ray diffraction), indicating, along with previous characteristics, deposition in a shallow, marine environment. The rock would have to be classed as "sandy"; the sections typically have 15-25% terrigenous grains—mostly well-sorted, equant, medium-sized sand quartz. Limonite, in part weathered from metallic trace minerals, and in part related to organic matter, is disseminated through the rock and imparts to it its yellow color. Modal analyses bear out the similarities in a particularly convincing way. Carbonate-quartz ratios vary very little and even the subcategorizations are extraordinarily similar (e.g. dominance of pellets over discrete fossil fragments, dominance of matrix spar over lime-mud, dominance of strained over unstrained and composite quartz).

In summary, both neutron activation analysis of the trace element compositions and petrographic study of the thin sections demonstrate that all nine reliefs were quarried from a very similar stone.

This reinforces the presumption that these reliefs once formed part of the same monumental complex.

The second stage of this investigation—pinpointing the quarry area in France and hence, the probable original site of the sculptural complex—obviously presents more difficulties.

Close examination of geological maps of France and of samples from French monuments in the collection of Monuments Historiques, consultations with French sedimentary geologists, and examination of both active and abandoned quarries narrowed the search to the present-day department of the Dordogne (roughly the area of old Périgord) in southwestern France.

The Dordogne, the third largest department of France in area, lies between the Massif Central to the east and the lowlands of the Aquitaine Basin to the west. The northeastern section of the department is made up of an extensive band of crystalline rocks that constitutes the southwestern margin of the Massif Central. In the southwestern regions of the department, tertiary sands, clays, and gravels cover a considerable area. The vast central plateaus, cut by beautiful valleys, are made up, for the most part, of Cretaceous limestones. The northern section of this central region, Périgord Blanc, takes its name from the frequent outcrops of chalky limestone that impart a whiteness to the landscape. In Périgord Noir to the southeast, bounded by the Vezère and Dordogne rivers, the limestone takes on a yellowish cast; the name of the area seems to derive from the greater density of trees that cover its plateaus. The region is known to archaeologists and tourists for the numerous rock shelters, dating from Paleolithic times, cut into the cliff walls of its river valleys. It is here that stone most closely resembling that of the United States reliefs is found.

This latter area, Périgord Noir, with extensions to the west, southwest, and south, became the focus of the first survey. The purpose of the survey was to characterize the general source area and to attempt to further localize the source. Samples were taken from abandoned and modern quarries at several locations in the region bordered to the northwest by Périgord Blanc and to the east and south by the extensive Jurassic formations of Quercy (fig. 10).

Analysis of these samples showed one of them, taken from a particular limestone formation in the region of Sarlat, to be most compatible with the stone of the reliefs (fig. 11). Except for iron and scandium, all of the concentrations for this sample lie within the 95% confidence limits for the elements reported for this group of reliefs, and neither iron nor scandium values greatly exceed these limits. The next step was to confirm these findings, and to attempt to further localize the source. With the aid of geologists from the Université de Bordeaux, and the Bureau de Recherches Géologiques et Minières, the limits of the particular formation under investigation were defined. The formation is marked by crosshatch in figure 10, and represents an area approximately twenty by twenty-six kilometers.

This particular formation, suggested as the possible source rock of the reliefs, was sampled throughout, and comparative quarry samples were taken from beyond the formation. A total of 111 samples was collected. Three individual quarries were sampled extensively: the remains of the abandoned quarries of Combe-de-Lama and Griffoul (indicated in the earlier study as close to the stone of the reliefs),

Table 1
Some Major, Minor and Trace Components in Related Limestone Reliefs

Relief I.D.	CaO (%)	Fe_2O_3 (%)	MnO (%)	Na_2O (%)	K_2O (%)	Rb_2O (ppm)	Cs_2O (ppm)	Sc_2O_3 (ppm)	ThO_2 (ppm)	Cr_2O_3 (ppm)	La_2O_3 (ppm)	CeO_2 (ppm)	Sm_2O_3 (ppm)	Eu_2O_3 (ppm)
Duke 45	38.6	0.240	0.0104	0.0193	0.291	11.4	0.366	0.813	1.44	17.3	4.60	11.6	1.47	0.211
Duke 50	39.2	0.225	0.0138	0.0204	0.278	11.3	0.346	0.710	1.20	17.0	3.70	9.0	1.28	0.174
Duke 146	40.6	0.257	0.0110	0.0215	0.326	12.5	0.415	0.782	1.24	18.1	4.27	10.5	1.56	0.213
Duke 148	38.7	0.225	0.0130	0.0197	0.281	9.3	0.328	0.828	1.22	16.7	4.16	10.2	1.48	0.190
R.I.S.D 41.045	36.9	0.268	0.0129	0.0249	0.322	12.9	0.439	0.857	1.36	18.2	4.12	9.8	1.17	0.207
R.I.S.D. 41.06	40.3	0.240	0.0137	0.0314	0.272	10.6	0.350	0.712	1.79	16.3	4.80	14.7	1.13	0.194
Rochester 43.35	39.0	0.257	0.0115	0.0204	0.287	12.4	0.396	0.804	1.58	16.8	4.69	11.5	1.33	0.206
Rochester 49.6	38.3	0.246	0.0130	0.0210	0.266	10.8	0.390	0.824	1.28	18.1	3.98	9.6	1.08	0.206
Smith 1973:12-1	36.1	0.289	0.0106	0.0266	0.344	13.5	0.496	1.025	2.00	18.7	5.28	13.6	1.32	0.243
Mean	38.6	0.250	0.0122	0.0228	0.296	11.6	0.391	0.817	1.47	17.5	4.40	11.2	1.31	0.205
Group Std. Dev(±)	1.4	0.021	0.0013	0.0040	0.027	1.3	0.053	0.093	0.28	0.8	0.48	1.9	0.16	0.018

Table 2
Similarities of Analytical Ranges Among Multiple Samples of a Single Relief[1] and those Within the Group of Nine Related Reliefs

Element Determined	Percent Spread[2] in Single Relief	Percent Spread[2] in Group of Reliefs
Calcium	3	4
Sodium	18	17
Potassium	13	9
Rubidium	12	11
Cesium	11	13
Scandium	14	11
Lanthanum	19	11
Cerium	23	17
Samarium	19	13
Europium	16	9
Thorium	29	19
Chromium	8	5
Manganese	9	11
Iron	6	8

Table 3
Cerium-Lanthanum and Thorium-Lanthanum Values for Three Quarries, the Reliefs and Six Samples from Outside the Formation

	Ce La	Th La
Griffoul	2.1	0.34
Combe-de-Lama	2.2	0.30
Les Combarelles	2.1	0.42
Reliefs	2.5	0.33
Ajat	1.1	0.068
Couze	2.2	0.38
Nazareth	1.5	0.11
Paussac	1.9	0.25
Grammont	2.0	0.28
Thenon	1.2	0.11

[1]R.I.S.D. 41.045 Relief

[2]Group Standard Deviations as Percent of Means

both to the south of Sarlat; and les Combarelles, a modern quarry approximately sixteen kilometers northwest of the city.

At les Combarelles, a relatively small quarry typical of the region, one gallery was sampled systematically. Petrographically there are significant differences between the lower horizon of the gallery (which is a fine-grained, pelletal limestone) and the upper horizon (which is coarse-grained, mostly a fossil hash); nevertheless, the elemental compositions proved relatively homogeneous. At Combe-de-Lama and Griffoul the stone is, in large part, exhausted. However, lateral sampling of both abandoned quarries shows considerable overlapping with each other and with les Combarelles (fig. 12).

Some distinctions can be made: for example, Combe-de-Lama has a higher range of values for iron; the values for europium at Combe-de-Lama tend to be on the high side as do the tantalum values at Griffoul. Griffoul and les Combarelles are perhaps more difficult to distinguish. Interestingly enough, petrographic examination of thin sections separates out the samples from les Combarelles, especially those of the lower horizon, as different from those taken at the Griffoul and Combe-de-Lama quarries. In general, however, on the chemical evidence, the differences between the three quarries were found to be of the same order as the spread within a single quarry; all three quarries are quite closely related to each other.

When the individual quarries are compared to the reliefs (fig. 13) the present-day samples from Combe-de-Lama, again, are not as close as the other two quarries. The smaller size of the Combarelles gallery is generally reflected in a tighter range, often of the same order or slightly larger than that of the reliefs. The spread at Griffoul, a larger quarry, is admittedly wide, but accepting that spread, the relation to the reliefs is rather close; the values of the reliefs fit quite well within the range of the Griffoul quarry.

Fig. 8. Angel, Memorial Art Gallery of the University of Rochester, N.Y.; 43.35.

Fig. 9. St. Peter, Smith College Museum of Art, Northampton, MA; 1937:21-1.

The ranges for the formation as a whole, including samples other than that from the three quarries mentioned, are, as one might expect, not as tight as those for the individual quarries (fig. 12). Nevertheless, average and standard deviation ranges for the formation correspond quite well to what would be expected from the three quarries combined.

The next question then, is whether samples from outside the formation are distinctly different. When six samples taken from outside the formation (Couze, Paussac, Ajat, Thenon, the stone of Grammont, Nazareth) are compared to the ranges for the formation as a whole (fig. 14), four of these immediately fall out on the excessively high or low absolute values of their trace element concentrations. (The stone of Grammont, which is quarried near villages of the same name within the Brive basin, has been suggested to be the material of the Brive sculpture discussed earlier in this study).

As would be expected, a study of correlations between elemental concentrations adds significantly to the interpretation of the analytical data. Illustrative is the high degree of correlation between lanthanum and cerium concentrations for samples from the Griffoul and Combe-de-Lama quarries. While, for instance in the Griffoul quarry

the group standard deviation for both lanthanum and cerium is about 32%, the group standard deviation for the lanthanum-cerium ratio is only 9.5%. Such correlation behavior may in itself be highly diagnostic.

Table 3 shows cerium-lanthanum and thorium-lanthanum ratios, both for the three major quarries, the reliefs, and the six samples from outside the formation under study. Quarries and reliefs show very close values for these ratios, but three of the six outside samples show much lower ratios for both the cerium-lanthanum and the thorium-lanthanum ratios.

Thus it becomes necessary to employ multivariate statistical techniques that take into account both absolute values and correlations. The data set was analyzed using the computer program "ADCORR," which calculates probabilities of group membership for individual samples on the basis of Mahalanobis distances.[8] It is interesting that, using these techniques, at a confidence level of 80%, the sequence of samples from the Combe-de-Lama quarry tends to separate out from all quarry samples as well as from the reliefs. Such definite distinctions cannot be made for the other two major quarries. The samples from outside the formation do not fit in any of the quarry groups

Fig. 10. Map of the Périgord region of France indicating first samples (□)
and (•) comparative samples from quarries outside the Sarlat formation.
Crosshatch delineates the limestone formation.

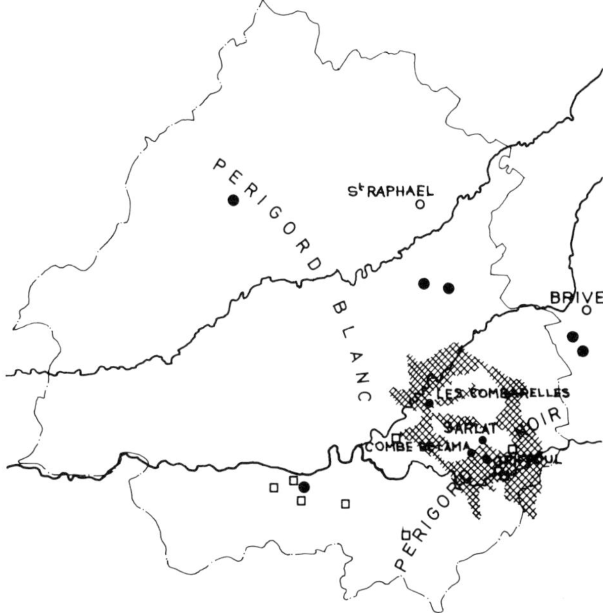

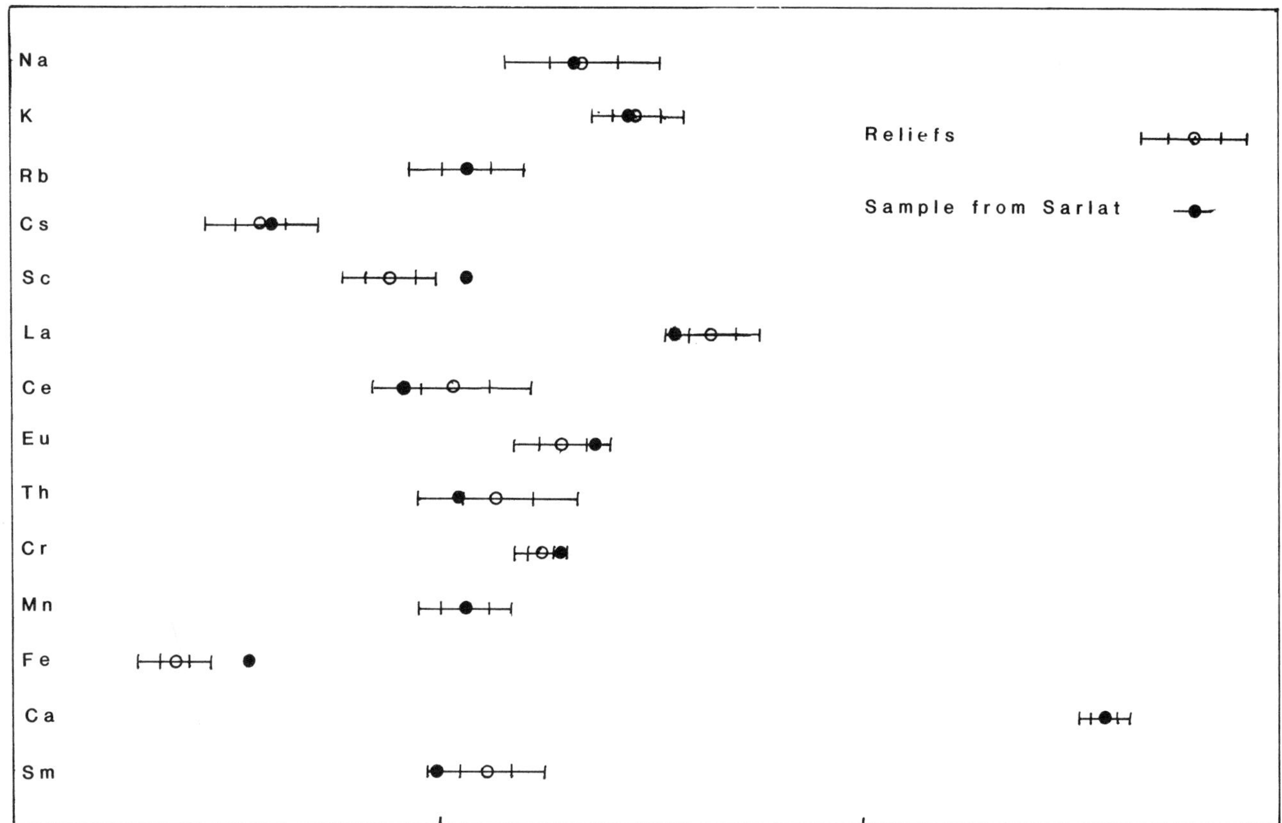

Fig. 11. Means with one and two standard deviation limits for the group of nine
reliefs, compared to a sample from near Sarlat.

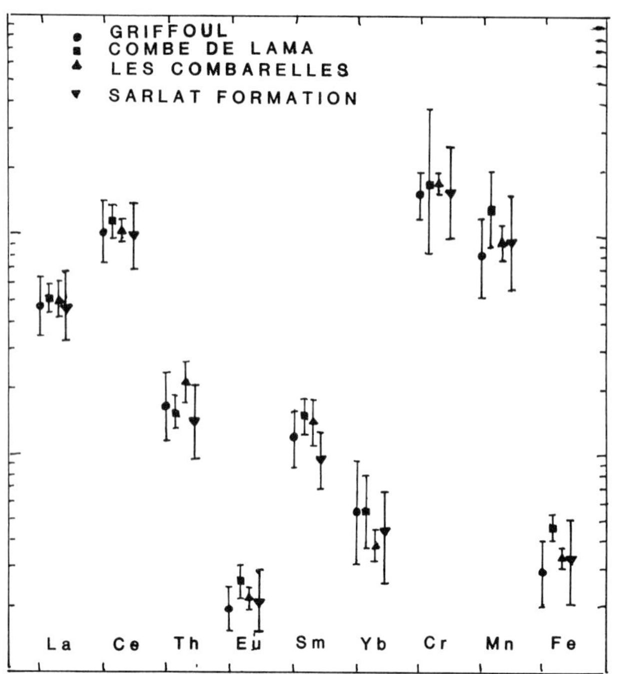

Fig. 12. Means and standard deviations for three quarries and for all samples taken from within the formation.

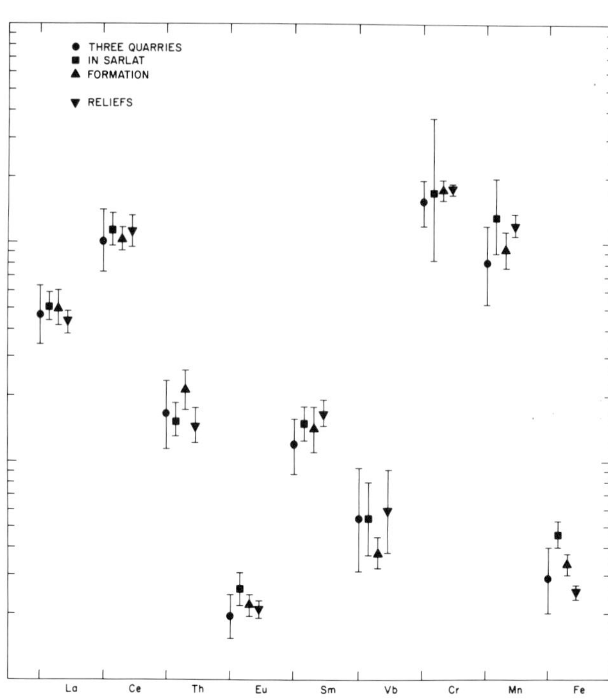

Fig. 13. Means and standard deviations for three quarries and the group of nine reliefs.

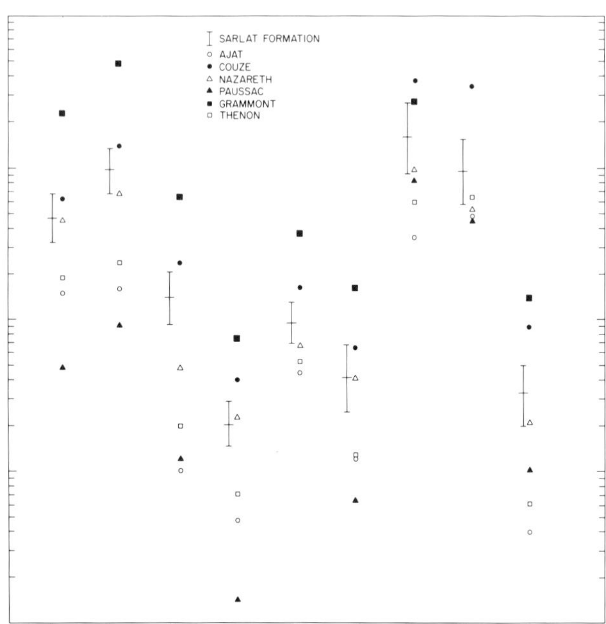

Fig. 14. Means and standard deviations for the formation, compared to six samples taken from outside the formation.

nor in groups formed by combinations of quarry samples.

When the reliefs are used as a core group, the number of samples admitted with a higher than 20% probability varies, of course, with the number and identity of the elements used in the calculation. Typically, however, samples from Griffoul, from les Combarelles, from le Pech de Giroux, and from several other locations within the formation are included in the core group for different combinations of elements. Chemical evidence indicates that the stone of the reliefs does originate from this formation; however, the exact location within the formation is not determinable.

On the other hand, when petrographic analysis of the samples is brought into the picture, a further refinement may be possible. For example, petrographic analysis of the thin sections tended to rule out the samples from both les Combarelles and le Pech de Giroux. When thin sections of all samples from the formation were subjected to petrographic analysis, a number of samples (primarily from Griffoul and Combe-de-Lama) were found to be "close" and "moderately close" to the reliefs; however, only certain samples from Combe-de-Lama and samples from the oldest section of Griffoul were found to be "virtually identical" with the reliefs.

Since the remains of these two quarries south of Sarlat are less than 2½ kilometers apart, it can be assumed that this particular stone was once quarried extensively in the immediate area. It can be best studied at Combe-de-Lama where the unquarried limestone is utilized as part of the foundation wall of a small twelfth-century

Fig. 15. (a) Mounting holes in the facade of the chapel of a local chateau. (b) Marks left on the side wall of the chapel.

church. Since the chemical composition of both quarries has been shown to be quite consistent with that of the reliefs, it may be possible to conclude that, through a combination of chemical and petrographic analysis, the source rock of the United States reliefs has been localized to a particular formation in the southeastern section of the Dordogne and, within that formation, tentatively to the immediate region south of Sarlat—the lower part of the Sarlat-Vezac-Vitrac triangle.

Possible corroboration of this source area came quite recently with the discovery of information regarding a large group of apostle reliefs, presumably from a local church, which was located until the early part of this century in the immediate vicinity of the Griffoul and Combe-de-Lama quarries. Two of the reliefs—a St. Peter and another apostle—were mounted on the facade of a small nineteenth-century chapel in the garden of a local chateau; the other rested on bases along the side of the chapel. The reliefs were bought by a French dealer and it is known that they were destined for the United States. The approximate date of the sale of the reliefs, the measurements of the mountings on the facade, and the measurements of traces from the reliefs along the side wall of the chapel suggest that this group of reliefs is the same as that which forms the basis of this study (fig. 15).

Finally, an analysis of two sculptures from the same geographic area provides another interesting illustration of the potential of trace element characterization in provenance studies of limestone objects.

There are very few objects in United States collections with a clear provenance to the Dordogne. However, a number of pieces in this country (at the Fogg Art Museum, Cambridge, Mass.; the Philadelphia Museum of Art; and at Williams College, Williamstown, Mass.) have been attributed to the church of Saint-Raphael near Excideuil in the Dordogne; of these a capital in the Williams College Museum of Art appears to have a firm provenance.[9] A second Williams College piece, a fragment of the torso of an apostle purchased two years before the capital, had been attributed to Saint-Raphael but with no further substantiating evidence.

Analysis of samples from the capital and torso demonstrates the virtually identical composition of the stone of the two Williams College pieces, which, however, is pronouncedly different from that of the apostle reliefs of the Sarlat formation to the southeast (fig. 16).

In this last example, trace element analysis has confirmed the previous assumptions of art historians and has helped to pave the way for further study of the monument. In the case of the larger apostle group, it has resulted in a previously unknown attribution: the localization to the immediate region of Sarlat of a monument with important links, stylistically, iconographically, and formally, to major sculptural complexes in south central and western France.

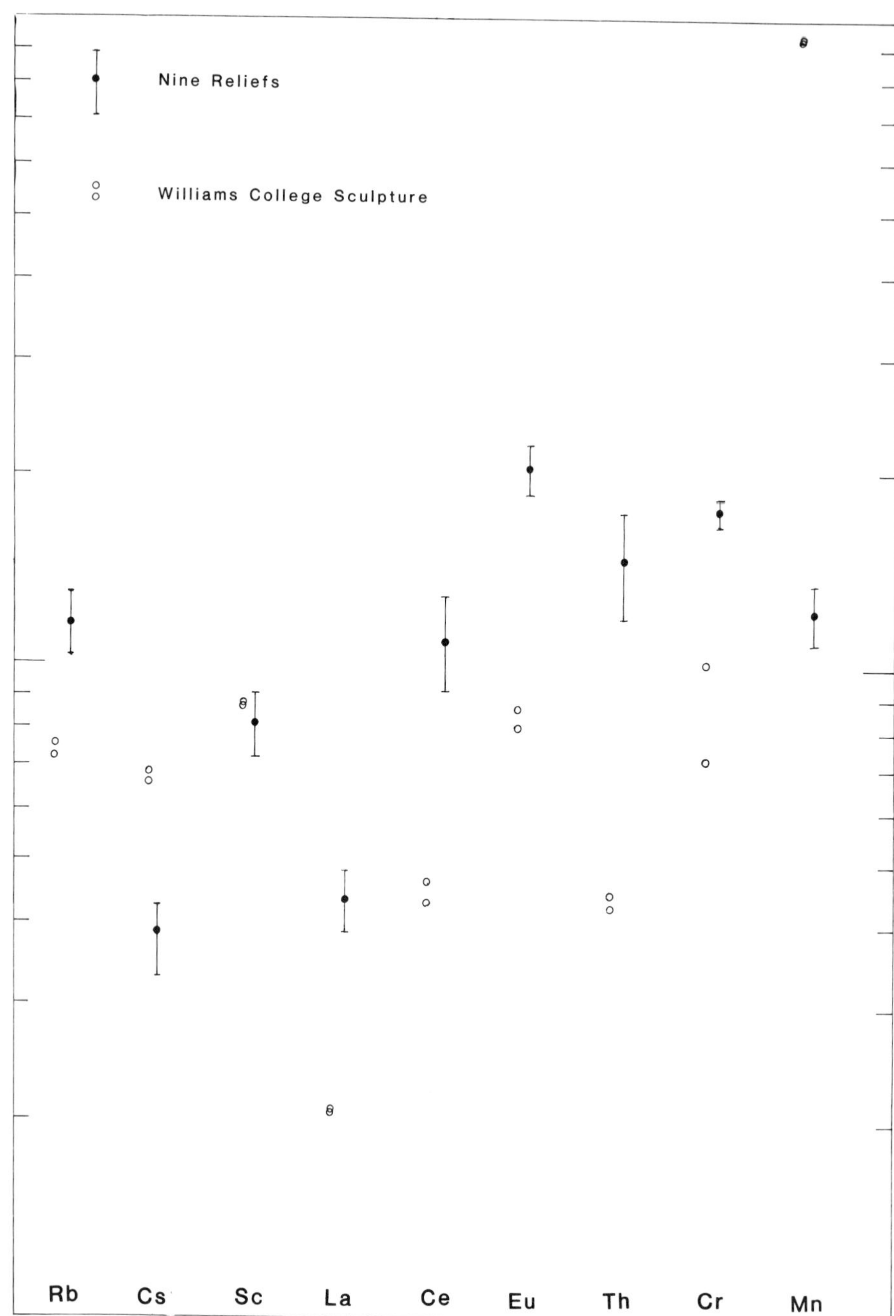

Fig. 16. Means and standard deviations for the group of nine reliefs, compared to two sculptures from Williams College, attributed to Saint-Raphael.

Acknowledgments

This investigation formed part of collaborative efforts between the Museum of Fine Arts, Boston, and the Department of Chemistry, Brookhaven National Labortory. Work at the latter institution was conducted under the auspices of the U.S. Department of Energy and supported by its Office of Basic Energy Sciences.

The research for this project was financed in part by grants from the American Philosophical Society, the National Endowment for the Humanities, and the Bard College faculty research fund.

The authors would like to express their gratitude to Dr. Judith Rehmer Hepburn who performed the petrographic analyses of the relief and quarry samples, and to Ms. Elaine Rowland and Dr. Ronald Bishop for their help in carrying out the analyses of the limestone. They would also like to thank the directors and staffs of the Museum of Art of the Rhode Island School of Design, the Smith College Museum, the Memorial Art Gallery of the University of Rochester, the Duke University Art Museum, and the Williams College Museum of Art; Mrs. Ernest Brummer for her kindness in permitting access to the Brummer files; Prof. Whitney S. Stoddard of Williams College; Mme. Annie Blanc of the Centre de Recherches sur les Monuments Historiques; Profs. Louis Humbert and Paul Lévêque of the University of Bordeaux; Mr. Jean Pierre Platel of BRGM (SGR Aquitaine); Mr. Fonquernie, architecte en chef des Monuments Historiques; Mr. Jean Beauchamps and Mr. Bernard Lourdon, architectes des Batiments de France et de la Dordogne; the late Mr. Roger Delmas, Conservateur du Musée des Penitents Blancs à Sarlat; and Mr. Alain Blanchard, without whose help this project would not have been possible.

References

1. E. H. Payne; *Bulletin of the Smith College of Art* 18-19 (June 1938), pp. 3-6.

2. R. C. Moeller, III; *Sculpture and Decorative Art*, (Raleigh: North Carolina Museum of Art, 1967), pp. 8-17.

3. R. C. Moeller, III; *College Art Journal* (1967-68) pp. 182-184.

4. R. C. Moeller, III; in *Renaissance of the Twelfth Century*, ed. Scher, (Providence: Museum of Art of the Rhode Island School of Design, 1969) pp. 50-58.

5. W. Cahn; *Gesta* 7 (1968) pp. 53-54; *Gesta* 8 (1969) p. 56; *Gesta* 10-1 (1971) p. 51; *Gesta* 14-2 (1975) pp. 67-68.

6. L. Bonnay; *Bulletin de la Société Scientifique, Historique et Archéologique de la Corrèze*, 1 (1878) pp. 223-238.

7. P. Meyers and L. van Zelst; *Radiochimica Acta* 24 (1977) pp. 197-204.

8. A. M. Bieber Jr., D. W. Brooks, G. Harbottle and E. V. Sayre; *Archaeometry* 10 (1976) pp. 59-74

9. A. de Roumejoux; *Bulletin de la Société Historique et Archéologique du Périgord* (1890), pp. 127-130.

NORMAN HERZ, SUSAN E. KANE, and W. B. HAYES

Isotopic Analysis of Sculpture from the Cyrene Demeter Sanctuary

Excavations carried out over ten years in the Sanctuary of Demeter, under the direction of Donald White, University Museum, Philadelphia,[1] have unearthed 724 catalogued pieces of sculpture. Included are large-scale statues, statuettes, heads, and reliefs that span the life of the sanctuary. Marble was the most commonly used sculptural material (89.1%) with the local limestone a distant second (10.9%). However, marble is not found in the region of Cyrenaica, and, therefore, all marble had to be imported, reflecting the wealthy economy of Cyrene. If the provenance of the marbles uncovered in the sanctuary could be determined, useful information would be gained on the commercial history of Cyrene.

Based upon a collaboration between a geologist, a classical archaeologist, and a statistician, this paper illustrates how a scientific technique (in this case stable isotopic signature analysis) and statistical treatment, coupled with stylistic analysis of marble sculpture from the Sanctuary of Demeter, can help to reconstruct some of the commercial connections of Cyrene.

Stable Isotopic Analysis

Until recently, a consistently accurate identification of the many visually similar classical marbles has been difficult.[2,3] Characteristics such as grain or color do not provide a secure basis for reliable "naked eye" identification because the physical characteristics of marble from any one quarry area are too heterogeneous. In order to determine the provenance of artifacts, various techniques were attempted including petrographic, petrofabric, and chemical analysis. Unfortunately, larger pieces are needed for the microscopic study of the first two techniques than any curator will allow to be taken from an artifact. Even then, the results obtained are not diagnostic because of the lack of an adequate data base for comparison and because of the above mentioned variability in mineralogical and petrographic characteristics. Trace element analysis has also been unsuccessful because of the large variability—commmonly with factors of over 100—within the same quarry.

An increasingly precise determination of marble provenance is possible using stable isotopic signature analysis.[4-7] Isotopic ratios of ^{12}C-^{13}C and ^{16}O-^{18}O are measured in each sample and expressed in terms of their deviation from a conventional standard, the Pee Dee belemnite, a carbonate fossil from South Carolina. This deviation, called δ values, expressed as $\delta^{13}C$ and $\delta^{18}O$ in parts per thousand (0/00 or per mil) forms the "isotopic signature." These values appear to exhibit a relatively restricted range for each quarry area or limited parts of a geological formation. For any signature technique to be successful, the signature values must be uniform over the dimensions of an artifact, should be uniform over a quarry area, and should show only small variations within the limits of a mining district. Variations in $\delta^{13}C$ and $\delta^{18}O$ within the dimensions of statuary blocks appear to be less than 0.5 0/00 and within an outcrop (or most quarries), less than 2 0/00.[8] Another important attribute, in the case of any type of destructive analysis, is the need for very small amounts of sample for analysis. Isotopic analysis consumes only tens of milligrams; a volume with the dimensions of a pencil point is sufficient.

The first use of isotopes to determine marble provenance was published by the Craigs.[5] They analyzed quarry samples which included some from Naxos, Paros, Penteli, and Hymettos and found that the samples from any one quarry fell into well-defined isotopic clusters. Manfra and his colleagues added to the Craig data base with analyses of samples from western Anatolia: Marmara. Ephesos. Aphrodisias, Denizli, and Afyon. This stable isotope data base for marble has been vastly increased in recent years by the work of the laboratories at the University of Georgia[9] and now includes all the known principal quarries of classical Greece and Rome.

Sample Analysis of Classical Quarries and the Demeter Sanctuary

The stable isotopic data base for classic marble now contains 234 samples, each analyzed for ^{18}O and ^{13}C. Most of the analyses were carried out in our laboratories, but the data also include analyses taken from the literature. The data base for Mediterranean marble sources, and the total number of analyses for each source, shown in parentheses, includes, Doliana (9), Mani (8), Paros, principally from the Lichnites mine (14), Penteli (12), Thasos, principally from the Aliki quarry but also the southern coast of the island (26), Naxos (29). Sounion (5), Ephesos (18), Afyon (7), Aphrodisias (8), Denizli (26), Marmara (14), Hymettos (10), and Carrara-classical Roman (34) and Renaissance (14) quarries. For this study, the samples analyzed from the Demeter Sanctuary were compared to the data bases for Paros, Penteli, Thasos, Naxos, Afyon, Aphrodisias, Denizli, Ephesos, Marmara, Hymettos, and Carrara. The data for Doliana, Mani, and Sounion were not used; these are smaller quarries, and, as far as is known, their marble was not exported over great distances.

Statistical analysis of the known quarry samples, and the correlation of the analyzed samples from the Demeter Sanctuary to this data base, was carried out on the Cyber 750 computer at the University of Georgia with the SPSS Discriminant Analysis (DA) program.[10] DA permits a statistical separation between groups of cases; for this study, each group was a quarry area and each case was a sample. Discriminating variables used were $\delta^{13}C$ and $\delta^{18}O$ values as determined for the known quarry samples. By forming linear combinations of the discriminating variables, DA calculates "discriminant functions" of the form $D_i = d_{i1} + d_{i2}z_2 \ldots + d_{ip}z_p$ where Di is the score on the discriminant function i, d is a weighting coefficient, and z is the standardized value for each discriminating variable used in the analysis. The discriminant functions can then be treated as axes for a geometric discriminant space into which the known groups are plotted. Polygonal regions of the discriminant space are thereby defined. The cases with unknown affinities (the Cyrene samples) can be categorized into one or another of these areas; the degree of certainty of this classification increases if the unknown is near the centroid of a polygon and decreases if it is near a polygon boundary. The probabilities of "most certain" and "second most certain" group membership are calculated by the program for each unknown.

Assignment of Samples by the DA Program

The DA results showed that certain groups of quarries could be better differentiated than others. Table 1 shows the results of analysis by quarry, and figure 1 shows the plots of each in discriminant space. For each discriminant function, the program calculated coefficients reflecting the contribution of each variable to the function. For function 1, carbon = 0.35962, oxygen = 0.94011 (shown as x-coordi-

Table 1
Discriminant Function Program Assignments of Known Quarry Samples Based on Oxygen and Carbon Isotopic Analysis

Shown are percent probability of assignment: high. average. low.

PAROS

N = 14	first choice		second choice		°₀ missed
Ephesos	7	82.3 75.4 67.5	1	18.8	57
Paros	2	39.0 36.5 34.0	7	26.1 21.1 16.1	36
Carrara	3	34.9 33.2 29.5	1	17.6	71
Aphrodisias	2	39.1 37.2 35.3	3	30.1 26.0 21.6	64

First Choices: Ephesos 50%, Carrara 21%; 14% each Paros and Aphrodisias.

Second Choices: Paros 50%, Aphrodisias 21%, 7% each Ephesos, Carrara, Hymettos, and Marmara.

PENTELI

N = 12	first choice		second choice		% missed
Penteli	9	62.0 51.2 34.4	1	40.8	17
Naxos	2	41.4 31.3 21.4	9	41.2 36.2 30.2	8

First Choices: Penteli 75%, Naxos 17%, Ephesos 8%.

Second Choices: Naxos 75%, 8% each Penteli, Aphrodisias, and Paros.

THASOS , ALIKI

N = 23	first choice		second choice		% missed
Thasos	21	70.8 43.9 31.1	1	27.5	4
Denizli	1	28.8	20	31.1 29.0 19.8	4

First Choices: Thasos 91%, 4% each Denizli and Paros.

Second Choices: Denizli 87%; 4% each Thasos, Marmara, and Hymettos.

NAXOS , APPOLLONAS

N = 29	first choice		second choice		°₀ missed
Naxos	11	52.8 47.0 33.5	5	40.6 27.4 23.3	35
Hymettos	8	42.9 35.2 30.3	2	27.4 26.7 25.9	66
Penteli	4	70.5 56.5 47.9	13	45.5 37.3 28.0	41
Denizli	4	66.3 54.5 44.0	3	32.1 24.8 20.2	76

First Choices: Naxos 48°₀. Hymettos 28°₀. Penteli 14°₀. Denizli 14°₀; 3°₀ each Carrara and Afyon.

Second Choices: Penteli 45°₀. Naxos 17°₀; 10°₀ each Denizli. Paros. and Aphrodisias; Hymettos 7°₀.

AFYON

N = 7	first choice		second choice		% missed
Afyon	2	41.7 41.2 40.7	2	36.0 33.5 31.0	43
Ephesos II	2	45.5 40.7 35.8	2	39.9 30.9 21.9	43

First Choices: Afyon 29%, Ephesos II 29%; 14% each Naxos, Aphrodisias, and Hymettos.

Second Choices: Afyon 29%, Ephesos 29%; 14% each Penteli, Denizli, and Aphrodisias.

APHRODISIAS I and II

N = 2	first choice		second choice		% missed
Aphrodisias I	2	96.3 64.0 31.7			0
Ephesos			2	29.0 16.0 3.0	0

N = 6	first choice		second choice		
Aphrodisias II	6	40.2 39.1 37.0			0
Hymettos			4	20.4 18.3 14.9	33

First Choices: Aphrodisias 100%.

Second Choices (combined I and II): Hymettos, 50%, Ephesos 25%, Afyon 25%.

EPHESOS I and II

N = 8	first choice		second choice		% missed
Ephesos I	7	81.2 64.7 40.7	1	30.1	0
Paros	1	40.3	7	38.0 26.9 16.7	0
N = 10					
Ephesos II	4	41.5 38.5 36.8	3	26.4 17.7 9.6	30
Aphrodisias I	2	88.4 83.8 79.1	1	32.1	70

First Choices: Ephesos 61%, Aphrodisias 17%, Naxos 11%; 6% each Paros and Afyon.

Second Choices: Paros 39%, Ephesos 22%; 11% each Afyon, Penteli and Aphrodisias; 6% Carrara.

MARMARA

N = 14	first choice		second choice		% missed
Marmara	6	27.5 26.1 24.7	1	19.4	50
Denizli	3	30.7 29.4 27.2	7	31.1 23.8 17.1	29
Thasos-Al	2	36.1 34.6 33.1	3	31.0 29.9 27.2	64

First Choices: Marmara 43%, Denizli 21%, Thasos-Al 14%; 7% each Aphrodisias, Carrara, and Paros.

Second Choices: Denizli 50%, Thasos 21%, Carrara 14%; 7% each Marmara and Ephesos.

HYMMETOS

N = 10	first choice		second choice		% missed
Marmara	3	23.8 23.3 22.3	0		70
Aphrodisias	3	40.3 36.8 33.9	2	30.3 26.4 22.4	50
Hymmetos	2	35.1 34.1 33.1	1	22.9	70
Carrara	0		3	18.6 17.4 16.8	70

First Choices: Aphrodisias 30%, Marmara 30%, Hymmetos 20%; 10% each Ephesos 7 Naxos.

Second Choices: Carrara 30%, Aphrodisias 20%, Ephesos 20%, 10% each Hymmetos, Naxos, and Afyon.

CARRARA

N = 34	first choice		second choice		% missed
Carrara	31	41.8 33.3 25.4	2	23.8 22.9 22.0	3
Marmara	2	26.3 26.1 25.8	19	25.1 21.0 16.1	38
Aphrodisias	1	36.6	6	29.6 21.7 16.7	79
Hymmetos	0		6	24.9 19.6 16.8	82

First Choices: Carrara 91%, Marmara 6%, Aphrodisias 3%.

Second Choices: Marmara 56%, Aphrodisias 18%, Hymettos 18%, Carrara 6%, Denizli 3%.

Fig. 1. Discriminant Space (DF) plots of the principal classical marble quarries. Cyrene samples shown as dots. Possible full DF field is shown for Cyrene-Carrara; other are plotted only within their relevant field segments.

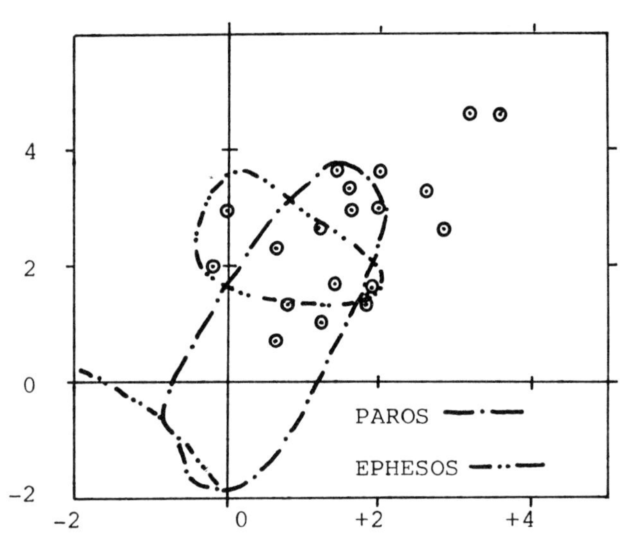

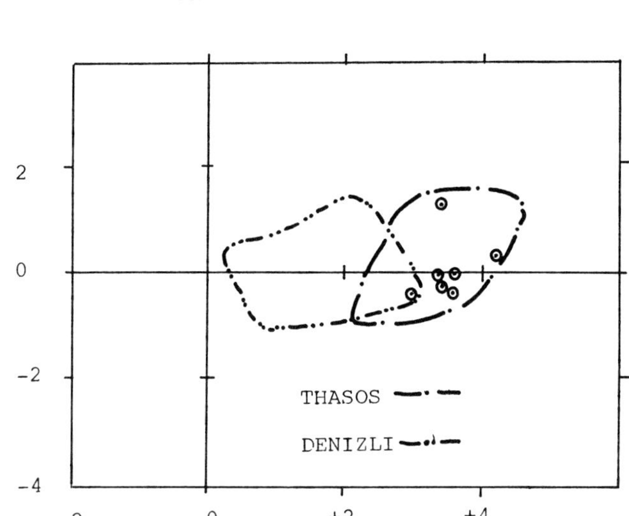

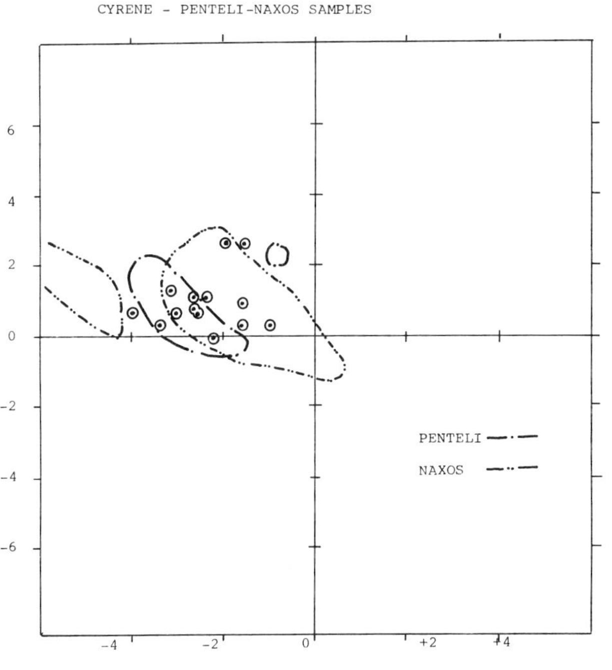

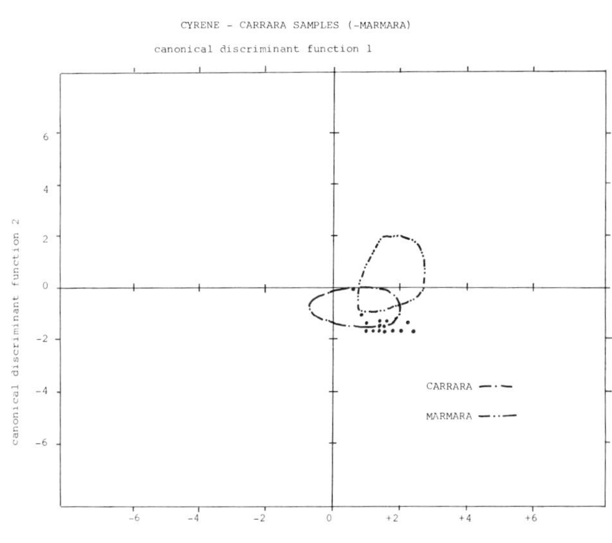

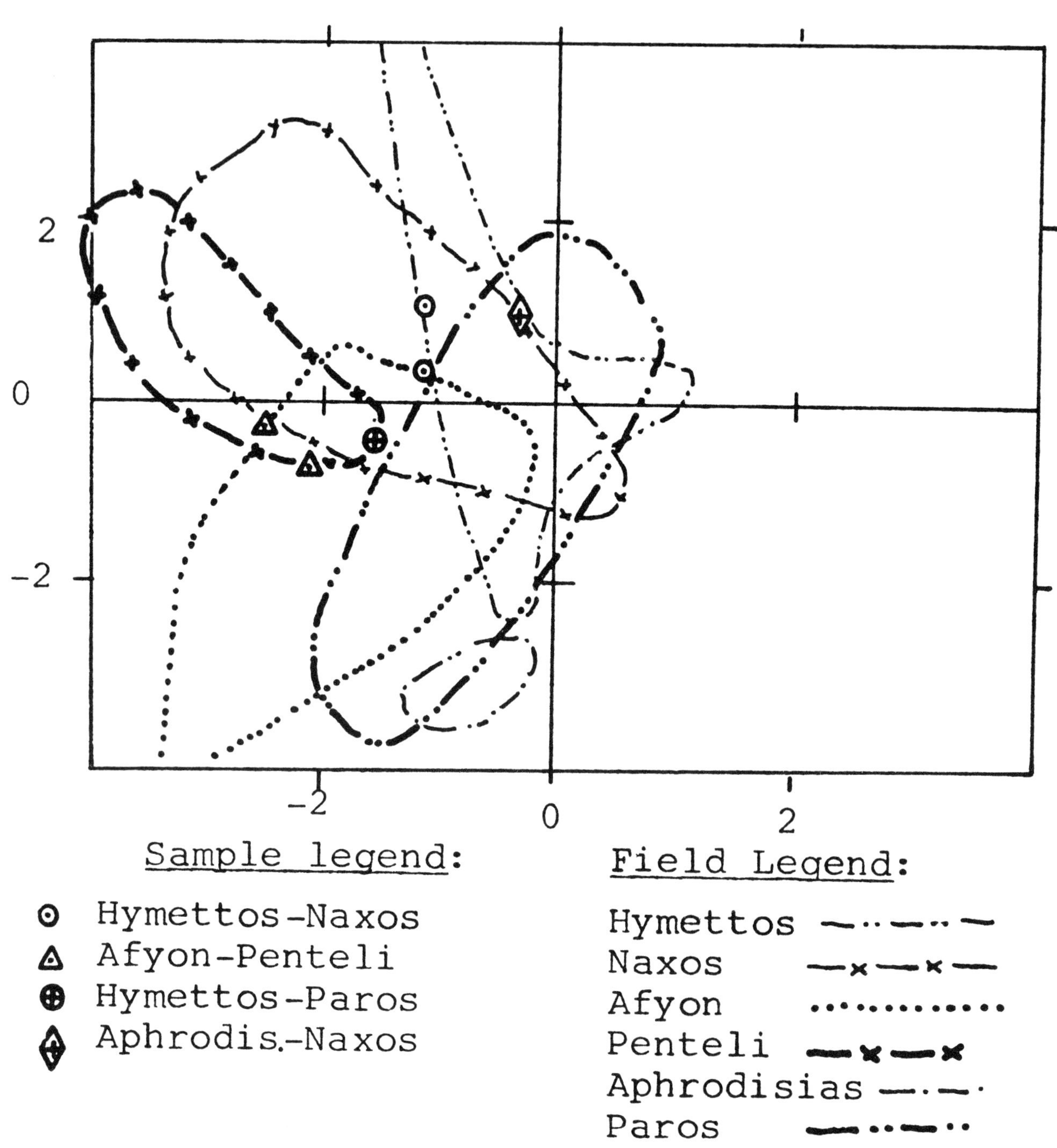
CYRENE - unknown samples
2
0
-2
-2
0
2
Sample legend:
Hymettos-Naxos
Afyon-Penteli
Hymettos-Paros
Aphrodis.-Naxos
Field Legend:
Hymettos
Naxos
Afyon
Penteli
Aphrodisias
Paros

Table 2

Cyrene Samples: Isotopic Analytical Data and Presumed Provenance

Sample #	$\delta^{18}O$	$\delta^{13}C$	‰ 1st Choice	‰ 2nd Choice
I. CARRARA				
69-292	− 1.04	+ 1.79	Carrara 48	Marmara 15
69-293	− 1.13	+ 1.82	Carrara 46	Marmara 16
69-360	− 1.89	+ 2.08	Carrara 39	Marmara 16
71-829	− 0.54	+ 2.37	Denizli 27	Carrara 25
73-263	− 0.64	+ 1.95	Carrara 39	Denizli 21
73-876	− 0.09	+ 2.10	Thasos 28	Carrara 27
73-978	− 0.19	+ 2.10	Carrara 28	Thasos 26
74-155 76-317	− 1.19	+ 1.76	Carrara 48	Marmara 15
74-958 78-655	− 0.48	+ 2.08	Carrara 35	Denizli 23
76-565	− 1.22	+ 1.96	Carrara 43	Marmara 17
76-1110	− 1.64	+ 1.98	Carrara 43	Marmara 16
77-947	− 0.82	+ 1.96	Carrara 41	Denizli 20
78-386	− 1.16	+ 1.93	Carrara 44	Marmara 17
F13-613/1977	− 1.44	+ 1.67	Carrara 48	Aphrodisias 15
UNK-59	− 1.07	+ 1.84	Carrara 46	Marmara 16
D14/E14Tr2St1	− 2.47	+ 2.68	Carrara 23	Hymmetos 21
II. PENTELI				
71-63	− 6.24	+ 2.51	Naxos 46	Penteli 45
71-500	− 6.75	+ 2.21	Penteli 49	Naxos 38
74-359	− 7.10	+ 2.76	Penteli 54	Naxos 42
74-931	− 6.50	+ 2.70	Penteli 48	Naxos 43
76-256	− 6.21	+ 2.60	Naxos 47	Penteli 44
77-160	− 6.34	+ 2.34	Penteli 45	Naxos 43
78-47	− 7.83	+ 1.99	Penteli 45	Denizli 33
79-2	− 7.03	+ 1.96	Penteli 45	Naxos 29
71-704	− 5.33	+ 2.75	Naxos 45	Penteli 28
74-94	− 6.09	+ 3.98	Naxos 48	Penteli 30
76-137	− 5.54	+ 2.03	Naxos 34	Penteli 28
76-460	− 6.43	+ 4.07	Naxos 50	Penteli 35
77-377	− 5.07	+ 2.43	Naxos 35	Penteli 21
77-940	− 5.89	+ 1.55	Afyon 40	Penteli 22
III. PAROS				
71-235	− 2.57	+ 5.58	Ephesos 82	Paros 16
71-304/76-618	− 3.12	+ 4.85	Ephesos 68	Paros 27
73-444	− 1.77	+ 5.48	Ephesos 68	Paros 23
73-1288	− 1.37	+ 6.68	Ephesos 43	Paros 28
74-25	− 1.69	+ 6.68	Ephesos 43	Paros 28
74-180 -517	− 1.37	+ 6.81	Ephesos 93	Paros 6
74-962 -1064	− 2.42	+ 5.29	Ephesos 73	Paros 22
76-844	− 2.78	+ 5.21	Ephesos 75	Paros 21
78-90	− 3.11	+ 5.47	Ephesos 85	Paros 14
78-701	− 2.93	+ 3.28	Ephesos 83	Paros 15
71-297	− 4.54	+ 4.66	Ephesos 77	Paros 18
71-700	− 2.23	+ 4.03	Paros 37	Ephesos 19
71-830	− 1.77	+ 3.91	Paros 27	Marmara 23
78-270	− 4.30	+ 4.00	Ephesos 47	Paros 32
78-501	− 2.44	+ 3.60	Paros 33	Marmara 21
78-550	− 3.47	+ 4.49	Ephesos 60	Paros 32
78-700	− 2.83	+ 3.28	Paros 29	Hymmetos 21
UNK-61	− 2.93	+ 3.65	Paros 38	Ephesos 17
IV. THASOS				
73-109-142	+ 0.67	+ 3.58	Thasos 61	Denizli 25
73-1255	+ 0.54	+ 3.35	Thasos 57	Denizli 27
74-295	+ 0.05	+ 3.09	Thasos 44	Denizli 30
74-415	− 0.15	+ 4.49	Thasos 48	Denizli 27
78-196	+ 0.43	+ 3.54	Thasos 56	Denizli 27
GrRel	+ 1.12	+ 3.79	Thasos 70	Denizli 21
UNK-60	+ 0.74	+ 3.35	Thasos 60	Denizli 25
V. UNKNOWN PROVENANCE				
76-39	− 4.67	+ 2.57	Hymmetos 32	Naxos 25
78-503	− 4.44	+ 2.58	Hymmetos 37	Naxos 19
76-543	− 5.00	+ 3.05	Naxos 37	Hymmetos 25
78-694	− 5.26	+ 1.43	Afyon 38	Ephesos 15
73-1288	− 1.37	+ 5.14	Hymmetos 36	Paros 27
71-841	− 4.77	+ 2.05	Aphrodisias 23	Naxos 19

Fig. 2. Head of a native Libyan. Parian marble. Exc. Inv. No. 73-1288.[11]

nates in figure 2); and for function 2 carbon = 1.237551, oxygen = -0.34142 (shown as y-coordinates). Thus function 1 is primarily influenced by $\delta^{18}O$ variation and function 2 by $\delta^{13}C$ variation.

The discriminative possibilities of the DA program are highly variable from quarry to quarry. Certain quarries proved to be easily distinguishable on the basis of their DA functions, but others overlapped in their average values for both discriminant functions. For all quarries, a distinctive pattern for both DA functions was sought by examining the manner in which known samples were assigned to groups. In most groups, the program assigned known samples to the correct first choice. In others, many samples were assigned as a second choice to the correct group, but in some groups, samples were incorrectly assigned both as the first and second choice (table 1). Consistent patterns were found, in most cases, for first and second choice which could be used as a DA "signature" for the recognition of the group.

Thus, from table 1, unknown samples can be assigned to groups on the basis of the DA signatures that were determined by analysis of known quarry samples. In some cases, assignments can be made with the highest degree of confidence, as for Carrara and Thasian marbles. For these two groups, DA showed 91% of samples correctly identified as a first choice, and with a consistent second choice: for the known Carrara samples, 56% of samples showed Marmara as a second choice, and for the Thasian samples, 87% showed Denizli. In the worst possible case, Hymettos, the DA program missed 70% of the samples for either first or second choices. Clearly Hymettos marbles cannot be distinguished solely on the basis of oxygen and carbon isotopic analysis.

One of the most important classical marbles is Parian. DA showed that 50% of the known samples were called Ephesos as a first choice, and 50% Parian as a second. However, from classical sources, it is known that the principal sculptural marble throughout classical times was Parian and that the Ephesian was not important. These two marbles can also be distinguished on the basis of their physical characteristics so the overlap in their isotopic signatures is not critical.

When the sixty-one samples from the Demeter Sanctuary were compared to the data base by the DA program (table 2) and plotted on the discriminant function diagrams (see dots on figure 1 plots), only six could not be assigned a provenance with a reasonable degree of confidence. An important factor in the assignments was the knowledge that Naxian marble was not extensively exported during classical times. Naxos was a common choice in the DA program and was used as part of the DA signatures.

The unknown pieces have been assigned to five groups on the basis of the DA classification. Sixteen pieces are from Carrara: fourteen (88%) showed Carrara as a first choice and two showed it as a second choice; eight (50%) showed Marmara as the second choice. This compares favorably to the pattern showed by the known samples (table 1). Fourteen pieces were assigned to Penteli. Of these, thirteen have patterns that are identical to the known quarry samples: Penteli or Naxos as a first or second choice. The odd sample did show Penteli as a second choice and plotted very close to the others on the DA Penteli diagram.

Of the eighteen pieces assigned to Paros, thirteen have distinctive Parian DA patterns: Ephesos first choice and Paros second. The five remaining samples showed Paros as the first choice. Because these pieces are sculptural, and clearly made of a high-quality and translucent marble, this assignment can be made with a high degree of confidence. The seven samples assigned to Thasos have patterns identical to the known Thasian samples: first choice, Thasos, second Denizli.

Six pieces could not be assigned a provenance with any degree of confidence. For four of them, the DA program listed Naxos as a possible source, but this can probably be discounted on historical grounds. Hymettos was shown as a first or second choice for four samples, which is a good possibility. However, as discussed above, the DA signature for Hymettos is not distinctive enough to permit a positive identification. Two samples may be from Asia Minor.

Assignment of Samples by Stylistic Analysis

Of the sixty-one samples analyzed, eighteen (30%) are from Paros, sixteen (26%) are from Carrara, fourteen (23%) from Penteli, seven (11%) from Thasos and six (10%) are unknowns (these may include four (7%) from Hymettos and two (3%) from Asia Minor). The predominance of marbles of east Mediterranean origin, primarily from Greece and the Islands (74%, with unknowns included; 64% without) might be expected on the basis of the historical evidence. Aegean marbles are known for their fine quality and were frequently used for sculpture in antiquity. It is reasonable to believe that Cyrene was an importer of these marbles. Always a city of Greek character, even during the Roman period, Cyrene kept its cultural and commercial ties generally within the eastern Mediterranean.

A percentage distribution of these marbles, according to sculptural type or part is given in table 3. Not surprisingly, Parian, one of the most prized marbles for fine sculpture in antiquity, is the most frequently used material for heads, statues, and statuettes. An example of one such high-quality head, that of a native Libyan, may be seen in figure 2.[11] Pentelic, another quality Greek marble, ranks second for heads and third for statues and statuettes. Italian Carrara marble, however, was not suspected to be present at Cyrene before isotopic testing began and therefore its frequent presence is striking. It ranks first for reliefs and second for both statues and statuettes. It is puzzling that there are no heads carved from this marble among the sixty-one pieces tested, although nearly one-quarter of the statues and the statuettes (all of which are headless) are made of it. This may be the result of unrepresentative sampling or may be related to the fact that there are proportionately fewer heads than bodies among the statuary from the sanctuary.

Each sample has been dated through stylistic analysis.[12] This evidence, when correlated with these provenance findings, can be used to describe changes over time in the importation of marble from different quarries. Preliminary analysis suggests that marble from Paros was used most frequently in the sketchily represented Hellenistic period. The Roman period is better represented. While Parian was still used frequently, marbles from Thasos and Carrara appear in quantity. Carrara marble seems to have been used often from the second to third century B.C., particularly in the Severan period. The use of Pentelic marble for several works in the Antonine period corroborates stylistic connections with Athens in the same period.

One such Attic stylistic connection can be clearly seen in a small-scale statuette group of Demeter holding her daughter Kore in her

Table 3
Percentage Distribution of Marble Types by Sculptural Type or Part

	Paros	Carrara	Pentelic	Thasos	Unknown	Total
Total	30	26	23	11	10	
Statue	42 (5)	25 (3)	17 (2)	8 (1)	8 (1)	20 (12)
Statuette	36 (5)	28 (4)	21 (3)	7 (1)	7 (1)	23 (14)
Head	44 (4)	0	33 (3)	11 (1)	11 (1)	15 (9)
Relief	13 (1)	50 (4)	13 (1)	25 (2)	0	13 (8)
Fragment	15 (2)	15 (2)	38 (5)	8 (1)	23 (3)	21 (13)
Unknown	20 (1)	60 (3)	0	20 (1)	0	8 (5)

n in ( ).

Fig. 4. Togatus statue, Carrara marble. Exc. Inv. No. 69-292.[14]

Fig. 3. Statuette group of Demeter and Persephone, Pentelic marble. Exc. Inv. No. 76-460.[13]

lap (fig. 3). The Cyrenean group is one of four known examples of this particular type.[13] The three other examples were found in Attica, home of Eleusis, an important cult to Demeter and Kore. Two of the three other Attic groups are believed to be of Pentelic marble and of Antonine date, the same marble and date for the Cyrenean group. This artistic evidence is congruent with the close political contacts between Cyrene and Athens during the Antonine period. Athens may even have been the source for the actual statuette, not just the stone and the iconography.

Examples of imported large-scale (life- to over-life size) statues may exist in a group of toga-wearing male figures made from Carrara marble. Two such works may be seen in figures 4 and 5.[14] Carved from thin slabs of stone, they could perhaps be shipped in stacks like so many plates. The existence of duplicate togatus statues in Cyrene may also confirm this.[15] The stylistic parallels for these togati are Italian for the most part, so that they have an Italian source for both marble and style. In addition, the togatus is a Roman type of statue that was not as popular in Cyrene as the Greek type of himation-draped male. The two togati from the sanctuary mentioned above

Fig. 5. Togatus statue, Carrara marble. Exc. Inv. No. 71-829.[14]

fact that it has been so successful suggests that isotopic analysis, already a proven technique for assembly of fragments of sculpture and inscriptions,[6,9] should also be considered in provenance studies. In addition, these efforts are part of an ongoing program intended to provide new perspectives on the study of this sculpture.[16] Hopefully, further research on the Sanctuary of Demeter sculpture will continue to elucidate the position of Cyrenean sculpture both as an art form and as a commercial commodity within the ancient Mediterranean.

are from the Severan period (late second to early third century A.D.), when a North African-born emperor and his dynasty ruled in Rome. Perhaps the trade in Carrara marble was part of the Severan dynasty's interest in economically revitalizing its native North Africa.

Further research is yet to be done using these new data. Questions to be investigated include: possible connnections between quarries and specific time periods, and correlation of sculptures with artists and specific workshops, both native and foreign are still to be investigated. More isotopic analyses are planned, to include not only additional sculptural samples from the sanctuary, but also some taken from works found in other parts of Cyrene. A sampling of architectural marbles is also planned, whose results might show a different trade pattern from that of the sculptural marbles.

Conclusions

On the basis of isotopic analysis for ^{18}O and ^{13}C, it has been shown that the sources for most statuary pieces from the Cyrene Demeter Sanctuary can be identified. This is the first time that this system of identification of marble has been tried on such a large collection; the

References

1. D. White, "Cyrene's Sanctuary of Demeter and Persephone: A Summary of a Decade of Excavation," *American Journal of Archaeology*, 85 (1981): 13.

2. B. Ashmole, "Aegean Marble: Science and Common Sense," *Annual of the British School at Athens*, 65 (1970): 1.

3. C. Renfrew and J. Peacey, "Aegean Marbles: A Petrological Study," *Annual of the British School at Athens*, 63 (1968): 45.

4. L. Conforto, M. Felici, D. Monna, L. Serva, and A. Taddeucci, "A Preliminary Evaluation of Chemical Data (Trace Element) from Classical Marble Quarries in the Mediterranean," *Archaeometry*, 17 (1975): 201.

5. H. and V. Craig, "Greek Marbles: Determination of Provenance by Isotopic Analysis," *Science*, 176 (1972): 401.

6. N. Herz and D. B. Wenner, "Assembly of Greek Marble Inscription by Isotopic Methods," *Science*, 199 (1978): 1070.

7. L. Manfra, U. Masi, and B. Turi, "Carbon and Oxygen Isotope Ratios of Marbles from Ancient Quarries of Western Anatolia and their Archaeological Significance," *Archaeometry*, 17 (1975): 215.

8. N. Herz and D. B. Wenner, "Tracing the Origins of Marble," *Archaeology*, 34 (1981): 14.

9. N. Herz, "Isotopic Analysis of Marble," in G. Rapp, Jr. and J. A. Gifford, eds., *Archaeological Geology* (Yale University Press, in press).

10. W. R. Klecka, "Discriminant Analysis," in N. H. Nie, et al., eds., *SPSS Statistical Package for the Social Sciences*, 2nd ed., (New York: McGraw Hill, 1975).

11. D. White et al., "Seven Recently Discovered Sculptures from Cyrene, Eastern Libya," *Expedition*, 18 (1976): 27.

12. S. Kane, "On the Sculpture from the Sanctuary of Demeter at Cyrene," to be published in the University Museum (University of Pennsylvania) Monograph Series.

13. R. Lindner, "Die Giebelgruppe von Eleusis mit dem Raub der Persephone," *Jahrbuch des Deutschen Archaeologischen Instituts*, 97 (1982): 303.

14. Fig. 2: D. White, "Excavations of the Demeter Sanctuary at Cyrene 1969. A preliminary report, *Libya Antiqua*, 8 (1971): 100 n.34; Fig. 3: D. White, "Excavations in the Demeter Sanctuary at Cyrene 1971: Second Preliminary Report," *Libya Antiqua*, 9 (1977): 189 n.82.

15. D. White, "Excavations in the Demeter Sanctuary at Cyrene 1971: Second Preliminary Report," *Libya Antiqua*, 9 (1977): 189 n.81.

16. S. Carrier and S. Kane, "Computer Analysis of Statuary Finds from the Demeter Sanctuary at Cyrene," *American Journal of Archaeology*, 86 (1982): 258; S. Kane, "Sculpture from the Cyrene Demeter Sanctuary in its Mediterranean Context," to be published in Colloquium Proceedings, "Society and Economy in Classical Cyrenaica", Cambridge, England.

RONALD L. BISHOP, E. V. SAYRE, and LAMBERTUS VAN ZELST

Characterization of Mesoamerican Jade*

*A collaborative research project between the Research Laboratory of the Museum of Fine Arts, Boston, and the Chemistry Department of Brookhaven National Laboratory.

One of the most prized materials in Pre-Columbian America was jade. The term *jade* refers not to a single mineral but to a number of colored stones. Jade has a long history of utilization being best known among the ancient peoples of Mexico, Central America, and China. At the time of the Spanish contact with the New World, several kinds of jade were distinguished and a summary of early references has been published by Foshag.[1]

Mineralogically jade consists of two principal minerals, nephrite and jadeite. Nephrite, the amphibole variety of jade, is a calcium magnesium silicate of widespread occurrence. Although tremolite and actinolite are known to occur in Mesoamerica, the presence of nephrite has not been demonstrated. The rarer variety of jade is jadeite; it and its mafic isomorph, chloromelanite, belong to the pyroxene group of minerals. Among the ancient Maya inhabitants of Southern Mexico, Yucatan, Guatemala, Belize, and Honduras, jadeite was subject to great artistic embellishment. While other stones such as albite, chloromelanite, glaucophane, and serpentine were also used, the greatest artistic elaboration was reserved for jadeite. Documentary sources and the frequency of jadeite occurrence in burials and tombs indicate that jadeite was a "status" material which, at least by Late Classic times (A.D. 600-900), was concentrated in the hands of an elite stratum of Maya society.

There is today but one known source of jadeite in the Maya area— the Motagua River valley of Guatemala. Given the number and distributional occurrence of jadeite artifacts, direct evidence for extensive Pre-Columbian exploitation of that source is surprisingly scant. The demonstration that the Motagua region was or was not a source for jadeite found throughout the Maya area would provide valuable data for investigations of raw material procurement and long-distance trade.

We have largely completed a multiyear investigation of jadeite that attempted to establish the characteristic properties of the Motagua source material, and to ascertain its relationship to jadeite artifacts from selected archaeological sites throughout the Maya zone. To place Maya jadeite in broader perspective, more limited analyses of jadeite and related minerals from Costa Rica—another area renowned for its jade carving—have been carried out. The characterization has been achieved through a determination of elemental composition by neutron activation analysis coupled with identification of mineralogical components by means of x-ray diffraction and petrographic thin-section examination.

Jadeite—the Mineral

Jadeite, a sodium aluminum silicate ($NaAlSi_2O_6$) is a member of the sodic pyroxenes. Chemically it is quite pure with almost no replacement of the silicon by aluminum in tetrahedral coordination and only limited replacement of Al by Fe^{3+} substitution.[2] Often additional minerals may be present, especially acmite and diopside.

Pure jadeite is white in color, tending to be darker by additional Fe^{3+} substitution. The preferred apple green color of the Maya jadeite is generally attributed to the amount of chromium oxide present, although some green specimens have been found to contain very little chromium.[3] When extensive substitution by heavier elements such as iron has occurred, one frequently encounters a mineral with an expanded jadeite structure, chloromelanite. This phase is characterized by intensely deeper color and a shift in position of x-ray dif-

fraction peaks to significantly greater d-values. Found widely in the Maya area, chloromelanite was subject to less artistic exploitation, commonly shaped into smooth stone celts.

Albite, ranging from blue green and green to white, is commonly associated with jadeite, often as a mineral surrounding mono-mineralic aggregates of jadeite. Other minerals occurring with jadeite include glaucophane, sphene, actinolite, and mica.

Jadeite is formed within a narrow range of limiting conditions of low temperature and high pressure indicating a mineral of low metamorphic grade. It is always associated with serpentinite bodies of blueschist metamorphism characteristic of suture zones and environments along tectonically active continental margins. These conditions have limited its known occurrence to only six areas in the world. Its presence in the serpentinites of the Motagua fault zone of Guatemala has been known since the publication of Foshag and Leslie,[4] and has been recently interpreted by Coleman utilizing experimental data and the theory of plate tectonics.[5] The source region lies in a boundary zone, containing peridotites and serpentinites, between the Caribbean and North American plates. The highly sheared serpentinites and tectonic blocks of blueschist and eclogite assemblages along the fault zone[6] indicate that the fault intersected a region of low temperature and high pressure made possible by the motion of the two plates. This motion resulted in the subsequent transportaion of tectonic blocks of jadeite upward through the plastic serpentinite.

North of the Motagua River, near the village of Manzanotal, Department of El Progesso, tectonic blocks of jadeite are encountered bounded by albite and protruding from the surrounding serpentinites. Other blocks have eroded from the surrounding material and have fallen or been transported into the river drainage where erosion has stripped away the softer material, often leaving mono-mineralic boulders approaching 0.5 meters in diameter. Subsequent participation in the sedimentary cycle results in the many cobbles and smaller detrital jadeite encountered along the alluvial terraces of the Motagua River.

Previous Investigations of the Motagua Source

The investigations of Foshag,[1,7] Foshag and Leslie,[4] McBirney et al,[8] Silva,[6] and Hammond et al.[9] have contributed to our knowledge of the Motagua Valley jadeite source zone. While the earlier workers concentrated on geological issues, Hammond and his associates focused directly upon the question of Maya utilization of the jadeite. Their investigation involved a ground reconnaissance of a restricted region of the Sierra de las Minas near the village of Manzanotal as well as a preliminary program of trace-element analysis. Arguing that since many artifacts might tend to be polymineralic in composition, it appeared justified to analyze a range of jadeite and jadeite-related minerals—material that might be called "cultural jade." It was hoped that the statistical analysis of data from a large number of samples would provide a basis for comparison of findings with some level of nondestructive analysis of the artifacts. Although they produced data on some seventy-five samples, which was of great value during the early stages of our project, their attempt to characterize the Motagua jadeite was inconclusive, as few actual jadeite specimens were analyzed.

Another investigation of jadeite in Mesoamerica was initiated in the mid 1970s by Russel Sykes and Charlotte Thomsen, stemming in part from the latter's interest in the sources of Olmec jade. Field reconnaissance in the Motagua River Valley and elsewhere in the Guatemalan Highlands was guided by an interest in relating jadeite to its host rocks. Actual jadeite sampling was limited to only six specimens from the Motagua source area. Sykes's and Thomsen's interest in jadeite established the initial Guatemalan contacts for our present project and provided jadeite source samples that served to orient our analytical program.

In light of this experience, as well as that of Hammond and his co-workers, our project orientation was to   cus on jadeite or jadeite-albite source material, which would allow us to more clearly characterize the composition of the Motagua jadeite. With this approach we hoped to provide a more direct level of compositional comparision of Motagua source jadeite with the Maya jadeite artifacts.

Jade is today being mined commercially from the Motagua source zone by the firm of JADES S.A. of Guatemala. In 1980 with the cooperation of the then-owners of the jade jewelry factory, Jay and Mary Lou Ridinger, Bishop visited five of their jadeite sources, sampling material up the smaller rivers extending north from the Motagua River in the Departments of Zacapa and El Progresso. Samples were taken from 0.3 to 0.5 meter alluvial boulders as well as from one large tectonic block. Although the latter had been largely removed to the JADES factory, remnant portions of the jadeite blocks remained, including portions where the jadeite was in sharp contact with albite. Additional samples from other locations in the Motagua source zone were obtained from the jadeite stockpiles at the JADES storage in Antigua. In all, more than 180 samples were obtained from the drainages of the Motagua River, covering an area of ridges approximately five kilometers north by some twenty-five kilometers along the river.

Artifact Sampling

Sampling of artifacts was first and foremost guided by object availability. At the onset of the project, the Peabody Museum, Harvard University, through the courtesy of its director, C. C. Lamberg-Karlovsky, allowed us to use jadeite objects such as beads and tubes in their collection while we were still establishing our method. Additional jade and jadeite specimens were made available to the project by the archaeological community resulting in the distribution of site coverage shown in figure 1. Although there is always a great temptation to focus on the jades of exquisite artistic expression on which a maximum amount of cultural information is available, our investigation is compositional and distributional and not, at this stage, primarily concerned with seeking patterns of stylistic and compositional covariation. With such notable exceptions as the jades from Altun Ha and Costa Rica, our sampling was, therefore, designed to restrict analysis to fragments or small beads. In all, 377 jade artifacts were analyzed with a resulting sample bias toward those sites relatively close to the Motagua source area and throughout Belize. The material from the Belizean sites provided a view of jadeite composition through time as these artifacts were associated with deposits ranging from approximately 2000 B.C. at Cuello through A.D. 1500 at Santa Rita. In addition to jadeite artifacts, the Maya material included

Fig. 1. Map of the Maya area showing locations of sampled artifacts and general Motagua jadeite source area with insert map of Central America.

forty-four albites, twelve chloromelanites, and thirty-three miscellaneous minerals.

Analytical Procedures

Our field specimens of jadeite were resampled to obtain fragments in the 100-500mg range. Those exhibiting lighter or darker phases, or coarse and fine textures were sampled using a diamond impregnated hollow-core drill, extracting a 60-100mg core from the target area. Large artifacts were also sampled with the hollow-core drill from an undecorated area, usually at the back of the object. The sampled artifacts were then filled, and the plugged areas were painted to match the surface color by the conservators at the Research Laboratory of the Museum of Fine Arts, Boston. Depending upon their shapes, whole artifacts of up to 10g, but more commonly in the less than 2g range, were analyzed in their entirety.

To remove surface contamination, including that which might have occurred during drilling, all samples were boiled for fifteen minutes in distilled water that had been rendered slightly acidic by the addition of hydrochloric acid. This was followed by three separate washings in triple-distilled, deionized water. Samples were then dried, and weighed to the nearest .01mg. Source samples received an additional step involving an etching with hydrofluoric acid prior to final washing.

Samples were grouped according to weight and shape for trace-element analysis. For standardization we used fired portions, of a size and shape approximating that of the specimens, of the Brookha-

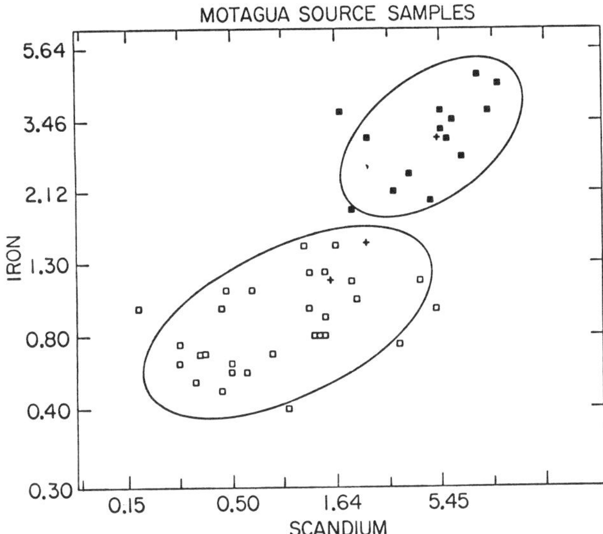

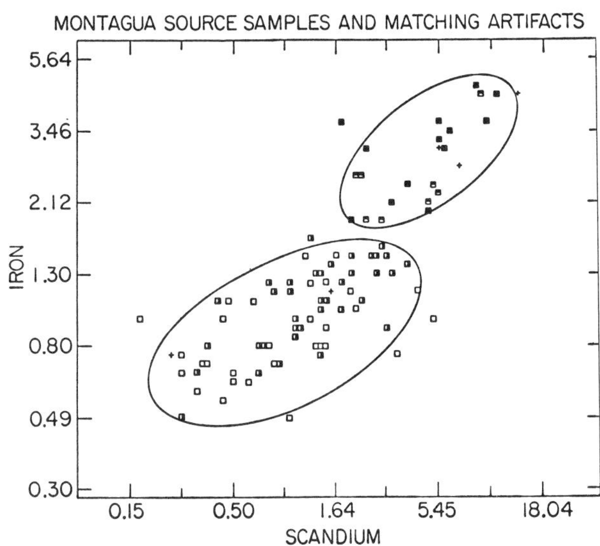

Fig. 2. Motagua jadeite source samples and matching artifacts compared to iron and scandium correlation plots. The ellipses in this and subsequent figures represent 80% confidence limits for specimen groups.

Symbols:
▫ Motagua light source specimens
◨ Artifacts matching Motagua Light
■ Motagua dark source specimens
◪ Artifacts matching Motagua dark
+ Positions of overlapping points

ven in-house standard Ohio Red Clay, which has been calibrated against USGS reference materials PCC-1, DTS-1, BCR-1, AGV-1, GSP-1, and G-2.[10] By replicating the general shape of the artifacts, the standard had the advantage of diminishing the geometric errors encountered during gamma counting.

Samples and standards were irradiated in a thermal neutron flux of 1.0×10^{13} neutrons cm^{-2} sec^{-1} for a length of time that was dependent upon the average sample weight, proportioned to the rate of 1.8 hours per 0.50g. This irradiation was found experimentally to induce no observable color shifts in the jadeite artifacts, and to result in less than one millirad per hour of residual radioactivity on contact after a six-month cooling; this low level of residual activity was an important consideration for the return of the artifacts to the lending institutions. Longer irradiations were possible, of course, for the source samples. Following a fourteen-day cooling period, gamma emissions were counted using a Princeton Gammatech Ge-Li detector (35cm crystal with 1.8 keV resolution on ^{60}Co) coupled to an ND-2400 4096 channel analyzer with magnetic tape readout. The gamma spectra were processed by the BRUTAL program. A short irradiation of approximately one second served to excite the short lived sodium and manganese isotopes. Reliable values of elemental concentrations were obtained for sodium, barium, scandium, europium, lutetium, hafnium, chromium, iron, cobalt, and ytterbium.

Sodium is a major constituent of jadeite and its large thermal neutron capture cross-section gave rise to potential problems that might arise due to self-shielding. In effect, the outer portions of the sample might capture a disproportionately high number of neutrons thereby lowering the neutron flux available for interior target nuclei. To assess this effect, an 8.8g jadeite bead with an outside diameter of

15mm and sides ranging from 5 to 8mm thick was prepared for standard irradiation including the placement of four iron wire monitors outside of the bead. Another wire was placed through the hole in the bead. Subsequent gamma counting indicated that the wires showed no significant difference. This was taken to mean that self-shielding was not a problem in jadeite analysis if samples were less than 8mm thick.

Data Analysis

The multicomponent characterization of the Motagua jadeite source area was carried out by first combining into a single group everything that could be considered jadeite, and eliminating only those few specimens that for one reason or another showed extreme chemical divergence. Pronounced divergence could be attributable to the inclusion of other mineralogical components but in some cases samples that were largely jadeite showed extreme concentration values. With a group thus formed, we calculated the variance-covariance matrix and group centroid and then determined the probability that indiviual specimens could belong to the group given their Mahalanobis distances from the group centroid. In this manner we found that most of the source specimens fitted into a single core group. As a further step of refinement, we separated the specimens into a light and a dark group, reflecting the tendency of the darker chloromelanite samples to deviate in a patterned manner from the majority of the lighter jadeite material. This phase separation is statistically significant at the 80% confidence interval and is illustrated by the elemental correlation plot in figure 2a.

Once the variance-covariance properties of the two Motagua groups were determined, the program ADSRCH was employed to

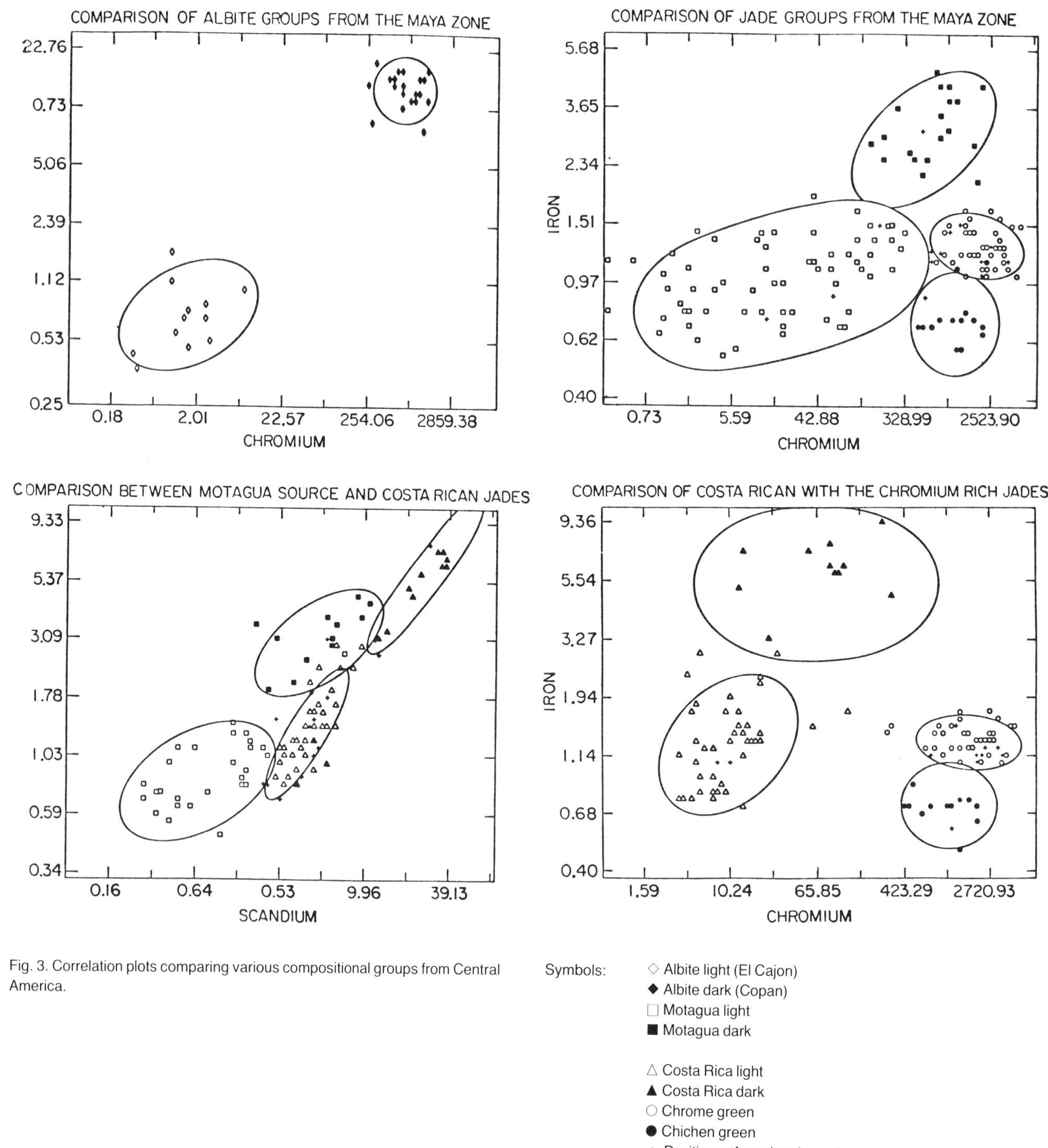

Fig. 3. Correlation plots comparing various compositional groups from Central America.

Symbols:
◇ Albite light (El Cajon)
◆ Albite dark (Copan)
□ Motagua light
■ Motagua dark

△ Costa Rica light
▲ Costa Rica dark
○ Chrome green
● Chichen green
+ Positions of overlapping points

Table 1

Comparative Means and Percent Standard Deviations

	Motagua light n = 43		Motagua dark n = 16		Chrome-green n = 54		Chichen-green n = 17		Costa Rica light n = 48		Costa Rica dark n = 14		Albite dark n = 24		Albite light n = 17	
Na	13.1	(16)	8.1	(31)	11.4	(12)	11.5	(25)	12.6	(8)	9.0	(24)	8.9	(26)	12.0	(11)
Sc	0.85	(146)	4.4	(68)	3.86	(40)	1.9	(70)	3.46	(45)	22.4	(58)	4.68	(86)	0.13	(109)
Fe	0.92	(46)	2.93	(42)	1.28	(13)	0.74	(23)	1.29	(46)	5.47	(43)	2.10	(28)	0.199	(29)
Co	2.31	(89)	20.6	(45)	7.09	(22)	4.30	(13.7)	4.39	(52)	17.4	(33)	14.6	(29)	0.73	(45)
Mn	278	(67)	867	(69)	504	(18)	285	(67)	207	(68)	1125	(56)	*	*	60.0	(37)
Cr	3.00	(421)	363	(126)	1580	(71)	932	(69)	8.0	(138)	55.0	(271)	671	(139)	1.00	(175)
Hf	1.14	(88)	210	(94)	0.33	(39)	0.57	(184)	2.7	(42)	4.17	(97)	0.62	(108)	0.37	(290)
Eu	0.028	(124)	0.053	(138)	0.146	(26)	0.036	(36)	0.168	(107)	0.948	(104)	0.054	(131)	0.013	(141)
Lu	0.018	(82)	0.063	(36)	0.097	(35)	0.0233	(42)	0.041	(166)	0.451	(296)	0.049	(76)	0.007	(62)
Yb	0.091	(84)	0.288	(52)	0.557	(29)	0.150	(33)	0.210	(148)	3.01	(258)	0.227	(76)	0.0035	(34)
Ba	75.0	(191)	118	(382)	192	(257)	105	(253)	32.0	(141)	770	(95)	48.0	(192)	83.0	(254)

*not determined

calculate the probability of membership of the artifacts in either of those groups. Using 20% as the lowest acceptable level of membership probably excludes some artifacts that might well relate to the Motagua, but at this stage of data analysis we wanted to establish the fact that there were a large number of artifacts that unambiguously related to the Motagua source area (fig. 2b). We leave for a future time the more detailed study of the borderline cases.

Group refinement of the residual artifacts revealed a major core group which, due to the color of the specimens and their elevated chromium concentration, we refer to as the chrome-green group. Initially, a subset of that group with a Copan provenience tended to diverge in like direction from the chrome-green centroid and was just within our level of group acceptance. Subsequently, x-ray diffraction revealed that group to be albite yet to differ significantly in chemical composition from albite specimens analyzed from the site of El Cajon, Honduras (fig. 3a). With the albite specimens removed and the group deviation thus tightened, an additional subset of the chrome-green group, all but one from the cenote at Chichen Itza,[12] was excluded (fig. 3b). We do not know if this group separation would hold up under more extensive sampling, for it would not require many samples of bridging composition to result in a single, larger chrome-green group. But for reasons unclear at present, the "Chichen-green" samples form a site-specific compositional pattern significantly different from the Motagua reference groups or the chrome-green artifact group.

To provide greater perspective on jadeite compositional varation, seventy-seven artifacts from Costa Rica were also analyzed. Similar to the Motagua samples, after removal of the non-jadeite specimens, the Costa Rican artifacts separated into light and dark groups that included most of the artifacts. A few of the artifacts showed compositional similarity to the Guatemalan Motagua source samples (fig. 3c). Compositional differences between the jadeite artifacts from Costa Rica and those of the chrome-green or Chichen-green artifacts were pronounced.

Discussion

While more extensive analyses of the trace element data remains to be carried out, the complexity of the data has become clear. The characterization of the Motagua source zone has resulted in the samples constituting two quite distinctive compositional groups that are differentiated primarily by the higher mafic component of the smaller group. When the artifacts are compared with the Motagua groups, forty-three specimens show unambiguous similarity in composition, including artifacts from Belizean sites (twenty-six samples) as well as from the Alta Verapaz. Yet almost an equal number of Belizean jades (twenty-eight) are included in the chrome-green reference group and the samples from Cuello, Belize are almost split between the two groups. Only the jadeite samples from the cenote at Chichen Itza show a tendency for strong site-specific patterning.

Some of the "cultural jade" albite artifacts from Copan and El Cajon segregate into distinct clusters and certainly those from El Cajon suggest the probable exploitation of a single source. The blue-green albite chips that make up the dark albite Copan-Focused group were obtained from a single location during turn-of-the-century excavations at that site and may actually represent artifacts from a common source rock. It is interesting to note that their composition is very close to that of several of our jadeite specimens and they were often included in a jadeite group during an early stage of data analysis only to be excluded during the statistical refinement of that group.

Only five of the Costa Rican artifacts show compositional similarity to our Motagua source groups or to the chrome-green Maya artifact group. The other specimens fall largely within the two derived Costa Rican groups that are clearly separable from the Maya jadeite.

We are left at this point with a great many Maya artifacts that show a significant lack of similarity to the Motagua reference groups or to the Chichen-or chrome-green groups. Doubtless many of the artifacts contain minor mineralogical components that are influencing the chemical composition. Some of the present groups must be expanded or merged to allow for more within-group chemical variation. While more inclusive compositional units may be formed during future data analysis, the significant compositional dissimilarity of the "ungrouped" artifacts to the analyses of the Motagua source samples is intriguing. Other areas of the Guatemalan Highlands or the Maya Mountains of Belize could have provided the sources of "cultural jade" but only the Motagua is known to be a source for jadeite.

Summary

Jadeite occurring in the Motagua River Valley of Guatemala has been characterized by neutron-activation analysis and forms two distinct, phase-related groups. Comparison of the compositional profiles of Mayan jadeite artifacts reveals many specimens with profiles matching those of the Motagua source. Of particular interest are the large number of jadeite artifacts that show internal similarity, yet have compositional patterns significantly different from the Motagua samples and Motagua-related artifacts. A few of the analyzed Costa Rican artifacts show patterns similar to those of the Motagua yet the vast majority fall within one of the two Costa Rican compositional groups. When considering the non-Motagua related Mayan artifacts, the analytical approach appears to be sufficiently sensitive to distinguish differences between the chrome-green and Chichen-green material. Even two Honduran site specific groups of albite—cultural jade—form distinct groups.

Given the demonstrated sensitivity of a trace-element approach to jadeite characterization, what are the implications for Pre-Columbian jadeite exploitation? Other sources of jadeite appear to be indicated but their proximity to the sampled area of the Motagua Valley is unknown. Perhaps a future step is to sort through collections of jade, isolating jadeite samples from the cultural jade of albite, diopside, serpentine, etc., and noting the abundance and distribution of jadeite. What is clear at present is that interpretations of jadeite exploitation of only the Motagua source are oversimplified.

Acknowledgments

This investigation is part of the Maya Jade and Ceramics Project of the Research Laboratory of the Museum of Fine Arts, Boston in collaboration with the Department of Chemistry, Brookhaven National Laboratory. Work at the latter institution was conducted under the auspices of the U.S. Department of Energy and supported by its Office of Basic Energy Sciences. Direct project support has been provided by Mr. Landon T. Clay of Boston.

We wish to acknowledge the many individuals and institutions who made this project possible, either by offering support or advice or by permitting objects under their care to be studied. It is an act of courage for a curator to allow his objects to be moved about, let alone to give permission for the removal of a portion—no matter how small—of the object for destructive analysis. In part, destructive sampling was made possible thanks to the fine conservation efforts of Merville E. Nichols and Jean-Louis Lachevre of the Museum of Fine Arts. Other individuals and institutions to whom we extend appreciation for their efforts on our behalf include: Jay and Mary Lou Ridinger, Robert Terzuola, David Sedat, Edwin M. Shook, and Dell Huelle in Guatemala; Department of Archaeology, H. W. Topsey -Archaeological Commissioner, and Mark Gutchen in Belize; Museo Nacional de Costa Rica, German Serrano P.-Executive President, Riccardo Monge O., Zulay Soto de Andrade, Raul Castallanos, Universidad de Costa Rica, Oscar Fonseca Z., Corporaccion Costarricense de Desarrollo (CODESA), Rolando Castillo M., in Costa Rica; Royal Ontario Museum, David Pendergast in Canada; Peabody Museum and its Director C.C. Lamberg-Karlovsky, as well as Wendy Ashmore, Arlen F. Chase, Robert Coleman, Elizabeth Kennedy Easby, David A. Freidel, James F. Garber, Norman Hammond, Kenneth G. Hirth, Tatiana Proskouriakoff, Frederick W. Lange, Robert J. Sharer, and Gordon R. Willey in the United States.

References

1. W. F. Foshag, "Chalchihuitl—A Study in Jade," *American Mineralogist*, 40, (1955): 1062-1070.

2. W. A. Deer, R. A. Howie, and J. Zussman, *Rock Forming Minerals*, 2A, 2d ed (New York: John Wiley and Sons, 1978), pp. 461-481.

3. G. R. Rossman, "Lavender jade. The Optical Spectrum of Fe^{3+} and Fe^{2+} — Fe^{3+} Intervalence Charge Transfer in Jadeite from Burma," *American Mineralogist*, 59 (1974): 868-870.

4. W. F. Foshag and R. Leslie, "Jadeite from Manzanal, Guatemala," *American Antiquity*, 21, (1955): 81-82.

5. R. G. Coleman, "The Natural Occurrence of Jade and its Bearing on Mesoamerican Jade Artifacts," Paper delivered at Mesoamerican-Central American Jade Conference (Washington, Dumbarton Oaks, 1980).

6. Z. C. G. daSilva, Studies on jadeites and albites from Guatemala, Rice University, M.A. thesis, 1967.

7. W. F. Foshag, "Mineralogical Studies on Guatemalan Jade," *Smithsonian Miscellaneous Collections*, 145, no. 5 (1957).

8. A. McBirney, K. Aoki, and M. N. Bass, "Eclogites and jadeite from the Motagua fault zone, Guatemala," *American Mineralogist*, 52, (1966): 908-918.

9. N. Hammond, A. Aspinall, S. Feather, J. Hazelden, T. Gazard and S. Agrell, "Maya Jade: Source Location and Analysis," in T.K. Earle and J. E. Ericson, eds., *Exchange Systems in Prehistory* (New York: Academic Press, 1977), pp. 35-68.

10. S. J. Yeh and G. Harbottle, *Intercomparisons of the Asaro-Perlman and Brookhaven Archaeological Ceramic Analytical Standards* (Department of Chemistry, Brookhaven National Laboratory, nd.).

M. HOURS

Toward the Unity of Culture

For nearly four centuries, the Louvre has been not only a palace but also a cultural meeting place. Since the seventeenth century, the academies have met there, gathering within the walls representatives of both the natural sciences and of the humanities. In the eighteenth century the royal collections were opened to the public. The encyclopedists founded the "Museum of the Arts" (the Louvre), one of the first accomplishments of the French Revolution. The eighteenth century encompassed an extraordinary group of scientists, physicists, and chemists who attempted to apply to the "arts" the fruits of their scientific discoveries. This is of such importance that Princeton University has dedicated a chair of History of Science to these scholars (Professor Charles Gillispie).

Thus it was in France in 1750 that was born the desire to understand not only the techniques of the artists, but equally to perceive their message by means of the new tools provided by science. The balloonist J. A. Charles is credited with the substitution of hydrogen for hot air to improve travel by balloon. The very same man installed a device for the study of works of art in his physics laboratory in the Louvre. This was the "megascope," the precursor of projectors and photographic enlargers. In the spirit of its inventor, it was destined to improve the perception of works of art. (This work took place between 1780 and 1790.)

A little later, in England, T. Wedgwood and H. Davy conducted chemical analyses of the paintings recently discovered at Pompeii. The expedition to Egypt conducted by Napoleon in 1798-99 brought together geologists, chemists, and archaeologists and was the first example of an interdisciplinary campaign directed at bringing the past to light, in this case in the history of Egyptian art. Since then, and throughout the nineteenth century, research was conducted in Europe, albeit in a sporadic manner.

It was toward the middle of the nineteenth century that a certain distrust of scientific development appeared in public opinion and among the authorities. This was partially due to the disturbances that the scientific developments had brought, as well as to the influence of German romanticism, which tended to elevate poetry at the expense of scientific rationalism. We should also acknowledge the effect of current politics, which tended to restrict the power of the encyclopedists, who had remained dedicated to the concept of revolution. These elements certainly restricted the institution of official bodies in the world of European museums. Nonetheless, the reverberations caused by the work of Louis Pasteur, dedicated to the study and chemical analysis of paintings (1863-1867), and that of the German scholar Roentgen who, at the end of the century, discovered x rays and performed radiography of a painting at the University of Munich in 1896, clearly demonstrated the constant interest of the European researchers into questions raised by the material properties of works of art.

The creation in 1888 at the Staatlische Museum in Berlin of the first museum laboratory affirmed the extent of the development of this work, which would come to fruition after the first world war.

Since then, work of this nature has continued to grow. It is here in Boston that laboratory work and the expansion of the collection of x radiographs were significantly extended. In particular, the work of Forbes and Burroughs was important. From 1930, the International Rome Conference has encouraged the large European museums to avail themselves of scientific services.

The first methods used were chemical and optical using the power of electromagnetic waves, from infrared to gamma rays. The importance of x rays, making invisible stages obvious, was immediately recognized: they contributed to the understanding of the condition of a painting (as did ultraviolet rays) and to the perception of stages in its creation. Thus awakened the interest, not only of technologists and conservators, but also of art historians and connoisseurs, all of whom were attracted by the emergence of hitherto invisible images.

The collaboration of photographers and radiologists, with help from optics, had until then not posed any problems for the staff of museums. The technologist provided documents whose relatively easy interpretation did not require his cooperation at the level of aesthetic judgments. After this, thanks to meetings of the ICOM and the IIC, the restorers could assimilate and utilize the work of chemists (for example those at the Doerner Institute in Munich, of Erner and Laurie, and of Gettens and Stout on pigments and media). These meetings allowed the resources of scientific research to wedge progressively into the work of "museum people."

Contemporary physics, by making available to laboratories new "non-destructive" analytical techniques, must break down the reservations and justified opposition to the taking of samples that is believed to interfere with the integrity of the work of art (for example, x-ray fluorescence and PIXE).

These new methods of examination have been more easily integrated into the archaeological disciplines than in the domain of visual art. The archaeologist is usually a "man of the earth," as much as an historian; he is used to collaborating with architects, photographers, and excavators. He is quick to see the value of new dating methods in developing a chronology of ceramics. The refinement and proliferation of these dating methods (carbon-14, fission-track dating, etc.) are of incontrovertible help to him. Palynology, as well as the new analytical methods, has transformed prehistory, and yielded new information on the archaeological object. The storage and use of analytical results has opened new windows on the sociology and commerce of antiquity. In the realm of the visual arts, the resources of physics, chemistry, and optics have been more reluctantly accepted. Directors and curators are usually historians and scholars, individualistic humanists accustomed to trust intuition as much as erudition, and not accustomed to integrate scientific measurements and objective results in a discourse that is essentially subjective.

Slowly, however, the restraints and barriers that have been erected for a long time in the museum world against scientific examination must come down. These barriers were caused equally by curatorial concerns and by a great distrust of the progress of a spirit often alien to directors, to those in responsible positions in museums, to historians concerned with maintaining their preeminence. That progress could play a nasty trick on individualistic humanism in a world where it is no longer possible to master all knowledge.

As the constant concern of scholars shows, contemporary interests tend toward interdisciplinary approaches. At Harvard University and at Brown University in Rhode Island, interdisciplinary departments are multiplying, as they are in Europe. Colloquia, such as that of two years ago in Cordova and the one to be held next year in Sicily, demonstrate the necessity of team work. When in 1960, Saint John Perse, a French poet living in the United States, received the Nobel Prize, he commented on the interrelationship between poetic and scientific endeavors, "Dans l'équivalence des formes sensibles et spirituelles une même fonction s'exerce initialement pour l'entreprise du Savant et celle du Poête."

We agree with J. Rigaud that "the great cultural functions: creation and conservation, have more and more meeting points with science and technology and on such a level that it could be qualified as operational; the science-culture duo is no longer that of non-communication or of conflict but that of cooperation." At the end of a long career, almost half a century at the Louvre, I am convinced that we have to find a common language in spite of the opacity of the specialities. Above all, a major commitment must be made together with a willingness to partake of the results of discovery and to include it in one's reasoning and in the fruits of one's personal erudition. The art historian must be willing to challenge his intuitions, to enrich them with perceptions of the intangible as much as with physical data, never losing sight of the fact that the creation of a work of art cannot be reduced to scientific measurements.

I think that it is the task of our laboratories and even more of meetings such as this one to publicize the growing resources of scientific progress, to improve not only the conservation of works of art, which is accepted, but above all to strive toward the establishment of an intimate dialogue in order to discern the true from the false and to better perceive the message of the artist.

One of the greatest thinkers of our time, Teilhard de Chardin, said, "joint planning and unison are agents of progress and form a stairway to the mind." This text recounts very summarily that long march toward the unity of culture in Europe, and is a tribute to Bill Young who for many years had directed the same process at the Boston Laboratory.

References

1. A. Burroughs, *Art Criticism from a Laboratory.*(Boston: Little. Brown and Co., 1938).

2. M. Hours, *La Vie Mystérieuse des Chefs-d'Oeuvre — La Science au Service de l'Art,* (Paris: Réunion des Musées Nationaux. 1980).

3. M. Hours, *Les Secrêts des Chefs-d'Oeuvre,* (Paris: éd. Denoël, coll. Médiation, 1980).

4. Saint John Perse, *Oeuvres Complètes,* Discours de Stockholm, (Paris: La Pléiade, éd. Gallimard, 1975). p. 443 et svtes.

5. J. Rigaud, "La Culture Scientifique dans le Monde Contemporain," *Science et Culture* (extraio) UNESCO-Scientia, Revue Internationale de synthèse scientifique.

6. France Culture, Colloque de Cordoue, *Science et Conscience, les Deux Lectures de l'Univers,* (Paris: Stock, 1980).

CATHLEEN A. BAKER

A Comparison of Drawing Inks Using Ultraviolet and Infrared Light Examination Techniques

The identification of old master drawing media especially in pen and ink drawings has always been a problem for the curator. collector, and conservator. There are three basic ink types that have been used over the past centuries: those based on carbon black pigments. usually lamp blacks: those based on gallo-tannic acids and ferrous or copper sulfate, commonly called iron gall inks: and those made from the extraction of water-soluble compounds found in wood soot. commonly called bistre. Unaged, these inks can look quite similar: either black, grayish or bluish black in color. However, while upon aging the carbon black inks retain their color, the other two inks tend to change and become brown. The two brown inks. iron gall ink and bistre, are often confused. For example, in recent research. some Rembrandt drawings identified as bistre have now been correctly identified as iron gall ink.[1] These findings were made possible through the use of x-ray fluorescence instrumentation, equipment not normally found in most museums or conservation laboratories. Chemical or spot testing on various drawings may lead to destruction of the ink or the support.

This paper reports on uncomplicated examination procedures using ultraviolet and infrared illumination. The findings were corroborated by artifical aging tests using an environmental chamber and a fluorescent light bank.

For examinations in ultraviolet light, any number of commercially available ultraviolet source lamps are adequate.[2] To record the findings, black and white as well as color photography can be accomplished simply. Objects can be examined in the infrared using the FJW Find-R-Scope (FJW Industries, Mt. Prospect, Ill.). Good results can be obtained using infrared photography (in black and white only; color infrared film was not found to be useful for this purpose).

Inks were purchased, found, or made up according to established recipes. They were then drawn out with a quill pen and a brush onto a sheet of paper of known composition. The pen was used to simulate the drawing line found on many old master drawings, the brush was used for comparison. The sheet of paper was then cut into three pieces. One was used as a control, and the other two were artificially aged under different conditions. After aging, the pieces were joined together for examination and photography.

Experimentation

Four different types of inks were used for the research: two carbon black inks, one iron gall ink, and one bistre. The carbon black inks were: Higgins® Eternal Permanent Black Ink, No. 813 (Faber-Castell Corp., U.S.A.), and a Chinese ink stick that dates to the 1950s but may be older. The latter will be referred to in this paper as *sumi*. Both traditional Chinese and Japanese ink sticks (*sumi*) are made from lamp or soot black mixed with animal glue solution. For this project, the *sumi* was ground in water on an ink stone, and then diluted enough to be used with the quill pen.

The iron gall ink was made according to the recipe (after Eisler) in James Watrous's book, *The Craft of Old-Master Drawings*.[3] The ingredients used were: 1/2 ounce oak galls, 4 ounces water, 1/4 ounce ferrous sulfate, and 1/8 ounce powdered gum arabic. After being crushed, the oak galls were soaked overnight, and the resulting liquor was a very dark brown color. The solution was filtered through layers of cheesecloth until no particles were visible. The ferrous sulfate and gum arabic were added to this solution. Immediately, the ink

turned a dark greenish brown color. This is contrary to most observations concerning the making of this ink. A dark coloration of this ink should not appear for several hours. even days, after being exposed to the atmosphere in a shallow dish. The recipes also say, however, that the extract from the oak galls should be a "pale ochre color."[4] The reason for the immediate coloration of this experiment's ink may be that the oak galls were too rich in gallo-tannic acids and too deeply colored. However, after a day, the ink did become a very dark bluish black. The pH of this ink was 2. The yield was 4 ounces.

The bistre was made according to the recipe found in Watrous.[5] It was made from soot taken from the out-of-doors flue of an air-tight, wood-burning stove. The soot was ground with a mortar and pestle until fine, and was then placed in a flask. It measured 350ml in volume. Tap water was added to the 500ml mark, and the solution was brought to the boil. After simmering for forty-five minutes, it was strained, and boiled gently for another hour. After cooling. it was passed through filter paper twice. The yield was 25ml. This bistre was a dark greenish brown.

The paper used for the experiment was "Yale with Jute Fleck" made by Twinrocker Handmade Paper. This paper is made from 100% cotton fibers with a small addition of jute from burlap. The coloring additives are iron oxide, yellow iron oxide, and carbon black. The internal archival size is an alkylketene dimer.[6] It is a laid and chain paper with a slightly textured surface, and is a light beige color.

The inks were applied to the paper using a quill pen for the words and lines of gradated width, and a brush for the washes (fig. 1). The inks used in the washes were diluted 1:1 with water. The sample was made so that it could be cut into three equal sections with the gradated lines and washes each cut down through the center so that these could be juxtaposed again for easy comparison after aging.

The section on the left, labeled "Oven," was placed in a Humid-Flow combination temperature and humidity cabinet (Blue M Electric Co., Blue Island, Ill.) A temperature of 91($\pm$1)°C and a relative humidity (R.H.) of 56($\pm$3)% was maintained continuously for 232.5 hours.

The center section, labeled "Control," was placed in the dark in a room where the environment was controlled at 21($\pm$2)°C and 70($\pm$2)% R.H.

The section on the right, labeled "Light," was placed under a light bank consisting of eight General Electric 100 watt F48PG17-CW Power Groove Cool White fluorescent tubes. The sample was placed face up in a plastic tray lined with aluminum foil. The sides of the light bank were covered to keep out any external light while maintaining an environment of approximately 42°C and 20% R.H. The output of the light bank was about 2700 footcandles, and this section was exposed for 258 hours.

The photographic specifications were as follows. Visible light: Kodak Plus-X 4147 sheet film, at ISO 100. Ultraviolet Light: Kodak Plus-X 4147 sheet film, at ISO 1.5 with Kodak filter no. 8; ultraviolet source were two filtered high-pressure mercury vapor lamps. Infrared light: Kodak High Speed Infrared, 4143 sheet film, at ISO 64 with Kodak filter no. 87.

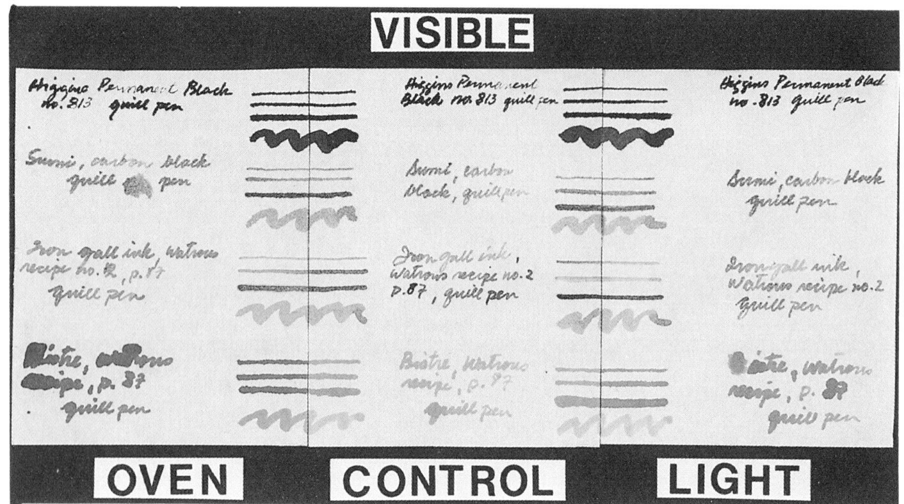

Fig. 1. The test samples. Visible light photograph.

Fig. 2. The test samples. Ultraviolet-visible fluorescence photograph.

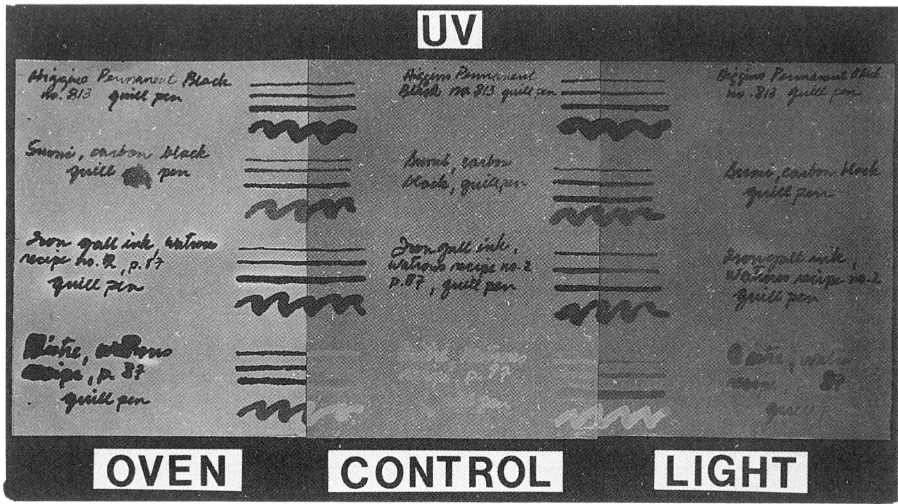

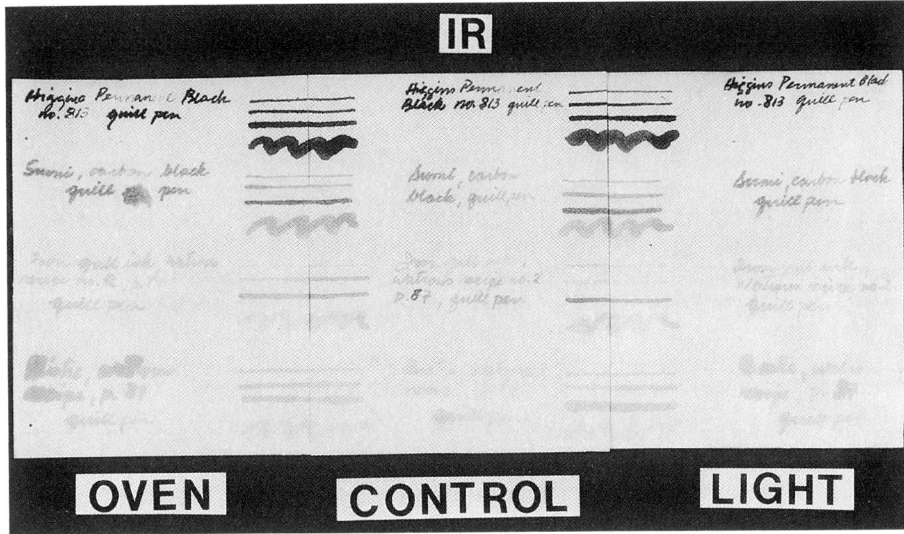

Fig. 3. The test samples. Reflected infrared photograph.

Observations

Following the aging tests, the samples were joined together and photographed under the above conditions. The following observations were made regarding each ink (Figs. 1-3).

The Higgins Eternal Permanent Black Ink remained unchanged under all conditions of control and aging. The ink does appear to live up to its name. However, it did not flow particularly well from the quill pen, and tended to feather slightly on the paper.

The *sumi* appeared to change after aging only when viewed under ultraviolet light. The *sumi* did fluoresce in the oven-aged sample, and this change is best seen around the edges of the heaviest line and on the verso. Like the Higgins ink, it did not flow evenly from the quill pen, and tended to feather slightly. It is a much lighter color than the Higgins ink, being a warm brown gray color rather than a deep black. Of course, value may be altered by dilution.

The iron gall ink did change in many respects after aging. Viewed in visible light, both of the aged ink samples are browner than the control sample's warm grayish black. Both the oven-aged and light-aged samples became crusty on the surface. The lightly drawn lines and the brush wash, although browner, also became slightly fainter. Viewed under ultraviolet illumination, a fluorescing halo appeared around the oven-aged iron gall ink lines; this halo could also be seen distinctly on the verso. No fluoresence on the recto or verso was seen in either the control or light-aged sample. When examined under infrared light, the iron gall ink samples also differed. In the oven- and light-aged samples, the inks became less absorbent in the infrared than the control sample. Physical damage was observed in the paper of the oven-aged sample; the paper split along the darkest pen line. The iron gall ink flowed more easily from the quill pen than did any of the other inks tested. It also did not feather out on the paper as the carbon blacks and bistre did.

The bistre underwent the most dramatic color changes upon aging. Viewed in visible light, the control bistre lines and words appeared gray, and the wash, a pinkish gray. In the light-aged sample, the bistre lines changed from gray to a more intense grayish brown. In the oven-aged sample, the bistre lines changed to an intense warm brown with no hint of gray. However, in each of the wash areas, there was still a hint of pink visible. Viewed under ultraviolet illumination, the control bistre fluoresced brightly especially in the areas of light application. In the light-aged sample, only the areas of the brush wash and the lightly applied ink lines still fluoresced. The oven-aged sample hardly fluoresced at all; there was a slight fluorescent halo around the darkest pen lines. (The fluorescent spot under the word "Watrous" in the oven-aged sample is a blotted smear of bistre.) When examined under infrared light, the aged bistre samples also exhibited changes; both became more absorbent than the non-aged sample in the infrared. However, the oven-aged sample became more absorbent in the infrared than did the light-aged sample. The bistre did not handle well in the quill pen; it tended to run uncontrollably from the tip, making blotches on the paper. The bistre feathered somewhat more than the iron gall ink, but less than the carbon black inks.

All in all, the iron gall ink performed best for use in the quill pen, and looked best on the paper as it did not feather. The bistre was the least satisfactory when applied with the pen.

Although not actually part of this experiment, the paper used in the samples underwent some noticeable changes after artificial aging. Viewed in visible light, little change in color between the sections was noticeable although the oven-aged paper appeared slightly darker than the other two. However, under ultraviolet illumination, the papers exhibited quite different fluorescences. Relative to the control, oven aging seemed to increase fluorescence while light aging seemed to reduce fluorescence. When examined under infrared light, no appreciable differences were seen.

Conclusions and Future Research

The limited number of samples and types of inks researched make it dangerous to draw definite conclusions about drawing inks in general. However, there are interesting observations to be made concerning the following: ease of use of the inks in a quill pen; simplicity of ink preparation and yield from raw materials; and certain color changes as well as physical changes upon aging. It is also notable that oven-aging produces different results from light-aging.

Observations made from this experiment confirmed those made on naturally aged drawings and archival materials, especially those done in iron gall ink. It has been noted by the author on several occasions that aged iron gall ink sometimes becomes almost transparent (non-absorbent) when viewed by infrared light. This property makes it easy to distinguish between dark iron gall ink and carbon black ink when they are present in the same drawing. In figures 4 and 5 one can see the difference in the infrared absorbency of the carbon black ink drawing and the iron gall ink inscription on a page from the late eighteenth-century John Vanderlyn sketchbook owned by the Albany Institute of History and Art, Albany, New York.[7] The absorbency of iron gall ink in the infrared may depend largely on the extent to which the ink has oxidized upon aging. Iron gall ink made with ferrous sulfate (green vitriol) could be expected to convert eventually to hydrated iron oxides (rust). Whereas ferrous sulfate has been observed to be very absorbent in the infrared, hydrated iron oxides are not. In addition, when iron gall inks are made up, either wholly or in part with copper sulfate (blue vitriol), it appears that the resulting copper compound may help retain the original black color of the ink.[8] This use of copper sulfate in recipes may account for the many differences in color, solubility in water, and acidity of various iron gall inks. There is also some confusion over the term "vitriol" in old treatises and recipes, and the terms green and blue vitriol may have been used interchangeably in the preparation of inks.[9]

In determining whether a drawing has been done in iron gall ink or bistre, it might be useful to view it by infrared light , as the bistre may be a much stonger absorber than the iron gall ink. Under ultraviolet illumination, the bistre may still have some fluorescence while the iron gall ink lines themselves should have none. Of course, if dyes or other ultraviolet-visible fluorescent or infrared absorbent materials have been added to any ink, the results under the different light conditions may also be different.

Another interesting observation from this research is the fluorescing halo around the oven-aged sample of the iron gall ink. Once that phenomenon was observed, a small number of naturally aged samples of iron gall ink were also examined under ultraviolet illumination and fluorescing halos were observed on about half of them. In most cases, these halos seem to be associated with inks that are decidedly brown, and have penetrated into the paper to the verso. In one case exhibiting this fluoresent halo, the ink had disintegrated portions of the paper completely, leaving a lace-work effect in the support. It may thus be possible to detect potentially damaging iron gall ink, and its susceptibility to conservation water treatments (without having to do dangerous spot testing) by first viewing the artifact under ultraviolet illumination. If a halo is observed, perhaps the degradation and resultant increased water absorbence of the paper immediately surrounding the ink may cause the ink to feather or bleed

Fig. 4. A drawing from a late eighteenth-century sketchbook by John Vanderlyn. Visible light photograph.

readily if the artifact is washed.

With respect to the changes in the ultraviolet-visible fluorescence observed in the Twinrocker papers used in the testing, it is impossible in the confines of this experiment to determine the aging mechanisms that might account for these differences. It is hoped that further work in this area will soon be done. Research by the author on the stability of iron gall inks prepared from different recipes is already underway and the results should prove to be quite interesting.

Acknowledgments

I would like to extend my appreciation to Mr. Dan Kushel, assistant professor in the Art Conservation Department, Cooperstown, N.Y., for all of his help in setting up the examination and photography of the samples, and in the editing of this manuscript. I would also like to thank Mr. Christopher Tahk, professor and director in the department for helping out with the environmental chamber and for his advice concerning this project. Ann Boulton, a student in this department, was instrumental in obtaining the oak galls used in this experiment, and I would like to thank her.

Fig. 5. Reflected infrared photograph of sketchbook.

References

1. G. W. Carriveau and M. Shelley, "The Study of Rembrandt Drawings Using X-Ray Fluorescence," *Nuclear Instruments and Methods*, 193 (1982): 297-301.

2. Some sources are: Stroblite Company, Inc., New York, N.Y.; George W. Gates & Co., Inc., Franklin Square, N.Y.; Spectronics Corp., Westbury, N.Y.

3. J. Watrous, *The Craft of Old-Master Drawings* (Madison: University of Wisconsin Press, 1975), p. 87.

4. Ibid., p. 71.

5. Ibid., pp. 87-88.

6. Twinrocker Handmade Paper, Brookston, Ind. Information concerning the furnish of this paper was provided by Howard Clark.

7. The photographs were taken by Joan London while a student in the department.

8. Watrous, *Craft of Old-Master Drawings*, pp. 72, 87.

9. M. dePas and F. Fleider, "History and Prospects for Analysis of Black Manuscript Inks," in N. Brommelle and P. Smith, eds., *Conservation and Restoration of Pictorial Art* (London: Butterworths, 1976) pp. 194-195.

M. SUSAN BARGER and WILLIAM F. STAPP

The Evolution of the Daguerreotype Art of Robert Cornelius: A Scientific Retrospect by Scanning Electron Microscopy

The first American portrait photographer, Robert Cornelius, took up daguerreotypy shortly after it was introduced to the United States in 1839, and continued his practice of the art until 1843. Contemporary reports acclaimed the artistic merit of his portraits, as well as his continual improvement of the daguerreotype process. In conjunction with the first major exhibition of his work since the nineteenth century, over half of his daguerreotypes were submitted for analysis by scanning electron microscopy. In the absence of written records left by Cornelius about his working methods, it was hoped that these analyses would provide chemical and microstructural data that could be used to reconstruct his working methods and the changes he made in his practice of daguerreotypy over the years. Not only was it possible to follow his improvements on the daguerreotype process, but it was also possible to assign dates to previously undated or loosely dated daguerreotypes on the basis of microstructural similarities with positively dated Cornelius daguerreotypes.

Background

Robert Cornelius (1809-1893), a manufacturer of high-quality lamps and lighting fixtures in Philadelphia, was a seminal figure in the early history of the daguerreotype in the United States. He began to experiment with Daguerre's process in the early autumn of 1839, just after the first descriptions of it became available in this country, and before the end of that year he began to collaborate with Dr. Paul Beck Goddard (1811-1866), a chemist, who shared with Cornelius his discovery that sensitization of the daguerreotype plate with bromine significantly increased its response to light. In the spring of 1840, Cornelius opened a commercial portrait studio—only the second of its kind in the world—that flourished until late in 1842, when he closed it because his other business had begun to demand his full-time attention. He continued, however, to make daguerreotypes into the late 1840s. He also taught the process to several of Philadelphia's most distinguished daguerreotypists.

Contemporary sources indicate that Cornelius was regarded at the time as one of the most technically advanced daguerreotypists particularly because of the exceptional finish of his plates. Comparison of his daguerreotypes to others of the period justifies this reputation. Cornelius's plates are distinguished by their superior polish, which became increasingly refined over the course of his career, and by their excellent tonal scale and contrast, especially in the later plates. These characteristics enhance and emphasize the particularly vital quality of his portraiture, which is in itself unique for the period. Surviving dated plates indicate that by the end of 1841, Cornelius was making daguerreotypes equal in technical and aesthetic quality to the best daguerreotypes of the early 1850s, when the development of the medium was at its peak.

As part of the research for an exhibition on Robert Cornelius organized for the National Portrait Gallery, Smithsonian Institution, fifteen daguerreotypes made by him between 1839 and 1843, as well as the earliest surviving America daguerreotype, a plate made by Joseph Saxton (1799-1873) in late September 1839, and an early (1839) portrait daguerreotype attributed to Walter Rogers Johnson (1794-1842), another early Philadelphia daguerreotypist, were analyzed at the Materials Research Laboratory of the Pennsylvania State University, using an ISI DS 130 scanning electron microscope with a Kevex energy dispersive x-ray analysis attachment. The purpose of this analysis was to trace, at the microstructural level, Cornelius's evolution as a daguerreotypist, and to determine whether these analytical results could be correlated with contemporary documentation to date the plates or assign them a position in Cornelius's chronology. It was also hoped that analysis would provide insights into his working methods, particularly into his polishing techniques and his use of bromine.

Introduction to the Analyses

A detailed description of the scientific findings of the Cornelius study, as well as the similarities and differences between daguerreotypes and conventional photographic materials, can be found in the catalogue to the National Portrait Gallery exhibit, *Robert Cornelius: Portraits from the Dawn of Photography*[1]. The raw data collected from scanning electron microscopic analysis were used to reconstruct Cornelius's working methods.

The primary concern for the scientific examination of any photograph should be the chemical and physical composition of the image, as well as what gives rise to the appearance or optical properties of the image. Because the optical properties are more often than not the direct result of the image microstructure, rather than image chemistry, the primary thrust of such examinations should center on the image structure. This approach puts the scientific emphasis where the most information can be gained. Chemical data adds to microstructural data, but when used alone, chemistry cannot provide adequate information for assessing questions concerned with processing or provenance.

This microstructural approach has been particularly fruitful for the examination of daguerreotypes. The importance of this particular study is derived from Robert Cornelius's position and skill as a pioneer daguerreotypist. Because he worked during the very earliest period of photography and because he, along with Paul Beck Goddard, was responsible for multiple sensitization (one of the only modifications made to the daguerreotype process), it is possible to view the progress of daguerreian practice through the examination of his work.

A brief description of the ideal daguerreotype microstructure and its relation to daguerreotype processing will serve as background for the following discussion. The appearance of a daguerreotype image is dependent upon the formation of a narrowly defined microstructure made up of image particles dispersed over a polished metal surface. These image particles have differing chemistries and distributions depending upon the amount of exposure from the original scene and the way in which they were processed. Ideally, highlight areas have several hundred thousand image particles per square millimeter ranging in size from 0.1 to 1.0μm. Shadow areas have from fifty to one hundred shadow particle agglomerates per square millimeter and these may be as large as 50μm in diameter. The dominant imaging mode for the daguerreotype is light scatter or diffuse reflectance in the highlight areas and specular reflectance away from the viewer in the shadow areas. When the daguerreotype microstructure does not match this optimal configuration, no matter what the cause, the image contrast range is compressed and the image may thus appear weak, faint, or faded.

The characteristic microstructure is formed during processing, and procedures that improve microstructure automatically improve image

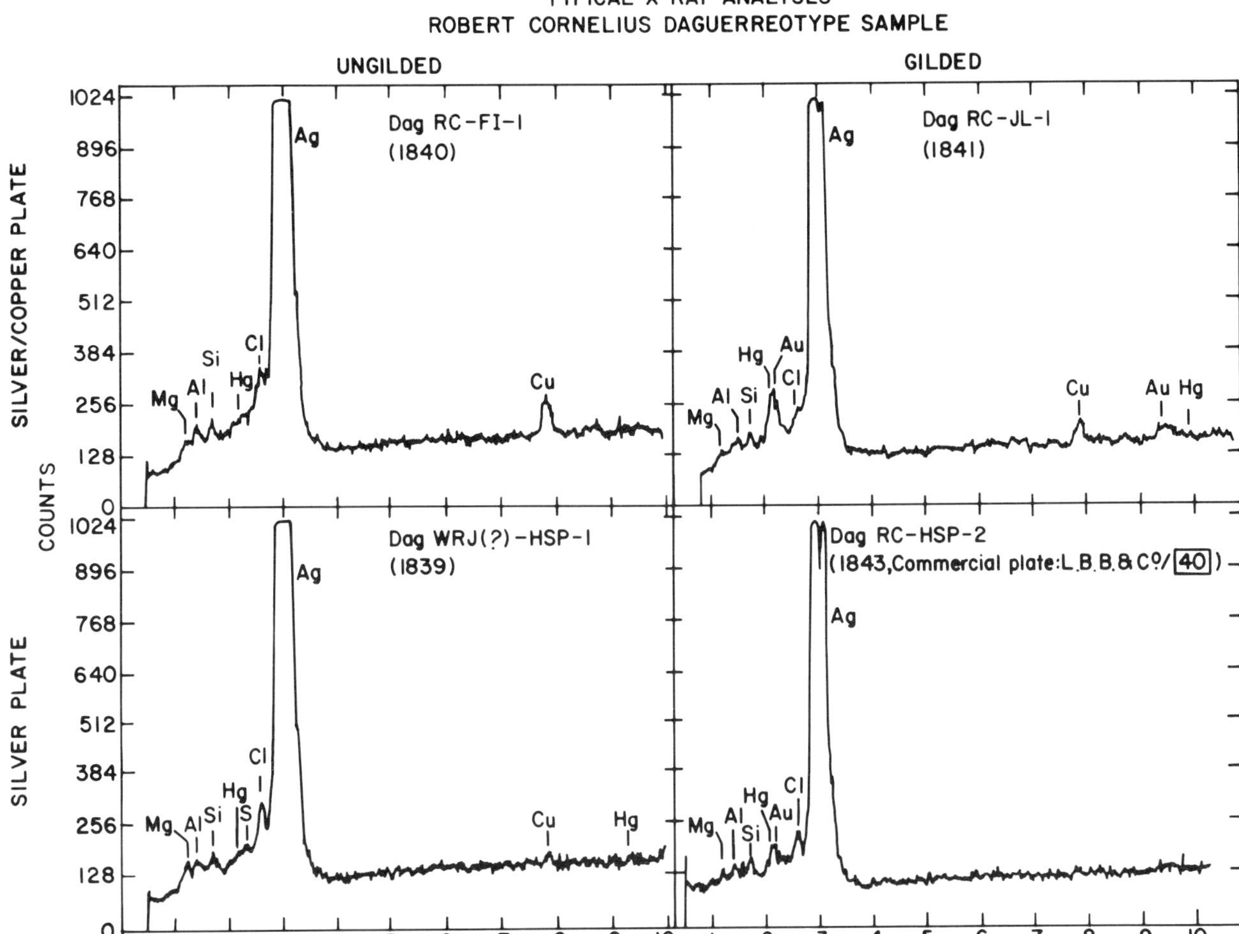

Fig. 1. Typical x ray spectra for both gilded and ungilded daguerreotypes, as well as daguerreotypes made on plates covered with pure silver and silver-copper alloy.

appearance. The original daguerreotype process introduced by L. J. M. Daguerre in 1839 involved the following steps. A highly polished piece of silver plate was placed over iodine vapor to form a light-sensitive layer. The "sensitized" plate was placed in a camera and exposed to light. The exposed plate was placed over hot mercury vapor until an image appeared (about 2-3 minutes) and the unexposed silver iodide was removed.

It has been generally assumed that the mercury used in the daguerreotype process acts as a photographic developer, i.e., it causes the reduction of exposed silver halide grains to silver to form the image. More recent work indicates that the mercury acts merely as a solvent that facilitates the formation of large (0.1- 1.0μm), crystalline image particles from photolytic silver formed by the photo-reduction of silver halide during light exposure. It does not appear that the silver of the underlying daguerreotype plate is involved in image formation. The exact details of this process are not pertinent here, but it is important to keep in mind that photolytic silver is required for this vapor-assisted crystallization and the formation of image particles.

The original processing procedure for daguerreotypes with its single sensitization step limits the amount of photolytic silver available for image particle production. The sensitization step was carried out in subdued light, forming some photolytic silver, but the primary source of photolytic silver in this processing mode is the camera exposure. As a result, the microstructure of daguerreotypes made in this way is made up of image particles that vary little in size, are few in number, and tend to be widely spaced. The images produced in this way have a compressed contrast range and ''flat'' tones and are often somewhat bluish or cold-toned and faint.

There were two additions to the daguerreotype process that greatly enhanced image microstructure and, thus, image appearance. One of these, multiple sensitization, was first routinely used by Robert Cornelius. Multiple sensitization was the adoption of several successive halogen treatments, usually iodine exposure followed by bromine, with a final exposure to iodine. This step does increase the photographic speed of the daguerreotype process somewhat. More importantly, because these steps were carried on in subdued light, a larger amount of photolytic silver was made available for particle production than in the single sensitization process. The resulting microstructure has more image particles and more image particle variety and distribution across the daguerreotype plate. The images that result have a larger contrast range and appear stronger than images made by single sensitization.

The second addition to the daguerreotype process was the gilding step. Gilding is a gold chloride treatment following the removal of unexposed silver halide. It improves both the mechanical stability of the image and the optical properties. During gilding, mercury in the image structure is replaced by gold. This causes a rapid hardening of the image and slightly increases the size of the image particles. This increase in image particle size improves the optical properties of the daguerreotype by increasing visible light scatter from the image. In other words, the light striking the image is more effectively scattered, and this is seen by the viewer as a stronger image with brighter whites. Although it has been claimed that a gilded daguerreotype can be recognized by visual inspection alone, positive verification of gilding is only possible with chemical data showing the presence of gold.

In addition to microstructural variations directly attributable to processing alterations, there is an intangible element, namely skill, involved in the production of daguerreotypes. The daguerreotype system is a complicated process of crystal growth involving a vapor phase. Its control is not trivial, especially with no analytical tools other than visual inspection of the final product to monitor progress. The skill element can be followed in microstructural examination by checking the ability of the daguerreotypist to produce discrete image particles of the correct size and distribution, as opposed to misshapen crystals or platelets.

In summary, the main features that are used to analyze daguerreotypes are microstructure and chemistry. The microstructural analysis compares the extreme highlight and the extreme shadow image areas (the areas of the highest and lowest image particle densities). This comparison is an efficient way to determine the average image particle size, shape, number, and distribution across the plate. It also gives an idea of how the daguerreotype was processed and of the skill of its maker. This type of examination must be done by scanning electron microscopy because the size of image particles is below the limits of resolution for a light microscope. General chemical information from energy dispersive x-ray analysis is quite sufficient to determine if a daguerreotype has been gilded and to give other information, such as plate composition. The exact chemistry of image particles varies from image area to image area, and from da-

guerreotype to daguerreotype, so specific information on image particle composition is not particularly helpful. In addition to checking the microstructure for how it was formed, it is also checked for signs of corrosion or cleaning treatments. Traditional cleaning treatments for daguerreotypes, potassium cyanide, and thiourea solutions cause characteristic alteration of the daguerreotype plate microstructure, as well as general etching of the underlying plate surfaces.

Results and Discussion

A summary of the analytical findings will be found in table 1. The dates in parentheses indicate dates that were assigned on the basis of the analyses. All other dates have contemporary documentation confirming the time at which the daguerreotypes were made. The spectra shown in figure 1 are typical of the daguerreotypes analyzed. These show general chemical data for gilded and ungilded daguerreotypes and for daguerreotypes made on plates plated with pure silver and those with silver-copper alloy. The secondary electron micrograph images shown in figure 2 are representative of shadow and highlight areas for each of the daguerreotypes. (The reader is referred to these data in the following discussion).

The primary concerns of this study were to construct a chronology of Robert Cornelius's work based on similarities (macro- and microscopic) between dated and undated daguerreotypes and to see if changes in Cornelius's technique could be monitored on the basis of analytical data. Once the analyses were completed, a chronology was constructed in the following way. The firmly dated daguerreotypes were used as the backbone of the chronology and the micrographs of these daguerreotypes were placed in order. The remaining daguerreotypes were compared and grouped by similar features within the ungilded and gilded groupings. These sorting procedures made it possible to group the analyzed daguerreotypes and to form the overall chronology.

All of the sorting was fairly straightforward. The difficult part was determining where each daguerreotype fit within each part of the overall scheme. The fine tuning of the dating procedure was done by looking for microstructural variations indicative of specific processing techniques and skill. For instance, Dag JS-HSP1, Dag WRJ(?)-HSP-1, Dag RC-APS-1, Dag RC-EH-1, and Dag RC-FI-1 are examples of single sensitization. In every case, these daguerreotypes have very little microstructure, which is made up of a relatively small number of image particles with very little variety in the size and distribution of the particles. Note that the difference between highlight and shadow areas is not pronounced in any of these daguerreotypes. Dag JS-HSP-1 is the oldest American photograph and its microstructure shows few image particle variations that would indicate control over the image formation process. Contrast this microstructure with that in Dag WRJ(?)-HSP1. In the second case, even though the microstructure is sparse, there is a difference in both image particle size

Fig. 2. Representative micrographs showing the shadow and highlight microstructure for each analyzed daguerreotype. The scale is indicated by the micrometer bar in the lower right hand corner of each micrograph.

JS-HSP-1
WRJ?-HSP-1
RC-APS-1
RC-APS-2
RC-EH-1
SHADOW
HIGHLIGHT
SHADOW
HIGHLIGHT
SHADOW
HIGHLIGHT
SHADOW
HIGHLIGHT
SHADOW
HIGHLIGHT

and number between the highlight and shadow area. Dag WRJ(?)-HSP-1 is a portrait and, as such, is not a first attempt at the process. Both the historic record and the microstructure bear this out.

The three remaining singly sensitized daguerreotypes were made by Robert Cornelius. Again, the sparsity of microstructure is clearly evident. The platelets seen in the highlight areas of these daguerreotypes indicate a further step in the evolution of good particle production. Platelets are a transition phase in the formation of mercury crystals from vapor and their presence shows some advance in the control of crystal growth. Platelets also indicate that there is room for improvement, i.e., the process needs to be driven through the transition phase. The similarity of the microstructure and the underlying plate surface of Dag RC-FI-1 to Dag RC-EH-1 suggests that these two daguerreotypes were made at the same time and possibly on the same plate stock.

The first of the multiply sensitized plates is Dag RC-APS-2. Because this plate differs from other plates made by Cornelius at the same time and also because it was made before he adopted the routine practice of multiple sensitization, it is thought that this was an experimental plate. The increase jn image particle variety and number is clearly seen in the micrographs. There is platelet formation in the highlight areas, but even the platelets have more substance than those found in the singly sensitized daguerreotypes from the same time grouping.

The next grouping of daguerreotypes are those assigned to late 1840: Dag RC-LC-1, Dag RC-WB-1, and Dag RC-MP-1. These are multiply sensitized and also appear to be made on the same plate stock. Platelets are still visible in the highlight areas, but they are fewer in number and there is an accompanying increase in the number of image particles found in these image areas. Both Dag RC-LC-1 and Dag RC-WB-1 had been previously dated 1841, but microstructural evidence, along with the lack of gold revealed by chemical analysis, led to the change in their dating.

Each grouping of daguerreotypes was evaluated in the way described above. It can be seen that over Cornelius's working period, not only do later plates have more image particles with a greater distinction between the highlight and shadow image areas, but there is also less and less evidence of platelet formation in the highlights. The microstructural evidence supports the visual evidence that his technique improved over time.

Some of the daguerreotypes have disturbed microstructures caused by previous cleaning treatment. The characteristic etching of the daguerreotype plate surfaces caused by traditional daguerreotype cleaners, as well as image particle rearrangement, can be seen in the micrographs of Dag RC-APS-1, Dag RC-APS-2, Dag RC-HSP-1, Dag RC-HSP-2, Dag RC-EH-2, Dag RC-EH-3, Dag RC-EH-4, Dag RC-MHS-1, and Dag RC-NZ-1. Those daguerreotypes cleaned with thiourea solutions (as opposed to cyanide solutions) show less damage (see Dag-RC-APS-1, Dag-RC-APS-2, and Dag RC-NZ-1). The catastrophic effects of cyanide cleaning on both image microstructure and plate surfaces can be seen especially in Dag RC-EH-3, Dag RC-EH-4, and Dag RC-HSP-2. These particular daguerreotypes may have been cleaned several times, judging from the severity of the microstructural alteration.

In addition to the chronology and evaluation of daguerreotype technique, there was a serendipitous discovery. It has been a strongly held belief that one of the contributing factors to the the flat tones and limited contrast range of early daguerreotypes was the use of a silver-copper alloy, rather than pure silver, for the silver layer of the daguerreotype plate. It can be seen from table 1 that all of the daguerreotypes analyzed, except for Dag WRJ(?)-HSP-1, Dag RC-EH-3, and Dag RC-HSP-2, were made on daguerreotype plates plated with a silver-copper alloy. Both visual and microstructural examination attest that Cornelius's technique improved in spite of the use of these plates. Re-evaluation of the historical statements indicates that all of the faults associated with the use of impure silver are attributable to plating flaws, rather than to the plating material itself. Examination of other scientific information about copper compounds shows that copper may be converted to copper halides that are light sensitive and form photolytic copper. However, because copper and mercury do not combine to form amalgams, any copper present will not be involved in the process of particle formation. Thus, it appears that copper would have no effect, either good or bad, on the process of image formation. Further, the presence of some copper in the silver layer would improve its mechanical properties and produce a better surface for polishing. Cornelius obviously reaped this benefit, for his polishing technique was highly acclaimed.

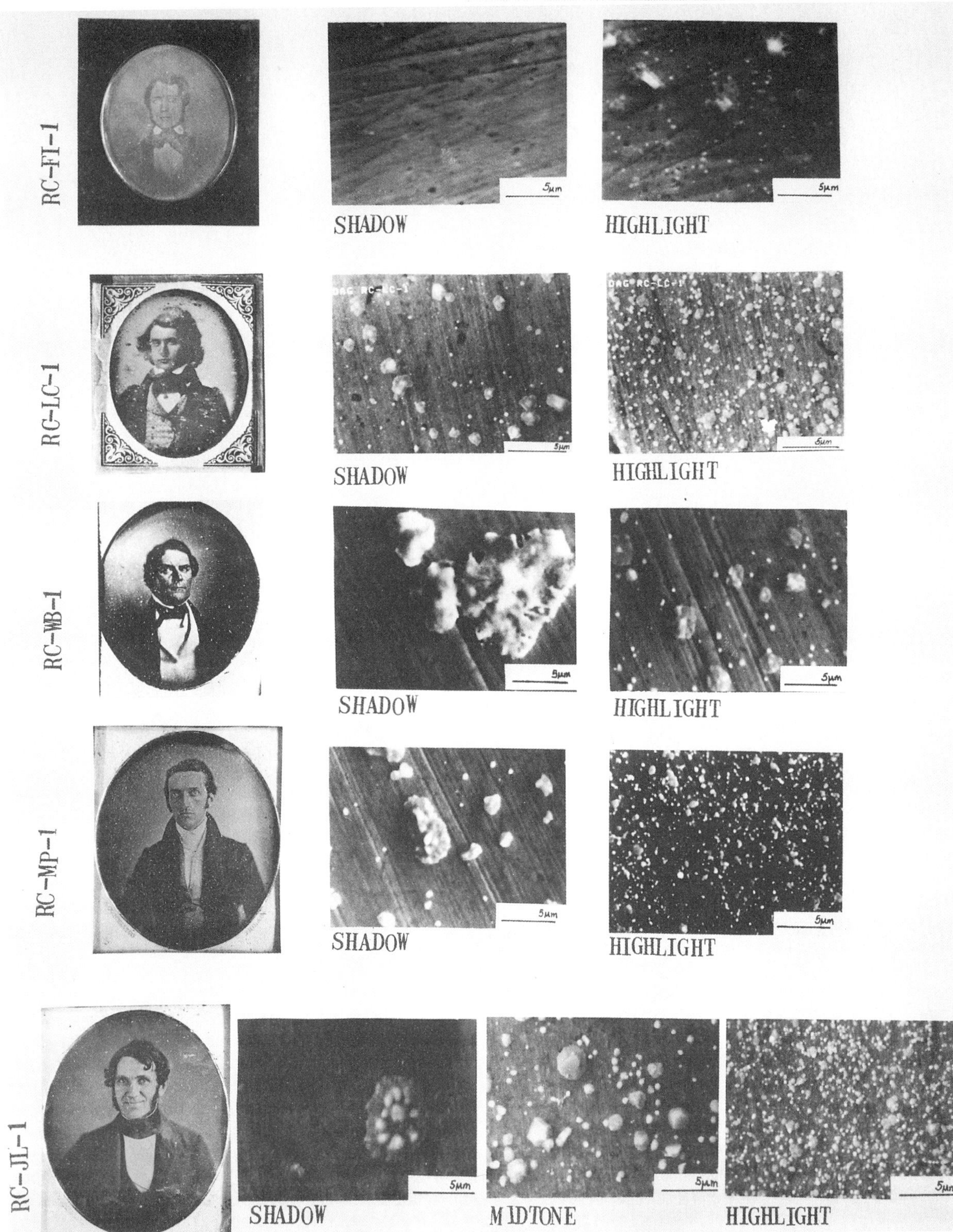
RC-FI-1
SHADOW
HIGHLIGHT
RC-LC-1
SHADOW
HIGHLIGHT
RC-WB-1
SHADOW
HIGHLIGHT
RC-MP-1
SHADOW
HIGHLIGHT
RC-JL-1
SHADOW
MIDTONE
HIGHLIGHT

RC–EH–4

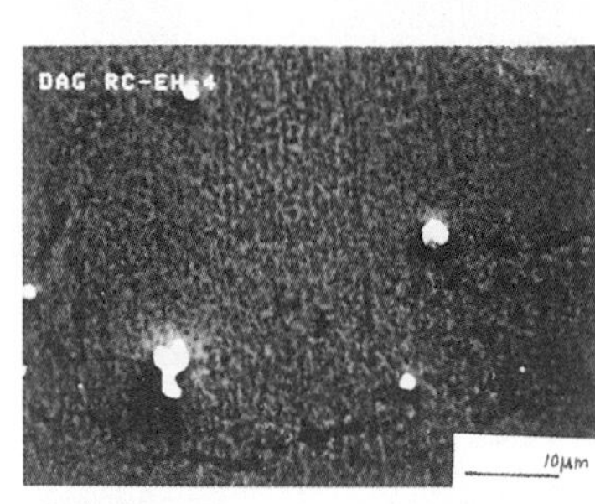

SHADOW

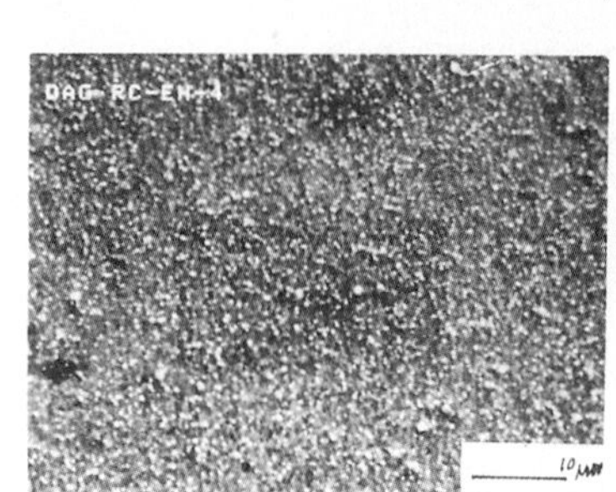

HIGHLIGHT

RC–HSP–2

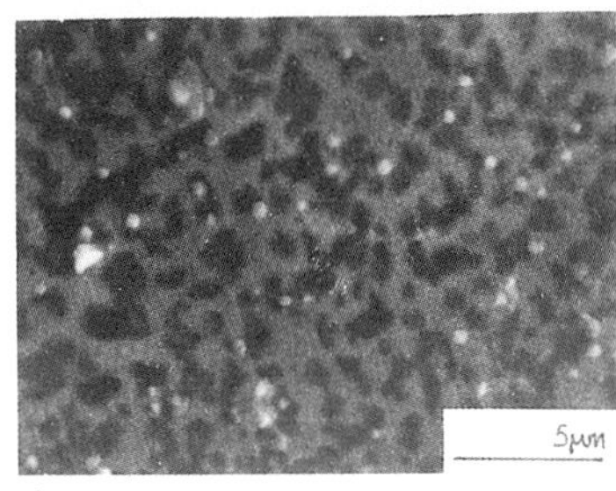

SHADOW

HIGHLIGHT

RC–EH–3

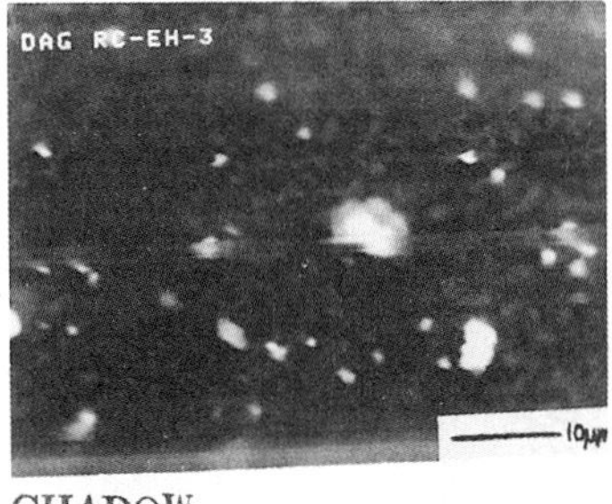
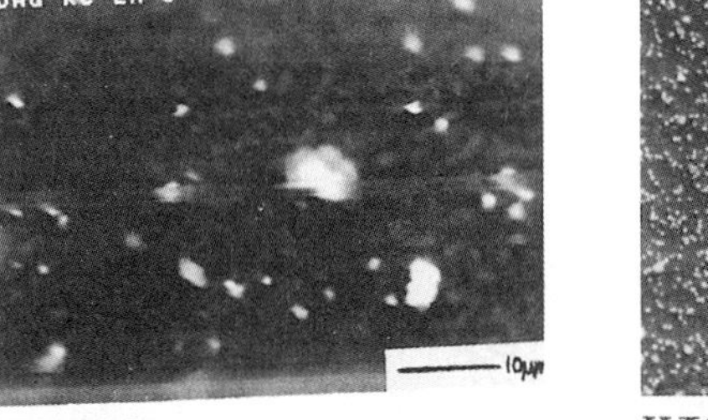

SHADOW

HIGHLIGHT

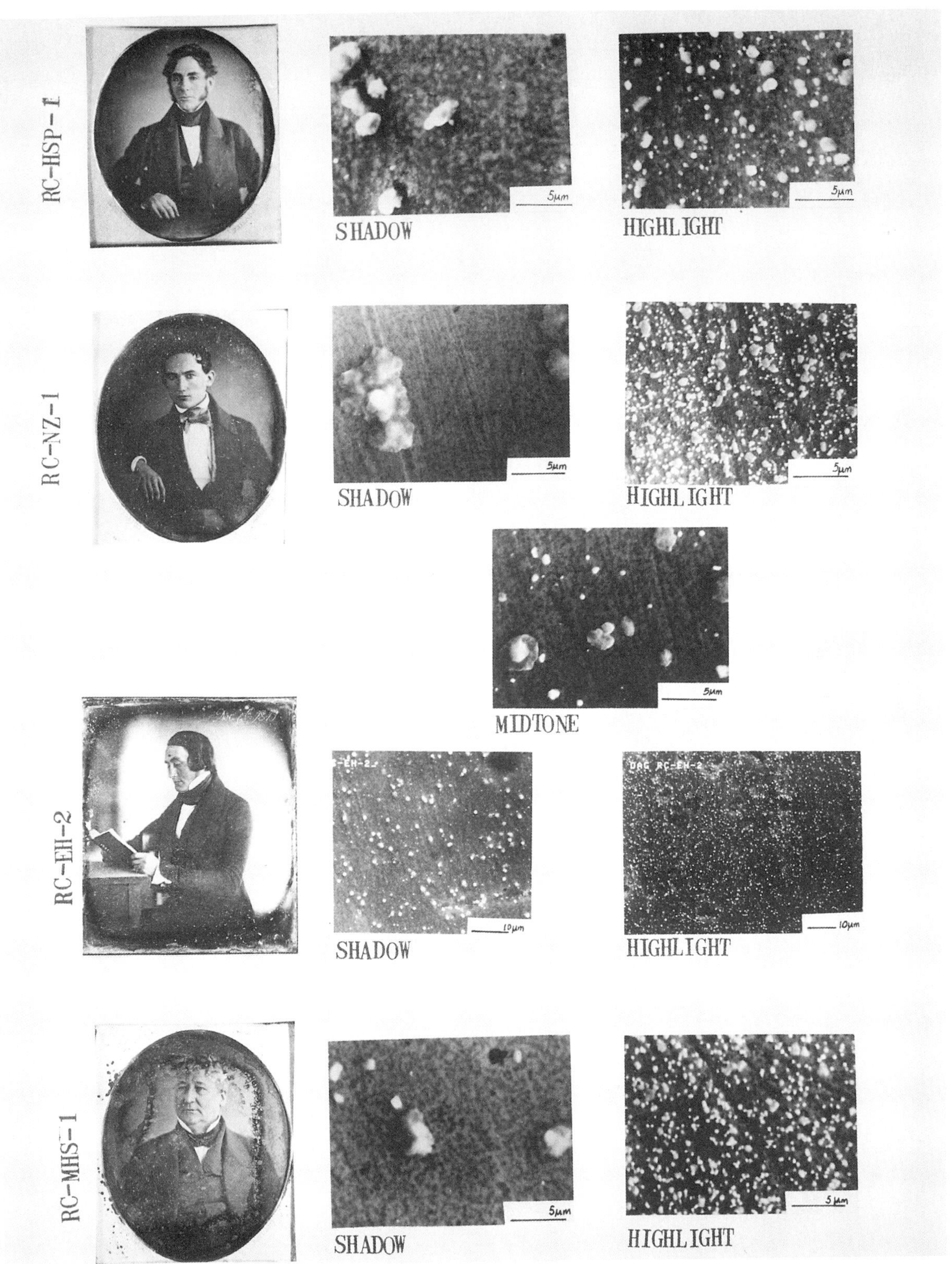
RC-HSP-1
SHADOW
HIGHLIGHT
RC-NZ-1
SHADOW
HIGHLIGHT
MIDTONE
RC-EH-2
SHADOW
HIGHLIGHT
RC-MHS-1
SHADOW
HIGHLIGHT

Table 1
Summary of the Experimental Findings

Analysis Number	Date*	Original Seal Intact	Plate Chemistry	Processing	Cleaned	
					Method	Noted in Records
Dag JS-HSP-1 View of Central High School Joseph Saxton (Historical Society of Pennsylvania)	(Sept. 25, 1839)		Ag-Cu	UG,S	-	-
Dag WRJ(?)-HSP-1 Portrait of Ezra Otis Kendall Walter R. Johnson (Historical Society of Pennsylvania)	Oct. or Nov. 1839	X	Ag	UG,S	-	-
Dag RC-APS-1 Portrait of Paul Beck Goddard Robert Cornelius (American Philosophical Society)	Dec. 1839		Ag-Cu	UG,S	Thio	Yes
Dag RC-EH-1 Portrait of Martin Hans Boyé Robert Cornelius (International Museum of Photography at George Eastman House)	(early 1840)		Ag-Cu	UG,S	-	-
Dag RC-APS-2 Portrait of Etiénne du Ponceau Robert Cornelius (American Philosophical Society)	May 15, 1840		Ag-Cu	UG,M	Thio	Yes
Dag RC-FI-1 Portrait of Henry Myers Robert Cornelius (Franklin Institute)	(early 1840)		Ag-Cu	UG,S	-	-
Dag RC-WB-1 Portrait of Samuel Bispham Robert Cornelius (William Becker Collection)	(late 1840)	X	Ag-Cu	UG,M	-	-
Dag RC-LC-1 Portrait of Henry Howard Houston Robert Cornelius (Library Company of Philadelphia)	(late 1840)		Ag-Cu	UG,M	-	-
Dag RC-MP-1 Portrait of Unidentified man Robert Cornelius (Miller-Plummer Collection)	(late 1840)		Ag-Cu	UG,M	-	-
Dag RC-JL-1 Portrait of Unidentified Man Robert Cornelius (Janet Lehr Collection)	(early 1841)	X	Ag-Cu	G,M	-	-
Dag RC-HSP-1 Portrait of Unidentified Man Robert Cornelius (Historical Society of Pennsylvania)	(early 1841)		Ag-Cu	G,M	KCN	-
Dag RC-NZ-1 Portrait of James B. Harrison Robert Cornelius (Nancy Zoza Collection)	(early 1841)		Ag-Cu	G,M	Thio	Yes

	Date*	Plate	Gilding/Sensitization	Cleaning	Solarized
Dag RC-EH-2 Portrait of Martin Hans Boyé Robert Cornelius (International Museum of Photography at George Eastman House)	Dec. 6, 1841	Ag-Cu	G,M	KCN	-
Dag RC-EH-4 Portrait of Martin Hans Boyé Robert Cornelius (International Museum of Photography at George Eastman House)	May 19, 1842	Ag-Cu	G,M	KCN	-
Dag RC-MHS-1 Portrait of Fielding Lucas Jr. Robert Cornelius (Maryland Historical Society)	(1842)	Ag-Cu	G,M	KCN	-
Dag RC-EH-3 Portrait of Martin Hans Boyé Robert Cornelius (International Museum of Photography at George Eastman House)	Dec. 1843	Ag (L.B.B. & Co./40)	G,M	KCN	-
Dag RC-HSP-2 Portrait of Martin Hans Boyé Robert Cornelius (Historical Society of Pennsylvania)	Dec. 1843	Ag (L.B.B. & Co./40)	G,M	KCN	Yes

*Dates in parentheses have been assigned on the basis of analyses.

UG = ungilded	S = single sensitization	KCN = cyanide cleaning	
G = gilded	M = multiple sensitization	Thio = thiorea cleaning	

Conclusions

In the absence of written records, scientific analysis of daguerreotypes can be a useful tool for dating daguerreotypes, as well as for
understanding how they were made and later treated. This study
was the first systematic, scientific examination of any group of photographs preparatory to their exhibition. Using the information gained
from extensive study of the material properties of daguerreotypes as
a "Rosetta Stone," it was possible to follow Robert Cornelius's work
as he increased his skill as a daguerreotypist. These findings lend
new importance to his position as a pioneer daguerreotypist and in-
novator of this imaging technique. Further, a deeper understanding
of how the daguerreotype process evolved in the earliest period of
photography has come about as a result of this study. An entirely
new insight into the role of copper in daguerreotype plates made with
silver-copper alloys has been arrived at from these results. It is
hoped that this study will set a precedent for future investigations into
the evolution of photography.

Acknowledgment

This work was made possible by grants from the Bara Foundation
and from the Secretary's Fluid Research Fund, Smithsonian
Institution.

References

1. M. S. Barger, "Robert Cornelius and the Science of Daguerreotypy," in W. F.
Stapp, *Robert Cornelius: Portraits From the Dawn of Photography* (Washington, D.C.: Smithsonian Press,1983).

2. M. S. Barger, R. Messier, and W. B. White, "A Physical Model for the Daguerreotype," *Photographic Science and Engineering* 26 (1982): 285.

GARY W. CARRIVEAU, LIN CHANG-SHAN, CHRISTINE GIERCZAK, and DAVID M. COLEMAN

The Laser Microprobe: A New Look at an Old Tool for Art Analysis

The laser microprobe has been used in qualitative elemental determinations of art objects for nearly twenty years, having been originally developed by Fred Brech and others at Jarrell-Ash in Boston.[1] This instrument uses a short intense pulse of focused laser light to sample the object, followed by excitation of this material by a pulsed electrical arc. Some of the radiation from the excited atoms and ions is collected and admitted to the entrance slit of a spectrograph. The resultant dispersed light can be measured to identify the components in the sample and, in some instances, to determine their concentrations. The instrument was a creative approach in combining the excellent sampling properties of a high-power pulsed laser with spark-like cross-excitation to enhance the normally weak spectral luminosity of the resultant laser plume.

For all the early enthusiasm associated with the introduction of the laser microprobe, its applicability has been surprisingly limited. There are a number of important factors that contribute to this situation. This manuscript describes work in progress that examines both advantages and limitations of the laser microprobe (as normally employed) and explores several new approaches to solving old problems.

Advantages and Limitations

For the study of art objects, utilization of a non-destructive method is generally called for—the laser microprobe approaches this restriction. Sample "spots" on the order of tens of microns in diameter are possible. Unfortunately small-diameter sampling sites have usually implied a corresponding lack of sensitivity. Thus cross-excitation is essential to improve signal-to-noise ratios and to minimize self-absorption and spectral line broadening. The energetic electric pulsed arc generally causes discoloration or surface damage unless several factors are carefully controlled; contamination by the auxiliary arc is also possible. Furthermore, to fully solve problems, quantitative analyses are needed that have the ability to detect major, minor, and trace constituents in a variety of sample matrices.

An excellent assessment of the laser microprobe's utility as a tool for quantitative analysis was reported by van Deijck, et al.[2] The authors expressed concern that both precision and accuracy will always suffer as a direct consequence of utilizing independent sampling and excitation methods. Their conclusions suggest that problems are so severe that further development of the laser microprobe is not warranted. Nonetheless, the many intrinsic advantages of the laser microprobe encouraged us to systematically study its utility and to make modifications to enhance its use in the study of art objects.

Instrumental Developments

The laser microprobe in use by the Detroit Institute of Arts is located in the Emission Spectroscopy Laboratories at nearby Wayne State University. The instrument is a Jarrell-Ash MARK II, Model 45-604 unit with a Q-switched, flashlamp-pumped neodymium glass rod laser. Energy output of the laser is adjustable from about 0.1 to 1.0 joules. Depending on experimental conditions, such as laser energy, degree of focus, nature of the sample, and quality and condition of the microscope lens, laser craters are created with diameters ranging from about 20 to more than 200 microns.

Laser output, which has a pulse duration of a few nanoseconds, is manually controlled by the operator and subsequent rotation of the laser Q-switch. Upon breakdown, laser energy is directed through the instrument head via a prism. The collimated light is focused by the microscope optics to the sample surface. The optical depth of field resulting from this arrangement is very shallow so "critical" focusing (i.e. within a few microns) is essential for efficient energy transfer.

The two ultrapure carbon electrodes are positioned close to the sample surface. These electrodes are capacitively charged and maintained at 1500-2000V DC. Material that is ablated from the sample surface is, in part, ionized due to the tremendous energy imparted to the gaseous plume during the sampling process. These charged species travel from the sample surface and "short out" the electrode gap resulting in the dumping of electrical energy into the laser plume and subsequent re-excitation of that material. The luminosity of re-excited material is often several orders of magnitude more intense than emission resulting directly from laser ablation only. Figure 1 details the arrangement of the Q-switched laser head.

Because our work involves art applications, attention was placed on achieving maximum spectral luminosity while sampling with minimum laser energy to minimize visible ablation from the surface. To this end, the standard electrode configuration was significantly altered. The angle between the electrodes was increased from approximately 90 to 150 degrees and the distance from the sampling site to the electrodes was reduced to about 1mm. This arrangement, coupled with careful optical matching to the 1.5-meter Wadsworth grating spectrograph increased spectral luminosity by one order of magnitude over standard conditions. The microscope lens is protected by a thin replaceable microscope slide cover to protect the optical surfaces from ablated material. Special airpath (non-cemented) objectives are essential; cemented lenses will fracture due to laser energy absorption.

The mechanical stability of the commercial instrument was not adequate to meet our stringent requirements of directing energy to specific sample sites. Since a major portion of our work has been to explore "depth profiling" a surface with the microprobe, it was essential that critical alignment be maintained in order to focus on the bottom of a crater left by a previous sampling event. Mechanical support stages were designed to accomplish this. Also added was an ultrafine adjustment microscope sample stage so that critical focus could be achieved. As a result, we have been able to "tunnel through" surfaces 1 to 2mm thick with thirty to fifty successive laser shots all following the same trajectory. With this improvement, coupled with development of sampling models derived from those introduced in this paper, we hope to be able to achieve selective elemental determinations of overlaid strata in complex objects such as paintings or oriental lacquerware.

Improvements in Sampling Conditions

Figure 2a illustrates the effect of one laser sampling event on a gold-plated (28-micron thickness) silver-copper alloy coin. It is obvious that "normal operation" is accompanied by severe surface discoloration and damage; the sampled area (crater) is easily detected (inside the letter U).

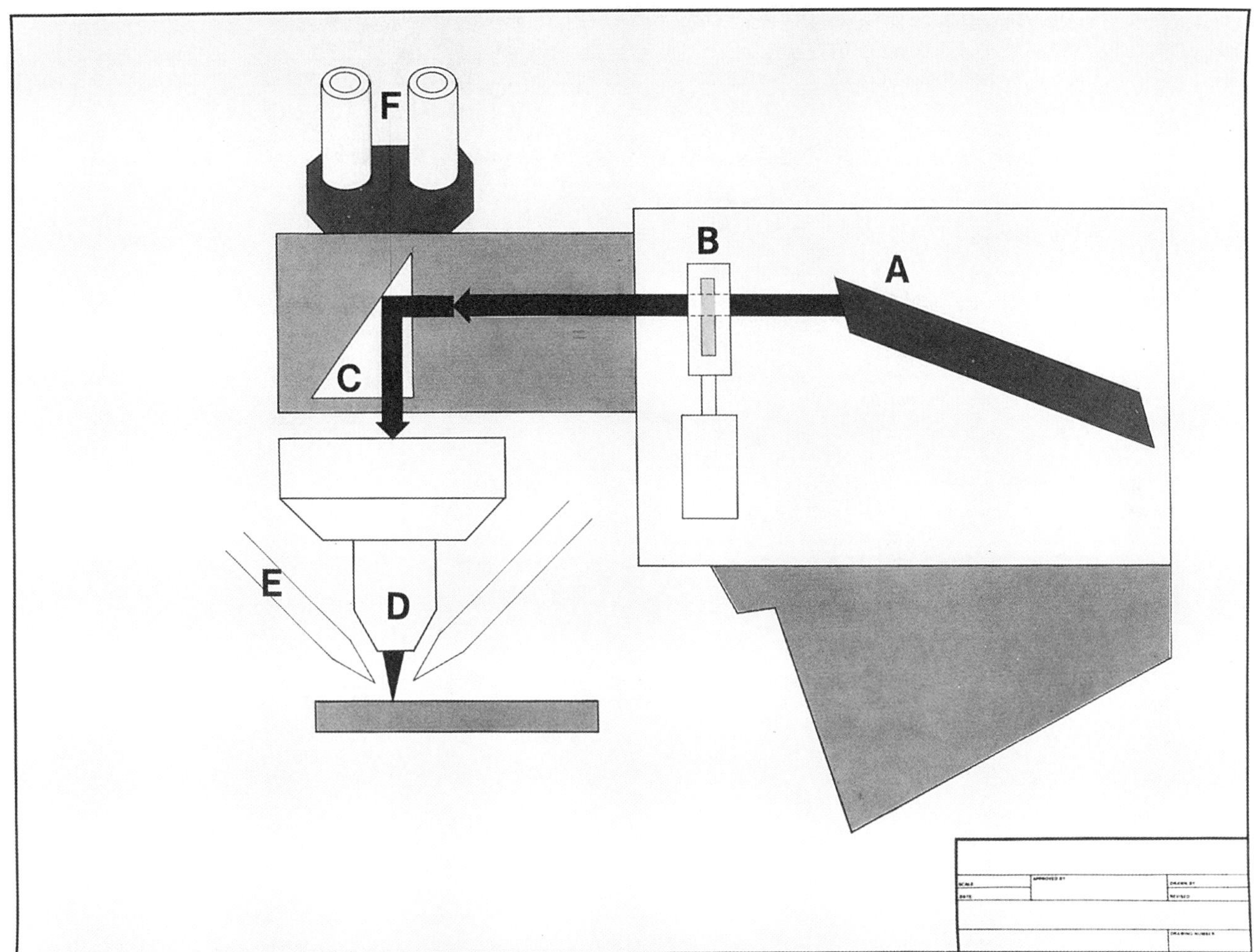

Fig. 1. Representation of laser microprobe elements. (a) flashlamp pumped neodymium glass laser rod, (b) rotating Q-switch, (c) deflecting prism, (d) microscope focusing objective, (e) re-excitation carbon electrodes, and (f) viewing eyepiece.

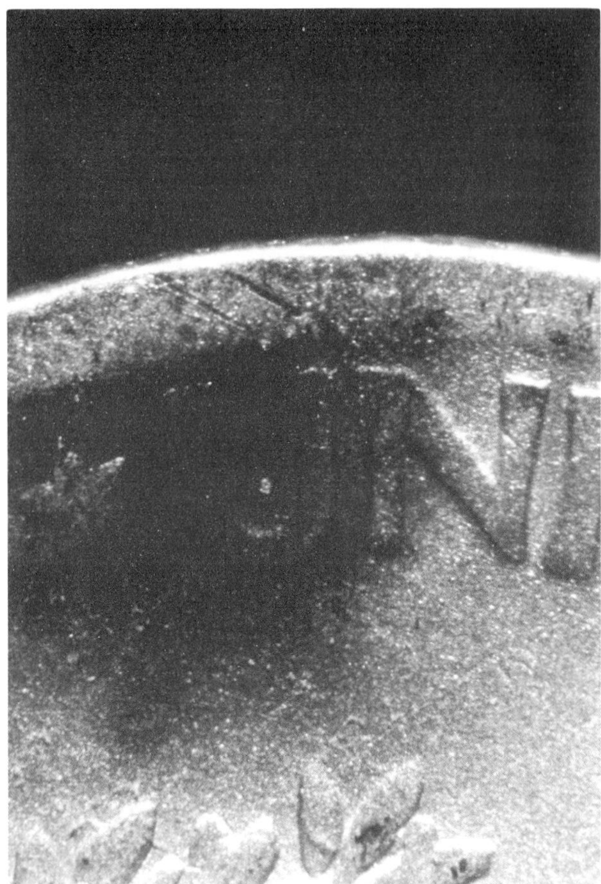

Fig. 2. Laser microprobe sampling of a gold-plated coin (a) without an organic protective coating, and (b) with organic protective coating.

A simple solution to this problem is the application of a thin organic coating (diluted rubber cement is one example) as a protective medium prior to laser microprobe sampling. Figure 2b illustrates corresponding results on the other side of the same coin after application of an organic coating. It is clear that damage and discoloration have been reduced; further the crater is not visible to the unaided eye (inside the letter O). While improvements may provide aesthetically acceptable results, changes in sampling or in spectral intensities must also be taken into account.

Figures 3 and 4 demonstrate typical performance. Shown are two sampling conditions with (fig. 3) and without (fig. 4) incorporation of a thin organic coating on an ultrapure (99.995%) copper surface. *Both* data sets were taken on the modified instrument and thus involve all improvements in electrode geometry, reduced laser ablation energy, critical focusing, and careful matching of the discharge emission to the spectrometer.

The inset microphotographs and corresponding spectral features show surface damage, crater characteristics, and integrated light intensity for both cases. The data result from a series of nine laser sampling events. In each case the laser was refocused at the bottom of the crater created by previous individual sampling events. Spectrum 8, for example, is emission from the eighth laser ablation only. The corresponding laser crater (8) is also shown. Exact location and subsequent refocusing was not possible with the instrument prior to the described mechanical modifications. The magnification for each photograph is approximately 1000X as observed. Note that crater damage is greater for the uncoated case and increases with the number of lases. The two intense lines in the illustrated spectrum are the copper 3247 and 3274Å resonance transitions. The integrated intensity of wavelength-dispersed light, recorded on film in the focal plane of the spectrograph, demonstrates that application of an organic surface coating results in enhanced line intensity and ultimately improved limits of detection. This enhancement may be due to improved arcing conditions brought about by the introduction of additional charged species in the ejected plume. Definitive experiments are underway to study the mechanism of this phenomenon.

An Empirical Model

A major thrust of this work has been to explore the feasibility of using the laser microprobe as a tool for depth-profile analysis of various surfaces. If available, this technology would be useful in such applications as analysis of plated surfaces, and layering of materials in paintings and decorated lacquerware.

The principle model that we have invoked for these studies deals with the effective trajectory of material following laser ablation from the surface. As noted in figure 5, the laser plume resulting from the first laser sampling event is probably hemispherical in nature. After a series of consecutive laser sampling events (into the same hole), a more conical expulsion pattern is expected due to confinement by the walls of the crater. Because of the change in plume density at the electrode position as a function of crater depth, this phenomenon is expected to have a major effect on observed spectral intensities.

This is seen for pure copper, which was chosen as a primary test case. Experimental data points for copper resonant transition emission are shown as a function of crater depth in figure 6. The general shape of intensity versus depth profiles are remarkably similar when going from material to material although the specific details are element and line specific. These shapes may also show a concentration dependency at trace levels. The dotted line corresponds to a third order polynomial fit of the experimental data points:

$$INT = b \pm mX \pm nX^2 \pm pX^3$$

where *INT* is the observed intensity, *b* the intercept, and *m, n, p*, the higher order polynomial coefficients associated with crater depth *X*. With a mathematical description of depth versus intensity information at hand, we view this as a first step to eventual deconvolution of emission intensities with an appropriate function in order to extract direct concentration data in practical experiments.

Figure 7 shows results of depth profiling through the silver-based gold-plated (28 micron) coin illustrated in figure 2. Note the experimental reproducibility for each data point collected (each point is the average of 10 *different* experiments) as well as the similarity of curve shapes for both the silver and copper components. Intensity due to gold emission is highest from the first "shot" and then falls off rapidly since the gold layer is rapidly penetrated. The diameter of the laser-sampled tunnel gets slightly larger as one continues to profile through the surface. As a result, spectral intensities resulting from gold transitions continue to be observed to a depth of about 400 microns. These and similar factors complicate a simplistic conical trajectory model and will have to be configured into our experimental description and model.

A final example (fig. 8) results from depth profiling through a 400 micron thick silver layer plated onto a pure copper surface. Penetration of the silver layer and subsequent detection of the copper layer are quite evident. In figure 8, silver data are plotted as both third- and as fourth-order polynomials. It is clear that the fourth-order polynomial gives an excellent fit to a depth of about 400 microns. As noted, neither polynomial provides a completely acceptable numerical fit at great depths. Research in this laboratory will continue to model surface depth intensities and will ultimately link this model in such a way as to allow one to calculate "corrected intensities" from complex surfaces.

The above observations indicate that we have not given up on the laser microprobe. Indeed, improved mechanical stability and optical considerations have resulted in increased spectral intensities and in more predictable sample-site selection. Incorporation of surface coatings reduces observed damage. Finally, we have completed first steps that encourage us to continue in the quest for improved "quantitative results" from complex layers and alloys.

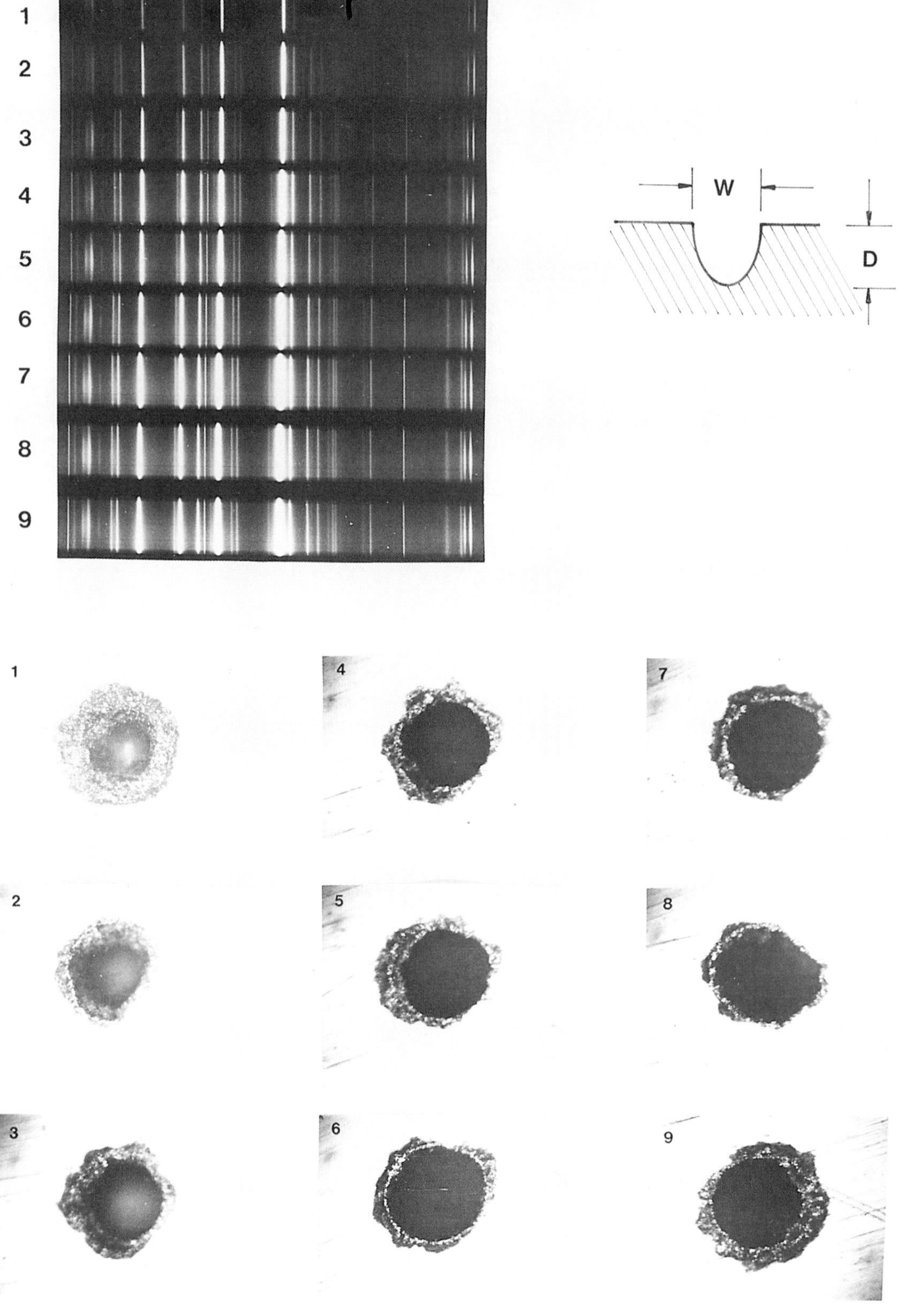

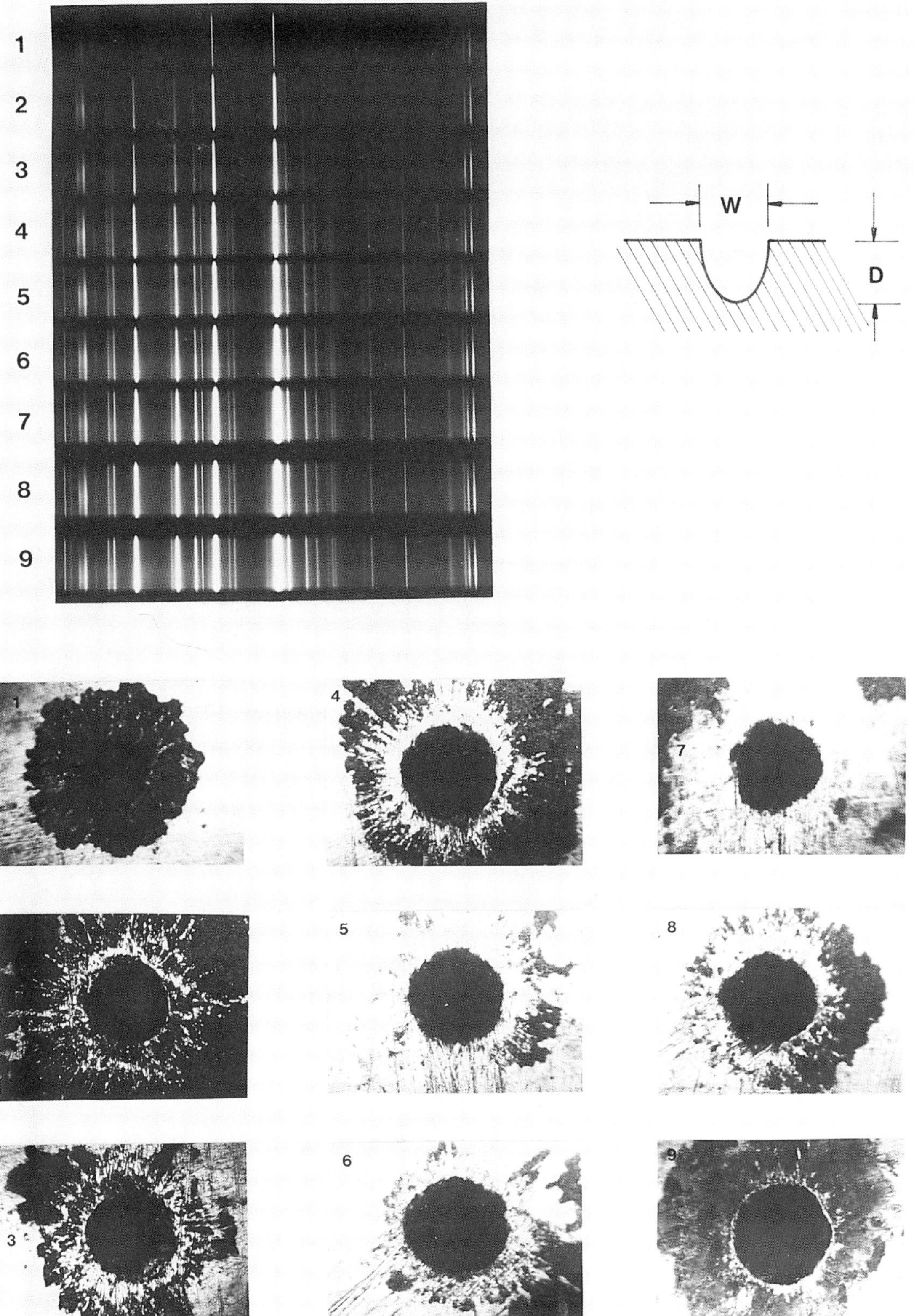
W
D

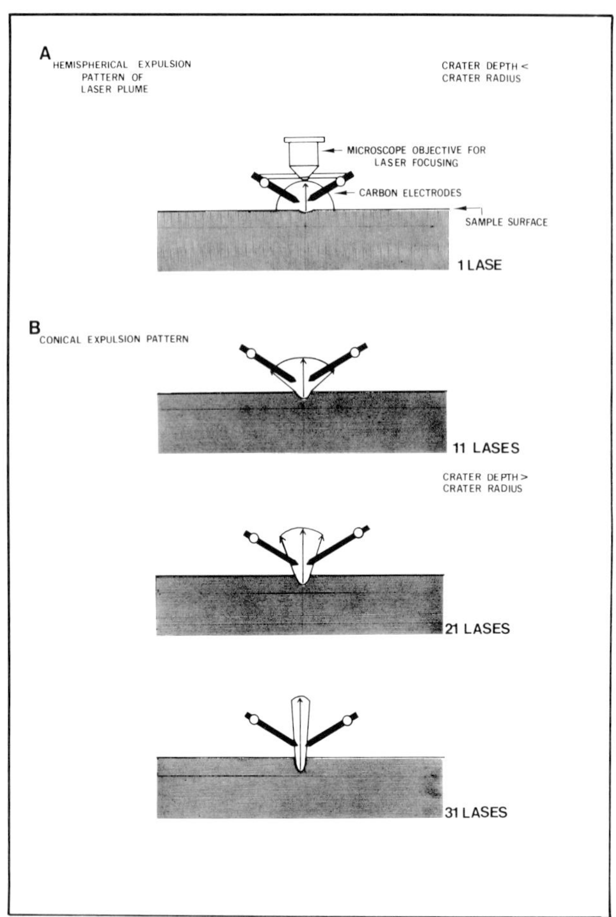

Fig. 5. Model of changing sample trajectories as a function of laser induced crater depth.

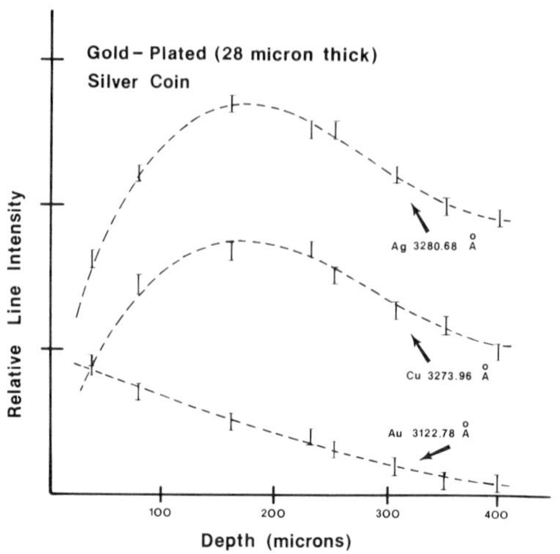

Fig. 7. Depth profiling through the thin gold-plated silver-based coin shown in figure 2.

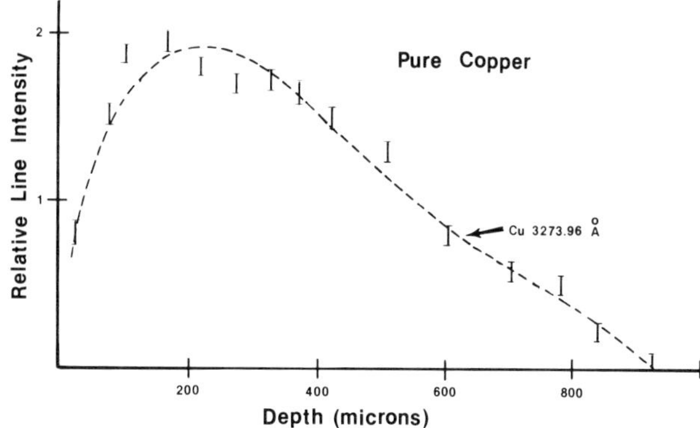

Fig. 6. Relative copper-emission intensity as a function of crater depth. Error bars indicate the range of experimental data for five individual experiments.

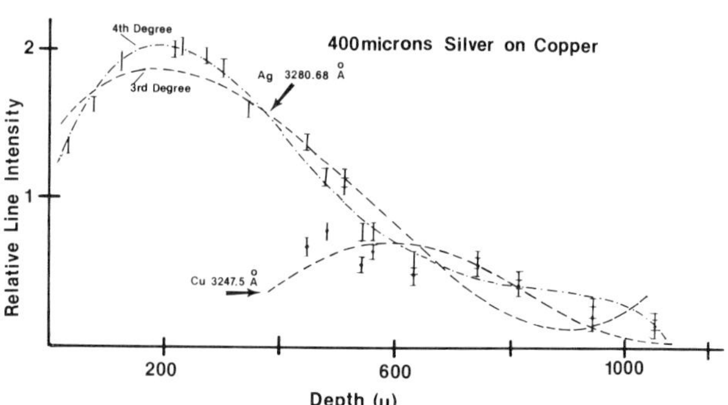

Fig. 8. Depth profiling through a thick (400 micron) silver coating on a pure copper surface.

Acknowledgments

Appreciation is expressed to Neil Lurie, Silvercraft, Inc., Detroit, Mich. for assistance with metal plating. Support for LC-S as a Leonard P. Woodcock visiting scholar is also acknowledged.

References

1. F. Brech and W. Young, "The Laser Microprobe and its Application to the Analysis of Works of Art," *Application of Science in Examination of Works of Art,* (Boston: Museum of Fine Arts, 1966).

2. W. van Deijck, J. Balke, and F. J. M. J. Maessen, *Spectrochimica Acta* 34B, 359 (1979).

YU-TARNG CHENG, JACQUELINE S. OLIN, ROBERT S. CARTER, MARTIN GANOCZY, CHARLES H. OLIN, and IVAN G. SCHRODER

Modification of the National Bureau of Standards Research Reactor for Neutron-Induced Autoradiography of Paintings

Neutron-induced autoradiography now makes it possible to examine paintings in a manner that complements x-ray radiography and often yields unique information that cannot be obtained by other means. After exposure to a field of thermal neutrons ($<$0.3eV), selected elements in the painting form radioactive nuclides that in turn decay at known half-lives with the emission of gamma rays and charged particles. The charged particles, mainly electrons, expose radiographic film that is held in close contact with the painting by a very slight vacuum. Unlike x-ray radiography, which involves detecting x rays that have passed through the painting to expose a film, autoradiography uses one activation with neutrons followed by sequential exposures of the painting to a series of films over a period of about two months to capture the images of activated elements that have different half-lives. The autoradiographs show the distribution patterns of the pigments in which the elements occur. These pigments may be on the surface and visible in the painting or they may be beneath the surface. Often autoradiography uncovers the preliminary sketch or underpainting preceding the rough blocking in of the subject and the subsequent paint layers that create the final image. This allows the study of the creative process from start to completion. X rays reveal almost exclusively the presence of pigments containing heavy elements; autoradiography identifies the presence of activated pigments and distinguishes the fine details of paint application as well as changes in composition.

Pioneering autoradiographic projects include a lengthy study of the works of Ralph Albert Blakelock at Brookhaven National Laboratory (BNL) in collaboration with the Sheldon Memorial Art Gallery and the Hecksher Museum.[1] Paintings for this study were also obtained from the National Museum of American Art, the Krannert Art Museum, and the Metropolitan Museum of Art as well as from private collections. More recently a study was performed at BNL on the works of Rembrandt, van Dyke, and Vermeer from the collections of the Metropolitan Museum of Art.[2]

The Conservation Analytical Laboratory of the Smithsonian Institution recognizes the potential of the technique for painting studies and is considering developing an autoradiography facility. Not all research reactors are suitable for such work, and certain criteria must be met for autoradiography to be successful. These criteria were established through analysis of the results of actual painting studies both at BNL and at the research reactor of the National Bureau of Standards (NBS) in Washington, D.C. The reactor should provide easy access to a relatively uniform thermal neutron field and the thermal neutron flux should be such that the duration of painting activation can be kept to a reasonable time. The short activation duration is important if one is to extract information of short-lived elements such as aluminum. Finally, it is important to minimize the intensity of high-energy neutron and gamma rays to which the painting is exposed. These serve no useful purpose in the painting activation process.

The National Bureau of Standards research reactor (NBSR) satisfies such criteria, including minimizing to 300 rads the dose of high-energy neutron and gamma rays received by the painting. However, at the present time the reactor is not available on a routine basis for autoradiography and the size of the painting that can be studied is limited to about three feet by three feet. Therefore, a design study,

funded by the Smithsonian Institution, has recently been completed to modify the thermal column facility.

The Facility Design Study

Given the aforementioned requirements for a neutron-activation autoradiography facility, a feasibility study was conducted to measure the inherent characteristics of the NBSR thermal column where the facility would be installed and to determine its suitability for modification. A thermal column is a prism of highly purified graphite stretching, in general, from the surface of the core to the edge of the outer shield of the reactor. The energetic neutrons leaking out of the core are strongly moderated but only weakly absorbed in this graphite column, so that a very pure thermal neutron field is created. In contrast to commercial nuclear power reactors, which operate typically in the 3000MW (million watts) range, the NBSR currently operates at 10MW and will be upgraded to 20MW in the near future. The neutrons released by the fuel elements in the reactor core are of relatively high energy ($\sim$ 2MeV) and must be slowed down (moderated) to thermal energies to maintain chain reaction in the reactor. This is accomplished by the heavy water (D_2O) in the reactor main vessel (fig. 1). To achieve as pure a flux as possible in the thermal column, the neutrons pass through an additional tank of D_2O and a six-inch wall of bismuth into the graphite region of the thermal column. The additional D_2O and graphite further moderate almost all of the high-energy neutrons, thus producing a very pure thermal neutron spectrum. The gamma rays from the reactor core are filtered out by the bismuth wall which allows the neutrons to pass through relatively unattenuated. Behind the bismuth wall a movable sheet of boron-loaded material (boral) serves as a neutron shutter. The thermal column is constructed of graphite blocks that comprise a column with a fifty-four inch by fifty-two inch surface area and thirty-seven inch depth. A movable door in the shielding provides access to the thermal column for activation of the painting.

A closely spaced grid of gold foils was used to measure the flux and the uniformity of the thermal neutron field over the entire surface of the thermal column. As expected, the thermal neutron distribution generally followed a cosine shape peaking in the center region and falling off toward the edges varying by a factor of four from the center of the column face to the edges. If the nonuniformity of the field is to be limited to within 50%, considered to be the minimum acceptable level, then the usable size of the field at the face of the reactor is reduced to a three feet by three feet area with a neutron flux of 6 x 10^9n/cm^2.s at the center. Next, an experiment designed to measure how the neutron field fell off at various distances away from the thermal column face, and to measure the radiation shielding requirements, was carried out. A four-inch wall of graphite was attached to the inside of the thermal column door to reflect back into the painting region some otherwise lost thermal neutrons. In addition, a one-half inch lead sheet was placed between the graphite reflector and the thermal column door to shield the painting from the intense gamma rays generated by neutron capture in the boral liner of the door. The graphite reflector enhanced the thermal neutron flux by more than one-third at all positions. Aided by the graphite reflector and the increased distance, the uniformity of the thermal neutron field at four feet from the thermal column face improved to $\pm$ 30% over an area

PLAN VIEW OF NBS REACTOR

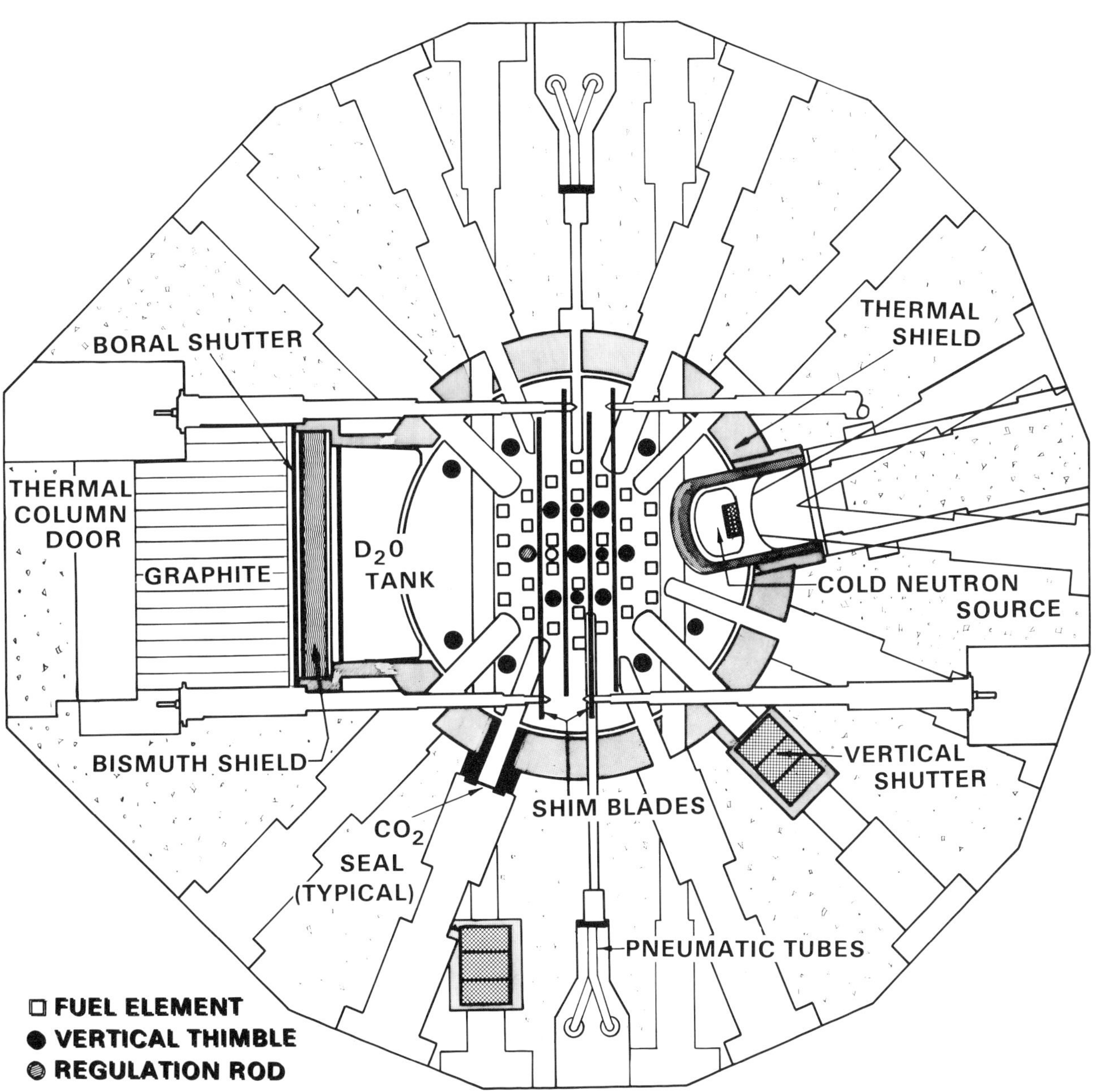

Fig. 1. The National Bureau of Standards Research Reactor.

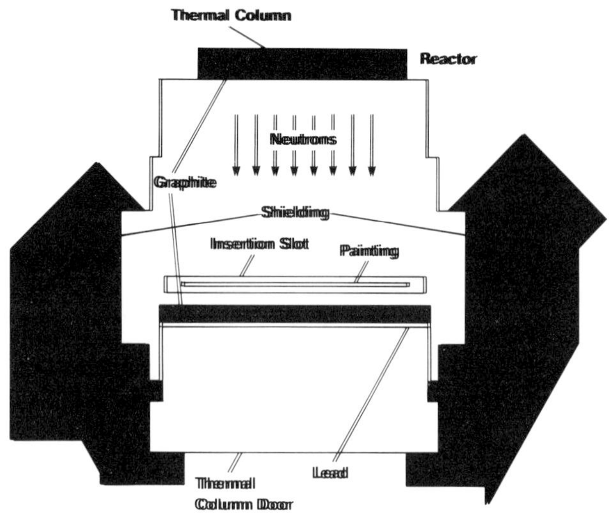

Thermal Neutron Flux:

5×10^9 n/cm^2 s

Size of Painting

Activation Area: 6 ft. × 7 ft.

Gamma Radiation:

1 Rad/2×10^{10} n/cm^2

Fig. 2. Plane view of the proposed painting activation port at the NBSR thermal column.

of six feet by seven feet. At this position, with the reflector in place, the measured flux was 2.5×10^9 n/cm^2.s. When the reactor power is raised to 20MW in the near future, the flux at this position will be 5×10^9 n/cm^2.s, which is adequate for making successful autoradiography studies of painting with just twenty minutes of activation. The activation duration selected is based upon a consideration of the half-lives of such short-lived nuclides as aluminum, which has a half-life of 2.3 minutes and commonly occurs in pigments used by artists. During these measurements, radiation dosimetry was used to monitor the radiation field. One major advantage of the NBSR thermal column is that the neutrons are so well moderated that essentially all high-energy neutrons have been removed, leaving only an intense thermal neutron field and a relatively low gamma-ray component. The absence of fast neutrons simplifies the measurement of the radiation field. Enriched ^{7}Li-based thermal luminescence dosimeters (TLD) were used to monitor the gamma rays. Each TLD was placed in a small individual plastic vial that was shielded from neutrons by one cm of ^{6}Li powder. The ^{6}Li powder reduced the thermal neutron flux entering into the TLD by a factor of 10^6. The plastic vial eliminated any possible contributions to the TLD readings by charged particles generated from neutron capture either in the ^{6}Li powder or other surrounding materials. The dosimetry result indicated the gamma field at the face of the thermal column was 1100 ± 250 rads/H, or about 1 rad per 2×10^{10} n/cm^2. This ratio did not change appreciably with the distance from the thermal column face within the region of interest. So, a painting exposed to a thermal neutron flux of 5×10^9 n/cm^2.s for 20 minutes would receive a gamma-ray dose of about 300 rads. The total energy being imparted into the painting materials

by all types of radiation from the NBS facility is calculated to be at a rate less than one-millionth of a watt. The dose of 300 rads is to be compared with 40 rads for a typical x-ray study of a painting. The actual number of atomic permutations from the neutron activation process in which the painting material has been permanently altered is of the order of a few parts in a trillion, well below the limit of detection.

The experiment described above showed that the NBSR thermal column facility could provide the following at a position four feet from the thermal column face: (1) a relatively uniform thermal neutron field covering an area up to six feet by seven feet; (2) a high thermal neutron flux of 5×10^9 n/cm^2.s at the anticipated reactor power level of 20MW; and (3) a field that is low in gamma radiation and is essentially free of fast neutrons. This provided the necessary input for a detailed design study. The new study was intended to design a painting activation port at the thermal column compatible with other usages for the thermal column; to provide quick, easy, and safe placement and retrieval of a painting in the irradiation position; and do so without undue radiation exposure to the personnel involved. To achieve these goals, the design calls for the thermal column door to be opened and closed on its track by a finely controlled hydraulic cylinder and a painting holder mechanism that provides simple and safe vertical movement of the painting. Figure 2 shows the plane view of the proposed painting activation port at the NBSR thermal column. The procedure for activating the painting at the proposed new facility is illustrated in figure 3. In step one, the painting is placed on the painting holder; in step two, the painting holder mechanism lifts the painting up through a slot in the shielding; in step three the thermal

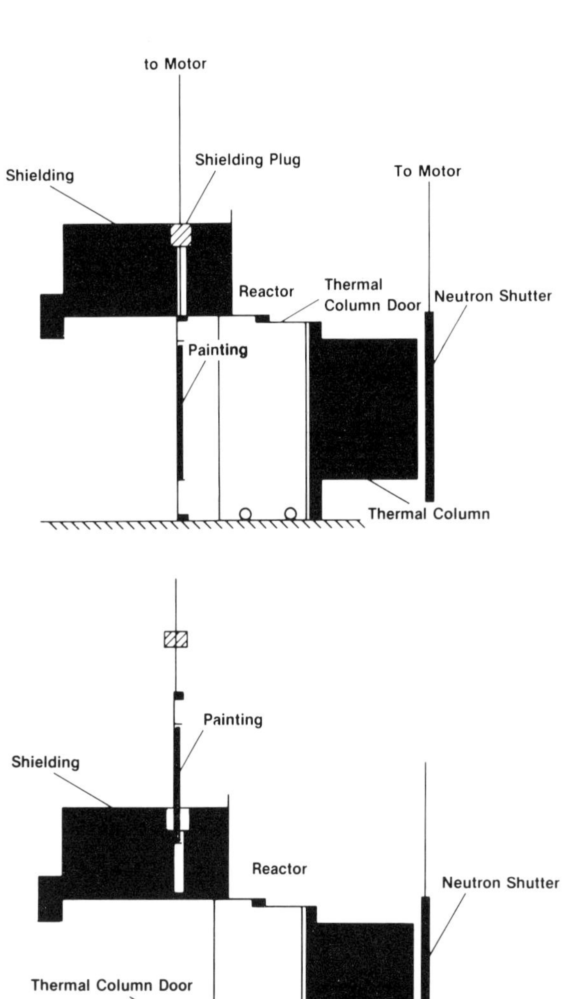

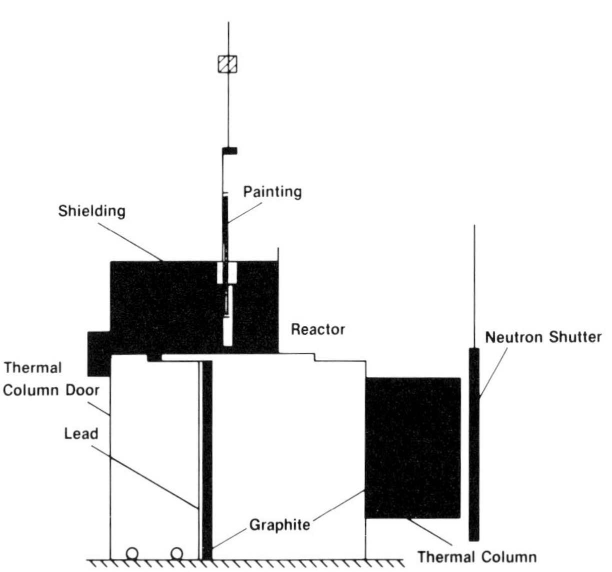

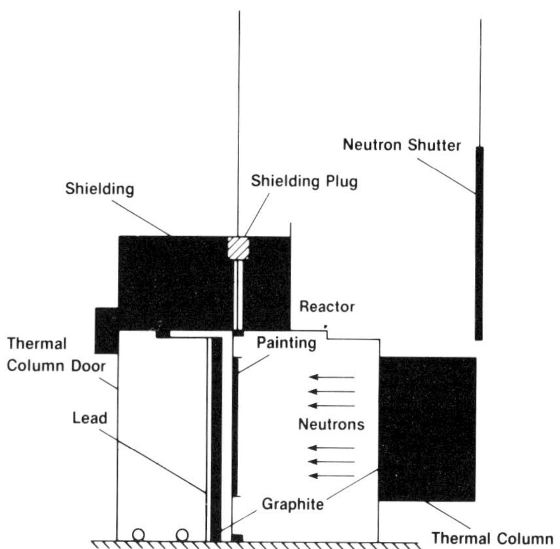

Fig. 3. Four steps of proposed painting activation procedure, counter-clockwise from upper left.

column door is moved into irradiation position; and in step four the painting is lowered, the neutron shutter is raised, and the activation process begins. The retrieval of the painting after the activation process is the reverse of the above procedure. After the activation, the painting would then be taken to a counting room for autoradiography and gamma-ray spectroscopy (fig. 4).

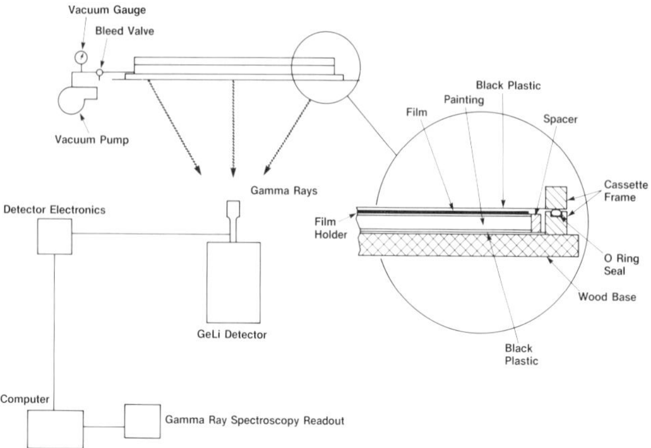

Fig. 4. Painting autoradiography and gamma spectroscopy equipment.

A Sample Painting Study

The NBS reactor has been used to carry out the study of several paintings by bringing the reactor up to power for the period of time necessary to activate the painting, and then immediately taking it down. This has been done by setting up temporary shielding at the thermal column door. As mentioned earlier, the present NBSR thermal column has an area of three feet square at the face of the graphite column that provides a relatively uniform field of thermal neutrons with flux at the $6 \times 10^9 \text{n/cm}^2.\text{s}$ range. A landscape painting signed *Ralph A. Blakelock* was examined by one of the authors (C. H. Olin) who had participated in the earlier autoradiography study of Blakelock's paintings. It was determined that an autoradiography study of the painting would be of interest in view of the comparative data available from the earlier study of the artist's work. The study would also help to establish a standard procedure and to identify specific needs involved in the process of exposing film to the activated painting within a vacuum cassette and obtaining the gamma spectra.

The painting was wrapped in 0.5 mil thick mylar sheet to prevent it from coming into contact with radiographic films and other materials. The painting with its mylar wrapping was then put in a polyethylene bag for further protection from possible contaminations and placed on an easel in front of the graphite column. After twenty minutes of thermal neutron exposure, the painting was removed to a painting counting room located on the lower level in the reactor building. The outer polyethylene bag was then taken off and the painting placed face-up into a light opaque plastic vacuum cassette with radiographic film positioned directly over the painting's mylar wrapping. A very slight vacuum that exerts less than 1/100 of the atmospheric pressure was applied to the cassette assuring close proximity be-

tween the radiographic film (Kodak direct contact film SO-445) and the painting surface. The top and bottom portions of the vacuum cassette, being made of pliable plastic, conform to the shape of the painting inside and exert a uniform pressure on both sides of the painting. While the film was being exposed to the electrons (or positrons) emitted from the activated painting materials, the gamma rays, which were also emitted by the activated painting, were continuously monitored using a Ge(Li) gamma-ray spectrometer. The registration of the electrons on the radiographic film indicates the spatial distributions of these elements while the gamma rays give an indication of the elemental (chemical) composition of the painting.

The exposure of film to the painting followed a set schedule (table 1) to optimize and separate the information of the various activated painting materials. The length of each exposure was determined by considerations of the elements anticipated to be present in pigments and other painting materials detected by concurrent gamma spectroscopy.

The painting is shown in figure 5. The painting (on canvas) was removed from its stretcher prior to the study. Figure 6 shows the x-ray radiograph of the painting; and figures 7 through 9 are respectively the neutron-induced autoradiographs 1, 6, and 9 of the painting. Associated with each autoradiograph is a table listing the most prominent elements responsible for the imaging of that particular autoradiograph. The relative strength of each element is normalized to the most prominent one in the same autoradiograph so one can readily follow the changes in relative strength of different elements from one autoradiograph to another. These autoradiographs demonstrate the unique capability of this technique; it allows one to see the spatial distributions of different elements separately and even to quantify them. Autoradiograph 9 is particularly interesting. Though zinc was the most prominent element that was still active at the time, the fact that images of the landscape do not appear before autoradiograph 9 indicates that it was mercury that was responsible for the image. Zinc has an isotope with a shorter half-life that was very active in the few earlier autoradiographs. Had it been zinc that was imaging the landscape scene in autoradiograph 9, one would have been able to detect it on earlier autoradiographs as well.

Fig. 5. *Landscape,* attributed to Ralph A. Blakelock, private collection, oil on canvas, 15 x 20 inches.

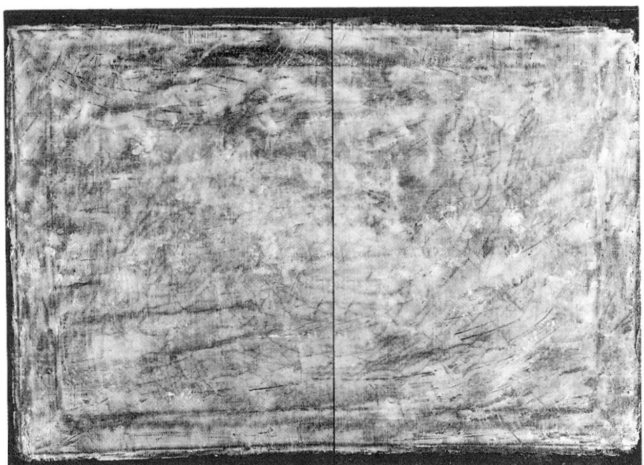

Fig. 6. X-ray radiograph of *Landscape*.

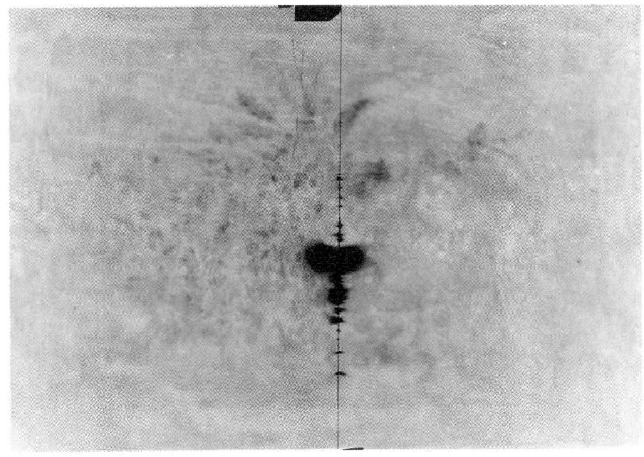

Fig. 7. Autoradiograph 1 of *Landscape*. Aluminum was not measured. The prime elements exposing the film were Mn (relative strength 100), Cu (31), Na (14), and V (4).

Table 1

Time	Event	Optimization of Autoradiographic Effects	
		Isotope	Half-life
$T_0 = 0$	End of neutron activation		
$T_1 = 5$ min.	1st film on for 5 min.	Ag 108	2.3 min.
	1st gamma spectroscopic count	Al 28	2.3 min.
	for 5 min.	Cr 55	3.6 min.
		S 37	5.0 min.
		Cu 66	5.0 min.
		Ti 51	5.8 min.
		Ca 49	8.8 min.
$T_2 = 15$ min.	2nd film on for 15 min.	Mg 27	9.5 min.
	2nd gamma spectroscopic count	Co 60m	10.5 min.
	for 15 min.		
$T_3 = 35$ min.	3rd film on for 35 min.	Sb 124m	21.0 min.
	3rd gamma spectroscopic count	Cl 38	37.3 min.
	for 35 min.		
$T_4 = 75$ min.	4th film on for 75 min.	(no new isotopes)	
	4th gamma spectroscopic count		
	for 75 min.		
$T_5 = 3$ hr.	5th film on for 3 hr.	Ba 139	85 min.
	5th gamma spectroscopic count	Mn 56	2.58 hr.
	for 3 hr.	Pb 209	3.3 hr.
$T_6 = 17$ hr.	6th film on for 17 hr.	K 42	12.5 hr.
	6th gamma spectroscopic count	Cu 64	12.8 hr.
	for 17 hr.	Na 24	15.0 hr.
		As 76	26.7 hr.
$T_7 = 2$ days	7th film on for 1 day	Au 198	2.7 days
	7th gamma spectroscopic count	Hg 197	2.7 days
	for 1 day	Sb 122	2.8 days
		Ca 47	4.7 days
$T_8 = 4$ days	8th film on for 3 days	P 32	14.2 days
	8th gamma spectroscopic count		
	for 2.25 days		
$T_9 = 11$ days	9th film on for 24 days	Hg 203	47 days
	9th film spectroscopic count for	Fe 59	45 days
	5.75 days	S 35	87 days
		Ca 45	153 days
		Zn 65	245 days
		Ag 110m	253 days
		Fe 55	2.6 yrs.
		Co 60	5.26 yrs.

Fig. 8. Autoradiograph 6 of *Landscape.* The prime elements exposing the film were Na (relative strength 100), Zn (65), As (17), and Mn (100).

Conclusions and Acknowledgments

The actual modification has not yet been undertaken; however, through the generous cooperation of the NBS Reactor Radiation Division staff, a limited number of paintings by the American artist, Thomas Wilmer Dewing, have been undertaken for irradiation.

We wish to acknowledge Mr. Conrad Little and express our appreciation for the opportunity to study the painting we have described. We also thank NBS guest worker Dr. M. James Blackman of the Smithsonian Institution Conservation Analytical Laboratory for his participation in carrying out the gamma spectroscopy measurements of the painting, and Ms. Cynthia Cox of the Eastman Kodak Co. for providing technical assistance and supplying the radiographic film used. Mr. Roland Cunningham, painting conservator at the Conservation Analytical Laboratory, assisted in the handling of the painting.

Fig. 9. Autoradiograph 9 of *Landscape*. The prime elements exposing the film were Zn (relative strength 100), Hg (5), and Cr (3).

References

1. M. J. Cotter, P. Meyers, L. Van Zelst, C. H. Olin, and E. V. Sayre, "A Study of the Materials and Techniques Used by Some Nineteenth Century American Oil Painters by Means of Neutron Activation Autoradiography," *Applicazione dei methodi nucleari nel campo delle opere d''arte,* (Rome: 1976).

2. M. W. Ainsworth, J. Brearley, E. Haverkamp-Begemann, P. Meyers, K. Groen, M. J. Cotter, L. Van Zelst, and E. V. Sayre, *Art and Autoradiography: Insights Into the Genesis of Paintings by Rembrandt, van Dyke, and Vermeer,* (New York: Metropolitan Museum of Art, 1982).

L. CONFALONIERI, M. MILAZZO, E. PALTRINIERI, and A. GALLONE

X-Ray Fluorescence, Infrared Reflectography, and Infrared and Ultraviolet Photography of *The Betrothal of the Virgin* by Raphael

Much recent research has been undertaken to mark the fifth centenary of the birth of Raphael. In collaboration with the Soprintendenza ai Beni Artistici e Culturali della Lombardia Occidentale, we have completed a detailed study of Raphael's *Betrothal of the Virgin* in the Brera Gallery, Milan, using the non-destructive techniques of x-ray fluorescence, infrared reflectography and photography, as well as complementary ultraviolet photography. Unfortunately, x-ray radiography was not possible because of a layer of lead white, which had been applied to the back of the panel in an 1856 restoration.

X-Ray Fluorescence Analysis

A portable system was used for the excitation and detection of the fluorescent radiation. A germanium crystal detector, cooled by a copper rod in direct contact with a small dewar of liquid nitrogen, was used. Excitation was obtained from a Am 241 source, after experimentation established that better data were obtained than when a Cd 109 emittor was employed. The results of the XRF analysis appear in the table and the analysed spots are indicated in figure I.

Infrared Examination

The underdrawing of the panel was investigated using black and white infrared photography, as well as infrared reflectography. For the infrared photography, Kodak high-speed IR 4143 estar thick-base film was used with a wratten 87 filter. The infrared reflectrographic work was based upon that of van Asperen de Boer.[1] This tells us that most of the pigments have a maxiumum absorption value for incident light in the wavelength range of 1.2-2 microns. This value has a minimal dependence on the ratio between the reflectances of the black underdrawing and its light background.

We used a Hamamatzu Camera C1000 equipped with vidicon type N 214.06. The sensitivity of this vidicon extended to 2.2 microns. A multilayer interference filter was used to eliminate wavelengths below 1.2 microns. The images were obtained on a twelve-inch black and white monitor (512 lines). Tungsten filament lamps were used for the illumination (Philips Argophoto 500W), controlled with a variac.

Before beginning examination of the panel surface. a series of color strips in egg, linseed oil, and glue medium were painted out on a chalk ground, concealing an identification number. It was found that the use of the filter improved the legibility of the numbers, with the exception of the modern pigment cobalt blue.

Observation of the Underdrawing

Figure 2 shows the hand of the Virgin receiving the ring, in the center of the panel. It is very clear that the ultramarine blue of the Virgin's robe is much more transparent to infrared radiation than is the priest's robe (copper green and ochre).

The hands of the girl in front offer one of the best examples of underdrawing in this study (fig. 3). In the fingers of her left hand can be seen the artist's preliminary sketching and pentiments.

The transparency of red lakes in infrared is well illustrated in figure 4, which shows the Virgin. This is in marked contrast to the areas of flesh, where lead white is the predominant pigment.

Figure 5, the face of the girl behind the Virgin, shows a change in the position of her eyes. Figure 6 illustrates the improvement in the infrared image, obtained when the filter is employed. The subject is Joseph's head.

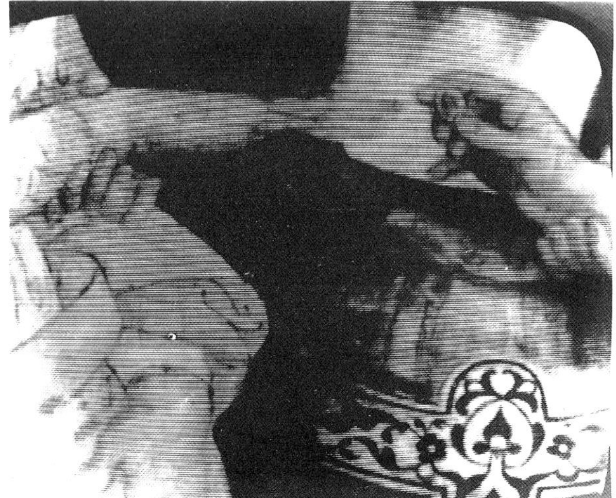

Fig. 2. Detail of infrared reflectography

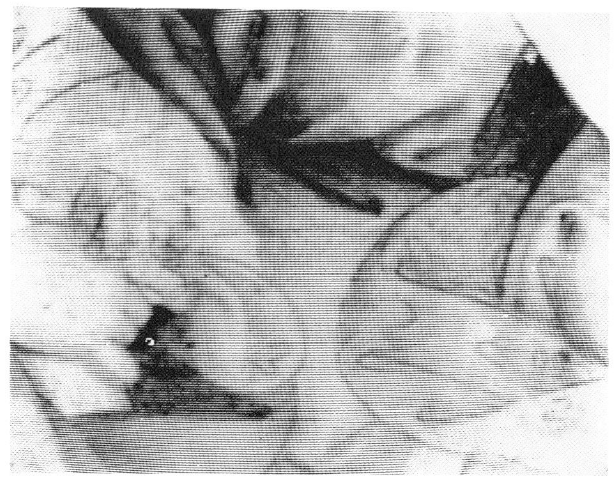

Fig. 3. Detail of infrared reflectography

Fig. 1. Allocation of seventy spots of XRF analysis of about 8mm in diameter
sited in different colors.

Fig. 4. Detail of infrared reflectography

Fig. 5. Detail of infrared reflectography

Fig. 6. Saint Joseph's head. Infrared reflectography without filter on the left.

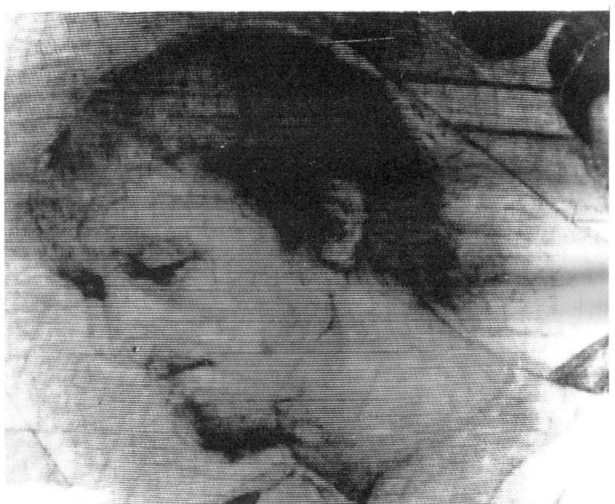

Table 1

Analyzed Zones: Reds	Identified Chemical Elements					Corresponding Pigments	Notes
Virgin's dress							
26 — bottom (light)	(Ca)	(Mn)	(Fe)	(Hg)	Pb	red lake (vermilion) (umber), lead white	
27 — bottom (shade)			(Fe)	Hg	Pb	red lake, vermilion, lead white	(1) (12)
28 — top (light)	(Ca)			(Hg)	Pb	red lake, (vermilion) lead white	
29 — cuff	(Ca)	(Mn)	(Fe)	(Hg)	Pb	idem	
first woman from left							
30 — dress	(Ca)	(Mn)		Hg +	Pb	vermilion, red lake (umber) lead white	(2) (12)
31 — dress	(Ca)	(Mn)	(Fe)	Hg +	Pb	idem	(2) (12)
third woman from left							
32 — arm		(Mn)		Hg	Pb	idem	(2) (12)
priest							
33 — chest	(Ca)		(Fe)	Hg +	Pb	vermilion, red lake (ochre), lead white	(2) (12)
man behind St. Joseph							
34 — shoulder			(Fe)	(Hg)	Pb	red lake, (vermilion), lead white	(12)
boy breaking the rod							
35 — right leg				Hg +	Pb	vermilion, lead white	
36 — left leg		(Mn)	(Fe)	Hg	Pb	vermilion, (ochre), (umber), lead white	
fourth woman from left							
37 — veil			(Fe)	Hg	Pb	vermilion, (ochre), lead white	
man at right rear							
65 — robe	(Ca)	(Mn)		Hg	Pb	vermilion, (umber) lead white	(4)
man at left rear							
66 — robe				Hg	Pb	vermilion, lead white	

Analyzed Zones: Yellows	Identified Chemical Elements						Corresponding Pigments	Notes
floor								
1 — bottom (left)(light)	(Ca)		Fe			Pb	ochre, lead white	(3)
2 — near St. Joseph's foot (shade)	(Ca)		Fe		Hg	Pb	ochre, vermilion, lead white	
3 — priest's foot (left)	(Ca)		Fe		Hg	Pb	ochre, vermilion, lead white	(5)
4 — boy breaking the rod (foot)	(Ca)	(Mn)	Fe			Pb	ochre, (umber), lead white	
5 — third woman from left (dress)	(Ca)		Fe			Pb	ochre, lead white	
St. Joseph's robe								
7 — shade	(Ca)	(Mn)	Fe		Hg	Pb	ochre, vermilion (umber) lead white	(3)
8 — light		(Mn)	Fe		(Hg)	Pb	idem	
stair								
59 — right (light)	(Ca)	(Mn)	Fe			Pb	ochre, (umber), lead white	
60 — left (shade)	(Ca)	(Mn)	Fe			Pb	idem	
62 — floor (yellow stripe)	(Ca)		Fe		(Hg)	Pb	ochre (vermilion), lead white	(5)
63 — beggar (near stairs)		Mn	Fe		(Hg)	Pb	ochre, umber, vermilion lead white	(5)
64 — man (rear left with red hat)	(Ca)	(Mn)	Fe	(Co)	(Hg)	Pb	ochre, (vermilion), (umber), lead white	(5) (9)

Analyzed Zones: Greens	Identified Chemical Elements					Corresponding Pigments	Notes
third woman at right (robe)							
19 — light	(Ca)		(Fe)	Cu –		Pb	a copper green. (ochre) lead white	
20 — shade	(Ca)	(Mn)	(Fe)	Cu –		Pb	a copper green. (umber) lead white	
priest's sleeve							
21 — light	(Ca)		(Fe)	Cu +		Pb	a copper green. (ochre) lead white	(7)
22 — shade			(Fe)	Cu +		Pb	idem	(7)
man behind St. Joseph							
23 — (hat)	(Ca)	(Mn)	(Fe)	Cu +		Pb	a copper green (ochre). (umber). lead white	(7)
boy breaking the rod (shoulder)							
24 — light	Ca	(Mn)	(Fe)	Cu		Pb	idem	
25 — shade		(Mn)	(Fe)	Cu +		Pb	idem	
grass field							
54 — at left of Temple	Ca	(Mn)	Fe	Cu		Pb	a copper green, ochre, (umber), lead white	(11)
55 — at right of Temple		(Mn)	Fe	Cu		Pb	idem	(11)
trees							
56 — at right of Temple	Ca	(Mn)	Fe	Cu		Pb	idem	(11)
57 — at left of Temple	(Ca)	(Mn)	Fe	Cu +		Pb	idem	(11)
58 — men (at rear left)				Cu +		Pb	a copper green, lead white	(8) (11)
69 — dome	(Ca)	Mn	Fe	Cu +		Pb	a copper green, ochre, umber, lead white	

Analyzed Zones: Blue	Identified Chemical Elements					Corresponding Pigments	Notes
Virgin's robe							
9 — fold (light)	(Ca)					Pb	natural ultramarine, lead white	(3)(13)
10 — fold (shade)	(Ca)	(Mn)	(Fe)		(Hg)	Pb	natural ultramarine, (umber) (ochre) (vermilion) lead white	
11 — knee (light)	(Ca)		(Fe)			Pb	natural ultramarine, (ochre), lead white	
boy breaking the rod							
12 — hat (light)	Ca	(Mn)	Fe	Cu		Pb	azurite, ochre, (umber), lead white	
13 — hat (shade)	Ca	(Mn)	Fe	Cu		Pb	idem	
second man from right							
14 — hat (left tip)	(Ca)	(Mn)	(Fe)			Pb	natural ultramarine, (ochre), (umber), lead white	(13)
15 — hat (central tip)	(Ca)	(Mn)	(Fe)			Pb	idem	
woman behind the Virgin							
16 — hat	(Ca)	(Mn)				Pb	natural ultramarine (umber), lead white	(13)
second woman from left							
17 — veil	(Ca)	(Mn)	(Fe)	Cu		Pb	azurite, (umber) (ochre) lead white	
St. Joseph							
18 — sleeve	(Ca)		(Fe)(Co)	Cu +	(Hg)	Pb	azurite, (ochre), (vermilion), lead white	
light zone of the mountain							
52 — at left of Temple	(Ca)	(Mn)	(Fe)	(Cu)		Pb	azurite, (umber), (ochre), lead white	
Sky							
53 — at left of Temple		(Mn)	(Fe)	(Cu)		Pb	idem	

Analyzed Zones: Browns	Identified Chemical Elements						Corresponding Pigments	Notes
	Ca	Mn	Fe	Cu	Hg	Pb		
priest								
40 — beard	(Ca)	(Mn)	Fe	(Cu)	(Hg)	Pb	ochre, (vermilion), umber, (a copper green), lead white	(5) (10)
St. Joseph								
41 — hair	(Ca)		Fe	Cu	(Hg)	Pb	ochre, (vermilion), a copper green), lead white	(5) (10)
priest's robe								
42 — light	Ca	(Mn)	(Fe)(Co)	Cu		Pb	a copper green, (ochre), (umber), lead white	(9)
43 — shade	Ca	(Mn)	(Fe)	Cu		Pb	idem	
second man from right								
44 — robe (shade)	(Ca)		Fe	Cu	(Hg)	Pb	azurite, ochre (vermilion), lead white	(3)
45 — robe (light)	Ca			Cu		Pb	azurite, lead white	
46 — tunic	Ca	(Mn)	(Fe)(Co)	Cu		Pb	probable copper green, (ochre), (umber), lead white	(9)
third man from right								
47 — hat	Ca	(Mn)	Fe(Co)	Cu	Hg	Pb	probable azurite, (umber) ochre, vermilion, lead white	(5)
48 — tunic	(Ca)		(Fe)	Cu		Pb	(ochre), unidentified copper pigment, lead white	
Priest								
49 — hat	Ca	(Mn)	(Fe)	Cu +	Hg	Pb	azurite, (umber), vermilion,(ochre) lead white	(5)

Analyzed Zones: Other Colors	Identified Chemical Elements						Corresponding Pigments	Notes
	Ca	Mn	Fe	Cu	Hg	Pb		
Black								
67 — man at left under porch				Cu		Pb	red lake, unidentified copper pigment, lead white	(12)
flesh								
38 — woman behind the Virgin	(Ca)	(Mn)			(Hg)	Pb	(vermilion), (umber), lead white	(10)
39 — Virgin's neck	(Ca)	(Mn)		(Cu)	(Hg)	Pb	(vermilion), ochre, (umber) (a copper green), lead white	(10)
lilac								
50 — third woman from left	(Ca)		(Fe)	Cu	(Hg)	Pb	azurite, red lake, (vermilion), (ochre), lead white	(5)(12)
51 — dress of woman behind the Virgin	Ca	(Mn)	(Fe)	Cu	Hg	Pb	vermilion, azurite, (ochre) (umber), lead white	(5)
White								
68 — robe of woman behind the Virgin	(Ca)	(Mn)	(Fe)	(Cu)		Pb	(a copper green), (ochre), (umber), lead white	(10)
61 — floor (top at left)						Pb	lead white	(10)

Notes:

1) The (+) symbol indicates a particularly abundant quantity of the element.

2) Elements in trace are indicated within brackets.

Notes

1. A small quantity of vermilion was detected here, the amount in the shaded areas being about ten times greater than in the highlights. This may be from a sketch of the Virgin's figure executed with light touches of vermilion directly onto the ground, before the application of the red lake.

2. In the shadows, red glaze is laid over vermilion.

3. In a number of shadowed areas the presence of mercury was frequently detected while this element is completely absent in the corresponding highlighted zones. This could possibly be attributed to the artist's peculiar use of vermilion in the execution of shades.

4. The outer portion of the robe consists of vermilion, the lining of red lake.

5. Vermilion is used in greater or smaller amounts together with other pigments (ochre, etc.) to obtain particular color tonalities.

6. In the center, azurite appears in a thin layer over a brown base.

7. The translucent aspect of the green layer, here also applied on a brown base, is typical of copper resinate.

8. The dress folds are achieved using azurite.

9. The traces of cobalt which appear at these points may be attributed to repaints, revealed at the same spots by UV examination.

10. A copper green is present, either mixed with other pigments (ochre, vermilion, lead white) or as a glaze in the shadows.

11. Malachite and copper green are undistinguishable using XRF.

12. Red lake is an organic pigment and thus is not detected with XRF. It consists, as is well known, of a transparent pigment on an inorganic substance and can be identified by its microscopic characteristics.

13. Natural ultramarine (lapis lazuli) mainly aluminium, calcium and sodium silicates) is not identifiable with certainty with the type of radioisotope which has been utilized. It has been identified indirectly on the basis of tradition and the lack of copper in the blue robe of the Virgin. Other blue zones in which copper is not detected may also consist, although quite improbably, of indigo — an organic pigment.

References

1. J. R. J. van Asperen de Boer, "Reflectography of Paintings Using an Infrared Vidicon Television System," *Studies in Conservation*, 14 (1969): 96-118.

ELIZABETH A. COUGHLIN, MICHAEL N. GESELOWITZ, and PHILIP M. RURY

Examination of a Spear from Late Iron Age Slovenia

The Peabody Museum of Archaeology and Ethnology at Harvard University contains the Mecklenburg Collection, which is the largest single collection of European Iron Age material in the world. This material was excavated at the turn of the century from what was then the Duchy of Carniola. The vast majority of the collection comes from an area that is today part of Slovenia, one of the six republics of Yugoslavia.

Slovenia in the Early Iron Age, ca. 750-400 B.C., was characterized by large population centers almost unique in Europe for that period of time. These sites are presumed to have been involved in an iron industry as well as other forms of commerce and craft production. The Mecklenburg material is mostly from grave groups associated with Stična and Magdalenska gora, the two best known of the Iron Age Slovene centers. Therefore, this material suggests itself as appropriate for an anlaysis that could address both the specific issues of the origin, dispersion, development, and organization of the iron industry in prehistoric southeastern Europe, and also the general archaeological issues concerning the interaction of technology and society.

The spearhead from Magdalenska gora Tumulus IV, Grave 25 (fig. 1) is in many ways representative of the numerous fine iron objects in the Mecklenburg Collection. Of the more than 300 objects of iron from Magdalenska gora in the collection, 28% are spearheads of varying sizes and forms. This particular spear was described by the late Hugh Hencken as possessing a "sharply angular blade."[1] The surface is relatively unweathered, and the haft contained a well-preserved segment of the original wooden shaft. When this wood was selected for botanical and radiocarbon analyses, an opportunity arose to take a sample of the iron for metallographic study. The metal was found to be highly corroded where it had been in contact with the wood, and the wood was metallicized where it had been in contact with the metal. The dimensions of the spear are indicated in figure 1, as is the location of the sample. The original weight of the spear was 120.3g. The sample weighed less than 0.1g.

The specific gravity of this sample was found to be 6, indicating a state of corrosion, as pure iron would be greater than 7, and steel higher still. Metallographic analysis showed the unweathered metal to be completely ferritic, and to contain several long, stringlike slag inclusions running parallel to the long axis of the spear (fig. 2). The grains themselves demonstrate no evidence of mechanical deformation. Hardness testing showed the sample to lie within the range for unworked pure iron.

Chemical analysis by means of electron microprobe confirmed these observations and further indicated that the metal was not carburized at all. It is pure iron, with only traces of phosphorus and nickel (fig. 3). The corrosion product (fig. 4) contains a number of impurities probably derived from the nature of the soil and the wood, both of which would have been active in the weathering process. The slag inclusions (fig. 5) are basically fayalitic, containing also calcium, phosphorus, magnesium, manganese, and a tiny amount of titanium. In the future it may be possible to identify at least the ore types, and perhaps the ore sources, by considering the compositions of the metal and the slag inclusions together.

The conclusion of the metallographic analysis is that the haft of this spearhead was hot-forged from pure bloomery iron, that no attempt was made to carburize this sample, to affect its structure by subsequent reheating or quenching, or to affect it by cold-working.[2] It must be recognized that the smith may have treated the blade separately from the haft. Without analysis of the edge, a final statement cannot be made about the level of technology that went into the production of this spear. The treatment of the haft was disappointingly simple. This is especially so since the grave in which this spear occurred was dated by Hencken to late in the sequence, either Certosa or Negau,[3] which would place it around 500-300 B.C.[4] In fact, within this span, spears of this sort were more numerous towards the end, in Negau 2.[5]

One fact can be ascertained about the construction of the spear from the orientation of the slag inclusions. The haft must have been first formed as a bar, then flattened along the long axis of the spear. This flattened piece would have been rolled to form the haft.

Numerous small iron concretions have been found in the interior of the wood and could perhaps support the idea that the haft was formed directly around the wooden shaft (fig. 6). Although these formations were first noticed as single nodules, larger, more amorphous, and more extensive occurrences became immediately evident (fig. 7).

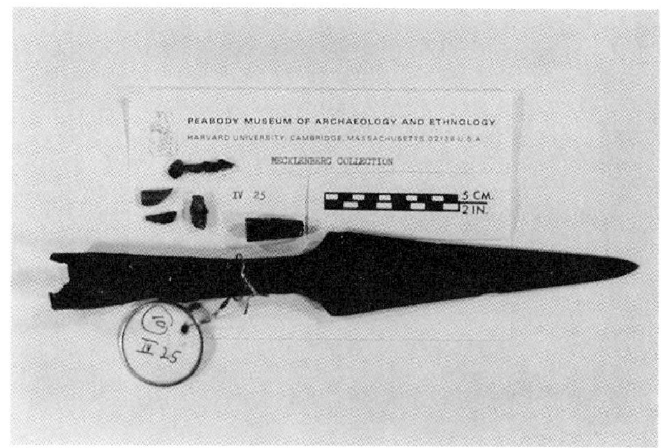

Fig. 1. Spearhead from Magdalenska gora.

Electron microprobe analysis showed the composition of these concretions to be 99% pure iron with the remaining 1% consisting of calcium, silicon, and sulfur (fig. 8). The extremely low percentage of these minor constituents of exchange and corrosion (silicon and calcium also occur in the slag) seems to indicate that these nodules were formed at the time of or closely following the actual making of the spear. However, since the spearhead was apparently hot-forged, there are many unanswered questions concerning whether the wood could have been inserted at that time without charring. It is more probable that the iron particles are indeed the result of chemical deposition as a consequence of the environment of burial throughout the yearly cycle of Slovenia, which includes wet springs and hot, dry summers.

Fig. 2. Metallographic section of the spearhead showing stringlike inclusions of slag.

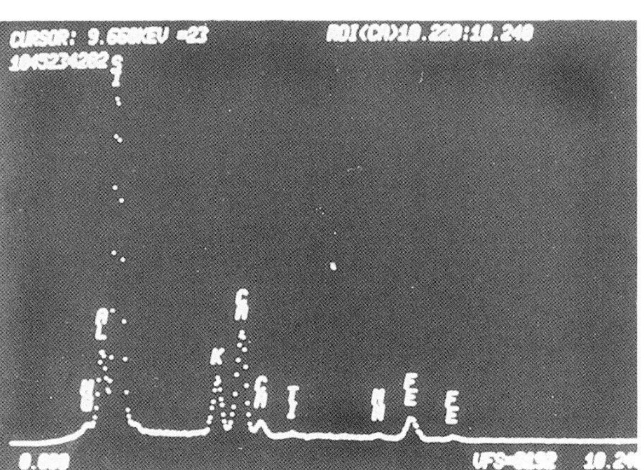

Fig. 5. Analysis of the slag inclusions of the spearhead.

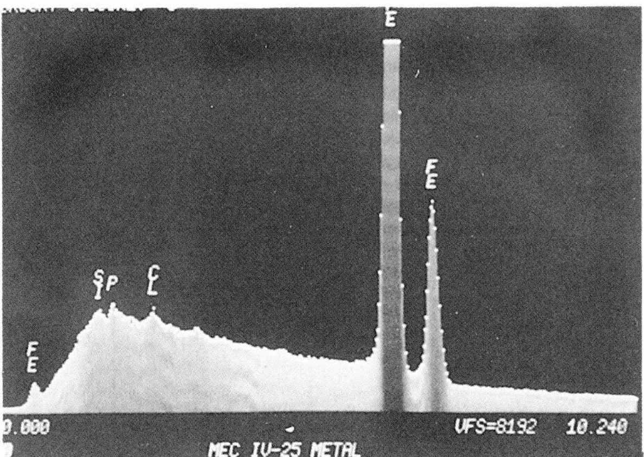

Fig. 3. Analysis of the metal of the spearhead.

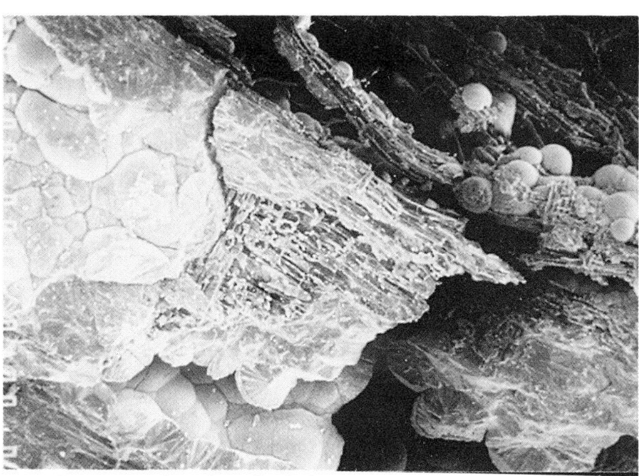

Fig. 6. Iron concretions, found in the interior of the wood.

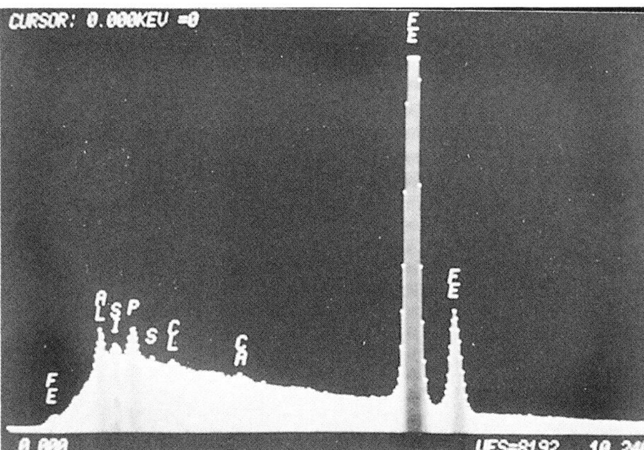

Fig. 4. Analysis of the corrosion layer of the spearhead.

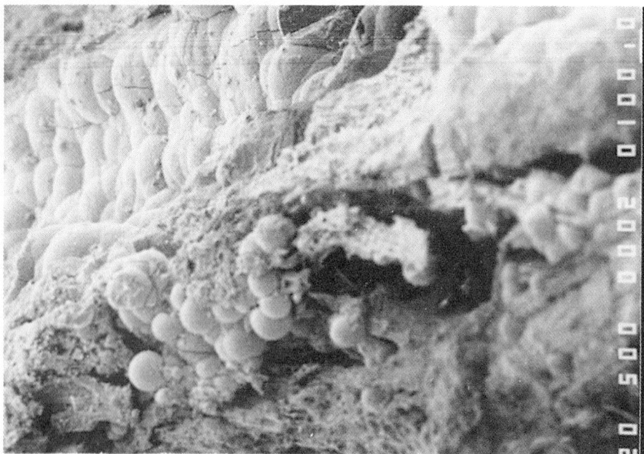

Fig. 7. Amorphous concretions in the wood.

Fig. 9. Internal structure of the concretions.

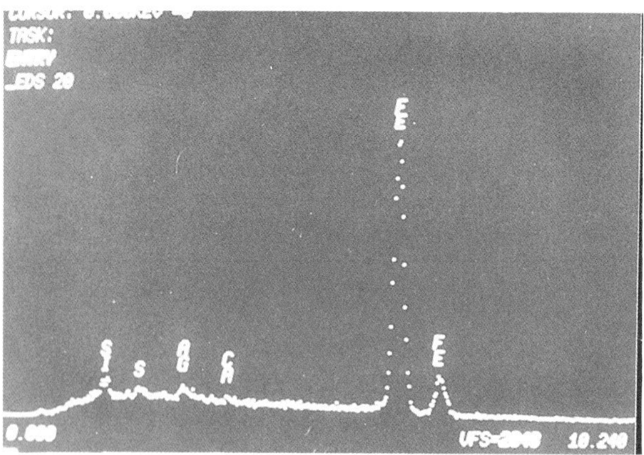

Fig. 8. Analysis of the concretions found in the wood.

These iron concretions, approximately 200 microns in diameter are, in fact, composed of goethite ($HFeO_2$) (limonite) routinely found as a result of the weathering of iron minerals under oxidizing conditions at ordinary temperatures and also as a direct precipitate in standing waters. The orthorhombic crystal structure, the stalactite mass, and the lustrous black surface characteristic of goethite formed by weathering are *not* found. Here it occurs as an amorphous array displaying a fibrous internal structure (fig. 9), with the fibers perpendicular to the surface, characteristic of a mineral gel precipitated from a colloidal suspension, indicating that these formations have occurred due to environmental conditions.

The fine outer surface (fig. 10) seems to indicate a completely undisturbed condition of formation or precipitations. Whether the source of the colloidal iron subsequently precipitated to form these nodules was the spear itself, or iron deposited in the shaft at the time of spear making, or iron environmentally available from groundwater and soil is not yet determined.

The wood comprising the shaft is extremely hard and fine-grained (fig. 11), with a specific gravity of 1.3, indicating a high-density wood. Portions of the wooden shaft in proximity to the basal end of the spear point's haft were metallicized and could be lifted with a magnet. However the portions of the shaft enclosed within the spear point were not metallicized and could be sectioned in a microtome.

The pith ran the length of the shaft and was surrounded by three to four growth rings of secondary xylem or wood. Hence, the shaft was from a trunk or branch that was at least three or four years old at the time of harvest. The anatomical diagnostic feature of scalariform

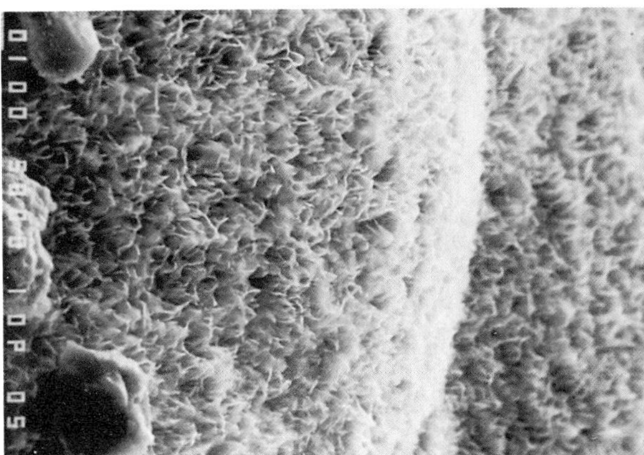

Fig. 10. Outer surface of concretion.

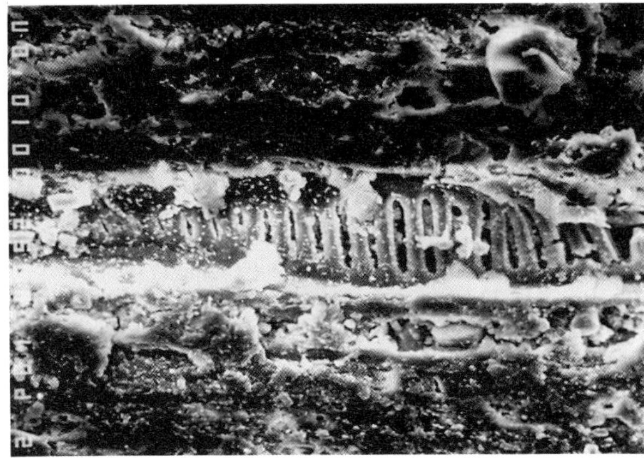

Fig. 12. Scalariform vessel perforation, diagnostic of *Buxus* or *Vibernum* genera.

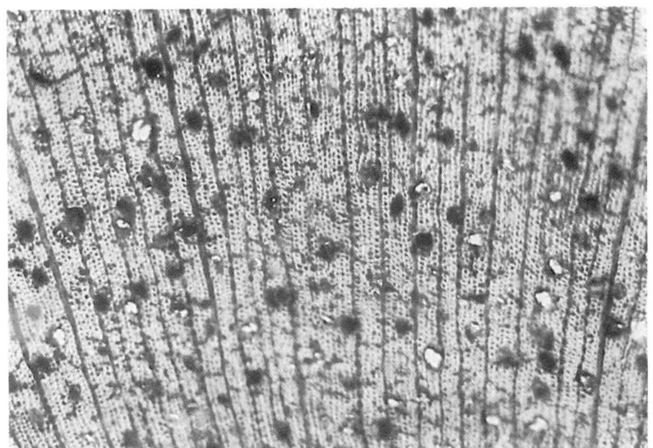

Fig. 11. Transverse section of the wood.

vessel perforation (fig. 12) present in this Slovene spear shaft as well as the physical properties of great strength and durability, are shared by the genera *Buxus* and *Viburnum* and both of these should be considered as equally probable identities.

Continued study of the Mecklenburg Collection holds the promise of addressing a number of important archaeological and archaeometric questions. This initial study has provided both answers and questions concerning not only the materials but also the technology of their manufacture.

References

1. Hugh Hencken, *The Iron Age Cemetery of Magdalenska gora in Slovenia*, American School of Prehistoric Research Bulletin 32, Mecklenburg Collection, Part II, Peabody Museum of Archaeology and Ethnology, (Cambridge: Harvard University, 1978) p.21.

2. For one discussion of these processes, see Radomir Pleiner, "Die Herstellungstechnologie der Germanischen Eisenwerkzeugeund Waffen aus den Brandgräberfeldern der Südwestslowaskei," *Slovenska Archeologia*, XXX-1 (1982): 79-101.

3. Hencken, *Iron Age*, p. 21.

4. Hencken, *Iron Age*, p. 11.

5. Hencken, *Iron Age*, p. 22.

JOHN E. DAYTON

Bronze Age Europe and the Discovery of Glass

It is only in the last twenty years that science has begun to have an impact on archaeological thnking. The first scientific breakthrough was the development of radiocarbon dating, which showed that the Bronze Age in Europe predated that of the Fertile Crescent and that tin bronzes were being produced ca. 2000 B.C. in areas where tin and copper occur—Sardinia, Spain, and Bohemia.[1]

Carbon-14 dating has revolutionized our picture of Central and Western Europe during the third millenium B.C. Spread over this large area from Denmark and Britain to southern Spain, and eastwards to the Elbe, northern Italy, Sardinia, Tuscany, and Sicily are the "Megalith Builders" who are peacefully associated with a group of people known as the "Beaker Folk." These twin cultures are densely located in the metal-rich areas; in fact no tin or copper deposit escaped their attention. Further, they were seafaring peoples whose works extend along the entire coastal area of Western Europe. How the Beaker Folk evolved into the later bronze workers of the Agaric Culture of Spain, and the Nuraghic Culture of Sardinia, both ca. 1800 B.C., is not yet clear, but their typical constructions of massive cyclopean masonry (in Sardinia there are some 7000 stone towers, or nuraghi, all associated with metal working) extend to Mycenaean Greece, Crete, and the Levant at Ras Shamra.

The second recent analytical tool is the use of lead isotope dating to fingerprint an ore deposit, and the artifacts derived from it; i.e. lead in bronzes, silvers, and glasses.[2-5] Some of these results, plotted in figure 1, are as follows:

1. A bronze spearhead of Amenophis III, ca. 1400 B.C. was made from Sardinian ore.
2. Blue copper frit from Fara (near Gaza) of the same date, was also made from Sardinian ore.
3. The ox-hide ingots of Haghia Triada, Crete, ca. 1600 B.C., come from Sardinia.
4. Blue frit from Mycenae, dated ca. 1400 B.C., is derived from Cyprus copper and indicates that the ores were being exploited by the Mycenaeans.
5. Silver from Tel Fara (near Gaza), ca. 1400 B.C., comes from Kremnica in Czechoslovakia.
6. An M.B.2 bronze notched axehead from Megiddo Tomb 911 comes from Tuscany.[3]
7. Yellow glass from Nimrud ca. 750 B.C. is from Bleiberg in Carinthia.
8. The lead block of Tukulti-Ninurta II from Ashur comes from Laurion.
9. Twelfth Dynasty kohls are from Uganda.
10. A cobalt blue bead from St. Veit has lead in it from Kremnica in Czechoslovakia.
11. A silver ingot from Khafaje and silver from the Death Pit at Ur are from Almeria in Spain.
12. A silver axe from Ur, PG 1422, silver from Shaft Grave IV at Mycenae, an Amuq figurine and a Jemdet Nasr lead cup from Ur are all from Kremnica, Czechoslovakia.
13. Silver from Abydos, Tomb 416, with Kamares Ware, is from Joachimsthal or nearby Schneeberg in the Erzgebirge.
14. Celtic tin money from Kent is from Bohemia, and not from Cornwall.
15. Wappenmünzen from Athens, pre 500 B.C. do not come from Laurion ore, whereas the Athenian 'Owls,' ca. 490 B.C., do. The contemporary 'Turtle' coins of Aegina are from Almeria.

The above few results give some idea of the revolutionary potential of the lead isotope method.

Sardinia

Sardinia is shown to have been an important center of metallurgy in the Bronze Age. It has rich deposits of copper and silver, and significant veins of cassiterite (tin oxide). The writer has shown by lead isotope analysis that Sardinia was the home of the Sherden and that bronzes and blue frit were reaching the Levant from Sardinia by 1400 B.C. and from Tuscany in 1600 B.C (fig.1).[3,5] An ox-hide ingot from Ilixi was of Sardinian copper mixed with scrap from either Spain or Tuscany, while the ox-hide ingots from Haghia Triada in Crete, inscribed with Linear A and dated to ca. 1600 B.C., are also made from Sardinian ore and cannot be confused with any other source (fig. 1). Ox-hide ingots have been found at eleven nuraghi, together with many molds and bun ingots. An El Agar-type sword, found in Sardinia, shows connections with Spain, but links are also clear with Tuscany and with the Remedello and Polada Cultures of Northern Italy. Silver has been found in the Ozieri Culture of Sardinia with copper, and in rock-cut tombs of the Beaker Culture ca. 1800 B.C., together with tuyères. A crucible was found with about ten kilogrammes (23 lbs.) of cassiterite in it, while a lump of pure melted tin was found at Lei (Sa Maddalena) weighing 700g.[6]

In recent years some hundreds of Mycenaean shards, mostly IIIC but with some IIIB, have been found on the island at Barumini, at Nora, and at Sulcia.[7,8] However, no glass or faience has been found on Sardinia, nor have lead isotopes yet linked any silver object with the island. What is important from the above is that strong contacts existed between the Levant and the western half of the Mediterranean as early as 1600 B.C. and before then, for the writer has found that silver ingots from the Diyala in Mesopotamia, dated to ca. 2500 B.C., were derived from the silver ores of Cartagena in southern Spain.[3] The postulated "fault line" between the eastern and western halves of the Mediterranean is thus disproved, while Sardinian influence on Crete, Cyprus, and the Levant was obviously considerable in light of the evidence of the ox-hide ingots, the other artifacts, and the Sherden mercenaries, whose existence is recorded in Egyptian annals for several hundred years.

Silver, Cobalt, and Bismuth

It is the appearance of silver in the Near East together with cobalt blue glass that gives us the clue to trade routes. Tin, another clue, has been discussed elsewhere.[1] Like tin and bismuth, cobalt is found in only a few places in the world. The main deposits are, as Stanton states, "one of the most spectacular and clearly defined of all ore types: the native silver-cobalt-nickel arsenide ores."[9] The classic deposits are those of the Erzgebirge of Upper Saxony, and of Sudbury, Ontario. Small sources of cobalt, also with silver, occur in Scandinavia, Cornwall, France, and Sardinia. (Traces of cobalt were found by Zwicker in the slag on an ox-hide ingot from Ilixi, Sardinia.[10])

It is the writer's contention, proved experimentally, that cobalt-blue glass was the slag *accidentally* produced from the smelting of a certain type of rich native silver and associated ores, occurring in a pure quartz gangue, with associated fluxes in the form of fluorite (CaF_2) and apatite $(CaF)Ca_4(PO_4)_3$ or $(CaCl)Ca_4(PO_4)_3$. Such an assemblage of minerals is found at Schneeberg in Saxony, the source of cobalt blue glass from early medieval times. Some tin and copper,

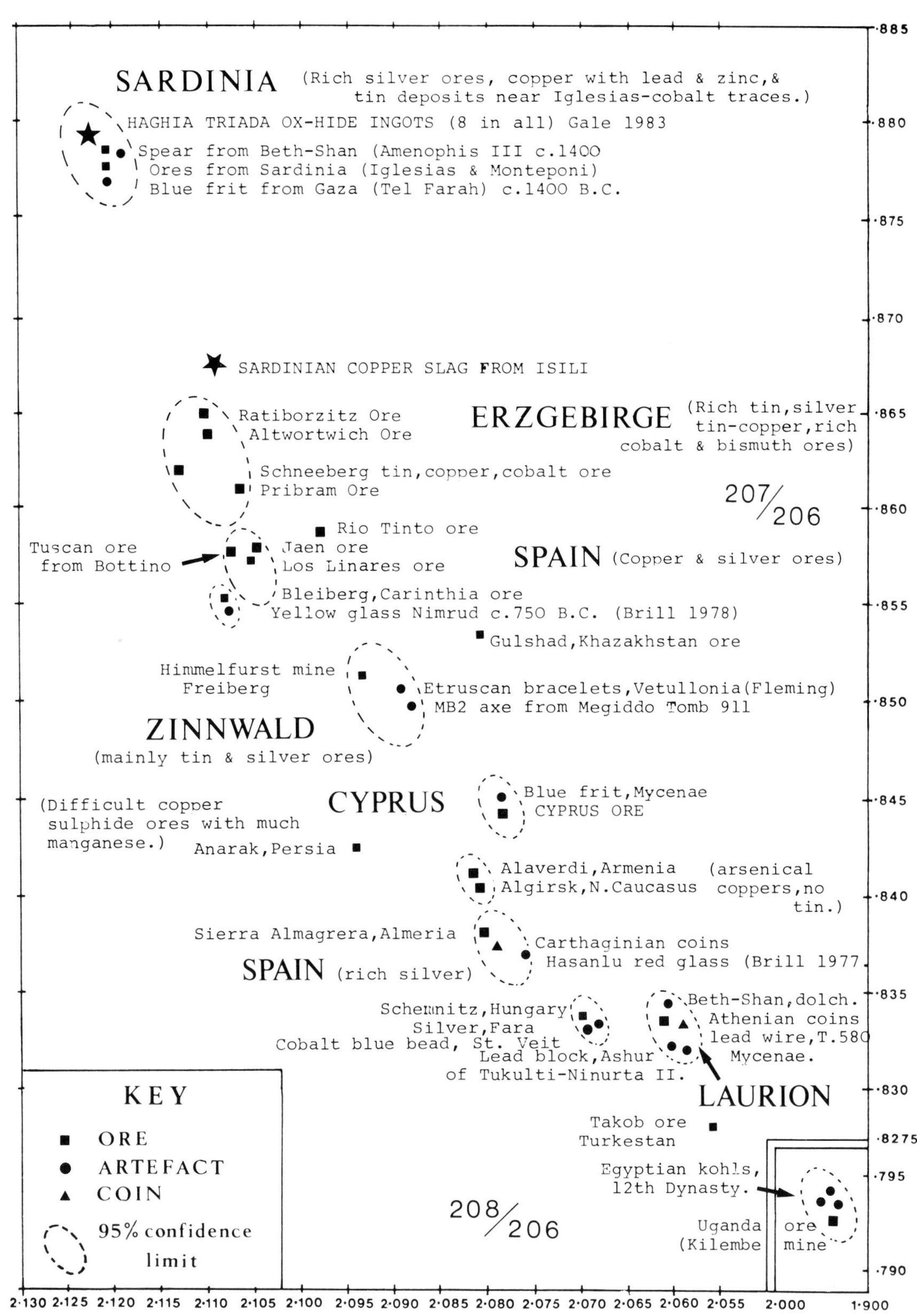

Fig. 1. Plot of lead isotope ratios of ores and artifacts, showing trade routes.

together with much bismuth are also found in this location, which must therefore be the source of the silver and cobalt glass that appear ca. 1650 B.C. at Mycenae. Amethyst and clear red jasper also occur in these deposits, which outcrop at the surface. Amethyst beads appear to be carried by the Aegean traders in the tomb of Rekhmir, and red jasper is a feature of Egyptian jewelry at this time.

Central Europe

The history of the production of glass in Central Europe goes back to the beginnings of bronze metallurgy in the Unetice Culture, ca. 1800 B.C. In the Unetice Culture, cobalt blue glass beads have been found at Nitra, north of Budapest.[11] These beads have been dated to ca. 1800 B.C., some 150 years earlier than those of the Shaft Graves of Mycenae. The bronzes of the Unetice Culture must have been made of tin and copper from the Erzgebirge, where, as we have stated, cobalt also occurs. A visit to the prehistoric museums of Munich and Hanover revealed an enormous number of bronze hoards—that from Luitpold Park, Munich, containing over 500 torques with a total weight of 85kg and dated to 1800-1500 B.C., slag and bun ingots form Obing and Pfakhofen (Regensburg), the Urnfield cemeteries of Munich-Unterhaching (thirteenth through twelfth centuries B.C.) and the cemetery at Kelheim (near Regensburg) with 259 graves containing swords, axes, and axe-molds. That the metal industry was local is proved by the quantity of molds from Ljubljana and Vučedol (ca. 2300-2100 B.C.) and Margarethenberg (ca. 1600 B.C.). An interesting mold for thirty spherical beads comes form Nitriansky-Hràdok and is dated to 1500-1200 B.C. (Deutsches Museum, Munich). This museum also has molds for typical Mycenaean ball-headed pins from Velem St. Veit, where antimony deposits are found, and which was later a center of the Hallstatt Culture.

Northern Italy, as Barfield has shown, is a logical outlet for the two cultural areas of Central Europe, that of the Postojna Pass to Trieste and Venice, while the rich metal-working areas of Austria and Bohemia had an easy route over the Brenner Pass.[12] It is no coincidence that glass beads are found in the Polada Culture of Northern Italy (ca. 1800-1450 B.C.) at Lucone, whilst Ledra has amber beads, a Vaphio-type cup, molds, crucibles, tuyères, bronze slag and ingots. At the later Bronze Age site of Frattesine, not far from Murano, lumps of blue and red glass for bead-making have been found together with metal-working slags. Also found in Northern Italy is silver, including a fine pectoral from Villafranca dated to ca. 2000 B.C. (Padua Museum). Nor must we forget the enormous bronze hoards from the Bologna region—one hoard with 14,838 scrap objects including 3,952 socketed axes and 3,200 fibulae. The uniformity of the Central European cultures and those of Northern Italy has been well documented, especially the sword types of Peschiera, Scamozzini, Monza, and Povegliano, with those of Rixheim (Rhine/Rhone) and Sauerbrunn in Hungary. A sword of the Rixheim type was found, significantly, at the Syrian site of Ugarit, with the cartouche of Merneptah (ca. 1234-1220 B.C.), who had to fend off the early attacks of the Sea Peoples.

Mycenae and the Mycenaean World

In spite of the remarkable architectural parallels between the western European world of the Megalith and Nuraghi builders and that of Mycenae, Tiryns, and Ras Shamra, no glass has been found in the western half of the Mediterranean. However, a feature of the Mycenaean world from the time of the earliest shaft graves (ca.1650 B.C.) is the abundance of dark cobalt blue glass, often covered with gold foil. This cobalt blue glass disappears at the end of the Mycenaean era (ca. 1200 B.C.) and does not reappear until Phoenician times (ca. 800 B.C.) This four-hundred year technological gap is one of the curiosities of archaeology. With the Mycenaeans there appears in the Levant true tin bronzes, silver, tin as a metal, amber, amethyst, and faience. The bronzes show strong traces of silver, nickel, cobalt, arsenic, zinc, antimony, and bismuth.[13-16] These trace elements give a clue to the source of the Mycenaean metals and artifacts, especially of the silver.

During the Mycenaean era (ca. 1650-1200 B.C.) the above metals and materials, in particular cobalt blue glass, are found in Egypt, and it is clear from the tomb scenes of Rekhmire (ca. 1435 B.C.) that these goods were brought into Egypt by Aegean traders who carried Sherden type swords, ox-hide ingots, Mycenaean rhtya, and strings of blue beads, as shown in figure 2. The writer suggests that the different groups of traders depicted represent Sherden and Syrians as well as Cretans and Mycenaeans.

The Phoenicians

In the history of glass we now have a curious gap of some 500 to 600 years until glass reappears in the Near East in Phoenician times. Again, most of this glass is colored with cobalt. Decoration is trailed in with yellow glass made with lead and antimony, white glass made from antimony, pale blue glass colored with copper, and green glass colored with lead and copper. Pliny credits the Phoenicians with the discovery of glass but in fact they were the importers of the material from the head of the Adriatic where the Hallstatt and later La Tène cultures had a flourishing glass industry.

The important publication of the Mecklenburg Collection by Hencken and Wells has shown the quality of glass that was manufactured in Slovenia and Hallstatt.[17,18] Wells notes that small glass beads, particularly of a blue color, are frequently found in the graves of the Urnfield period of the Middle Bronze Age.[18] Six graves at Stična of the Early Iron Age produced some 20,500 glass beads.[19,20] Stična, ideally situated in the Ljubljana Gap, was conveniently near the antimony deposits of St. Veit. Glass bowls were found in the graves of Hallstatt and Most-na- Soči.

Analysis of typical beads from the Mecklenburg Collection (by kind permission of Professor Lamberg-Karlovsky) showed that all of four typical dark blue beads contained cobalt, copper, a high percentage of antimony, and traces of tin. Three beads from Magdalenska Gora, grave 56, were dark blue (copper, cobalt and antimony), white (antimony only) and yellow (lead and antimony only, no iron or tin).

The cobalt and copper, as we have argued above, came either from Schneeberg or from Dobschau (Dobsina) in Slovakia. The antimony probably came from St. Veit or from Kremnica in the Nitra Valley. Lead for the yellow could have come from Bleiberg at nearby Villach.

We have here three interesting archaeological horizons: the introduction of cobalt-copper blue glass, the appearance of a white antimony glass, and the appearance of yellow lead-antimony glass.

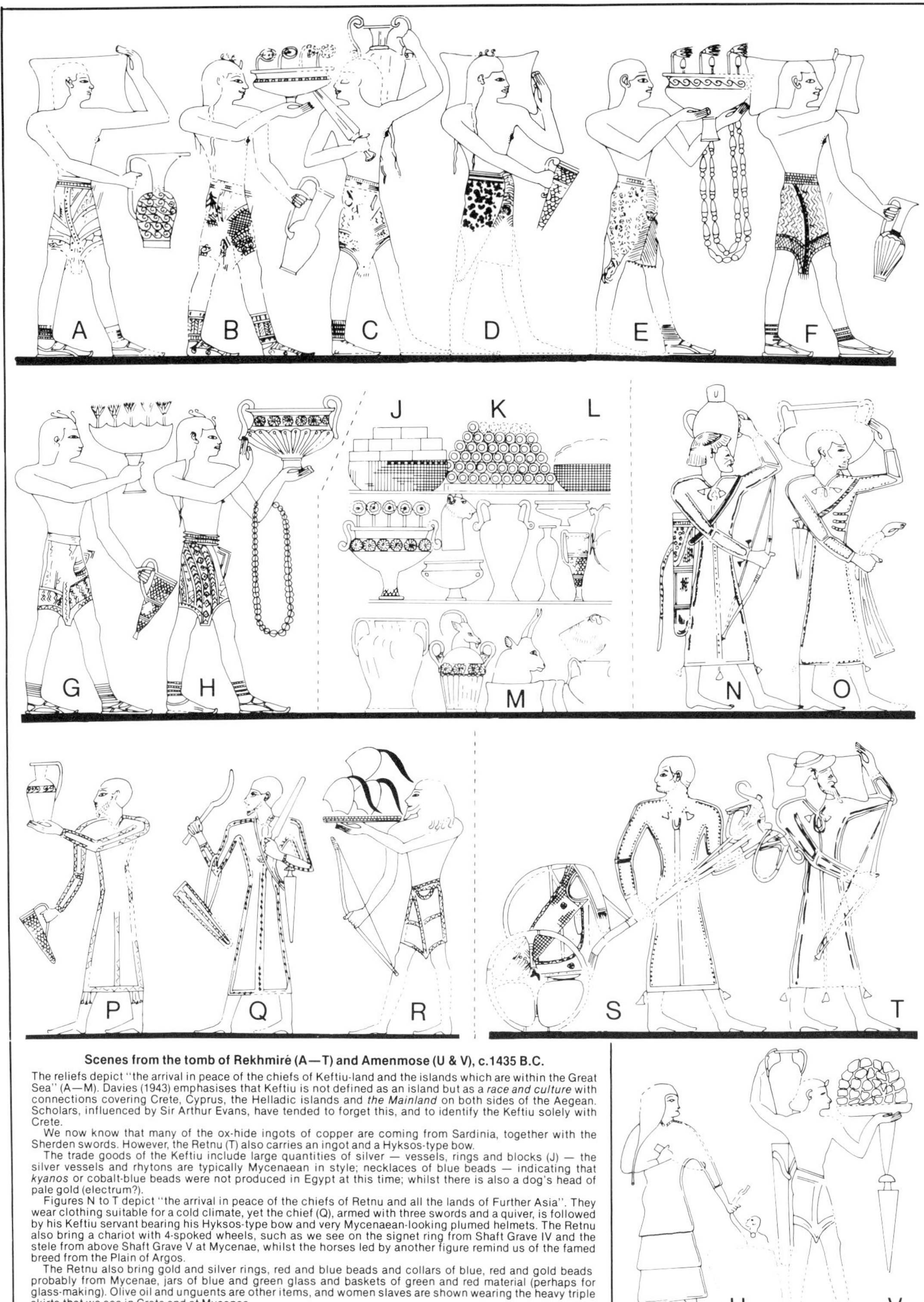

Scenes from the tomb of Rekhmirė (A—T) and Amenmose (U & V), c.1435 B.C.

The reliefs depict "the arrival in peace of the chiefs of Keftiu-land and the islands which are within the Great Sea" (A—M). Davies (1943) emphasises that Keftiu is not defined as an island but as a *race and culture* with connections covering Crete, Cyprus, the Helladic islands and *the Mainland* on both sides of the Aegean. Scholars, influenced by Sir Arthur Evans, have tended to forget this, and to identify the Keftiu solely with Crete.

We now know that many of the ox-hide ingots of copper are coming from Sardinia, together with the Sherden swords. However, the Retnu (T) also carries an ingot and a Hyksos-type bow.

The trade goods of the Keftiu include large quantities of silver — vessels, rings and blocks (J) — the silver vessels and rhytons are typically Mycenaean in style; necklaces of blue beads — indicating that *kyanos* or cobalt-blue beads were not produced in Egypt at this time; whilst there is also a dog's head of pale gold (electrum?).

Figures N to T depict "the arrival in peace of the chiefs of Retnu and all the lands of Further Asia". They wear clothing suitable for a cold climate, yet the chief (Q), armed with three swords and a quiver, is followed by his Keftiu servant bearing his Hyksos-type bow and very Mycenaean-looking plumed helmets. The Retnu also bring a chariot with 4-spoked wheels, such as we see on the signet ring from Shaft Grave IV and the stele from above Shaft Grave V at Mycenae, whilst the horses led by another figure remind us of the famed breed from the Plain of Argos.

The Retnu also bring gold and silver rings, red and blue beads and collars of blue, red and gold beads probably from Mycenae, jars of blue and green glass and baskets of green and red material (perhaps for glass-making). Olive oil and unguents are other items, and women slaves are shown wearing the heavy triple skirts that we see in Crete and at Mycenae.

Two unique types of figures are represented (N and T), both archers, one with an ingot, the other with a jar of olive oil. Note also that Retnu V has a Keftiu kilt, a quiver, a Sherden sword and a tray bearing lumps of ore or glass.

Conclusions

We thus have, on the geological and technical evidence, at least two active sea-faring groups of peoples in the Mediterranean from about 1800 B.C.—progenitors of the later "Sea Peoples" of historical record. One group obtained tin bronze, amber, cobalt glass, rock crystal, amethyst, blue anhydrite, jasper, and silver from Central Europe by way of the Adriatic. At this time the climatic evidence shows that the Brenner Pass was ice-free. We may reasonably assume these people to have been the ancestors of the Mycenaeans.

The second group obtained copper from Sardinia, Spain, and Tuscany. They also exploited the silver deposits of southern Spain and appear to have been associated with the megalith builders of western Europe including Los Millares and the builders of the nuraghi of Sardinia. They did not have glass, but the ores they exploited would not, on smelting, have produced a pretty, glassy slag. This western Mediterranean group of whom the Sherden appear to be the most important part have close connections, on the evidence of the Sardinian origin of the Haghia Triada ingots, with Minoan Crete. On the historical evidence from Egypt, both of these groups must have been in contact with the Hyksos rulers of Egypt, (ca. 1650 to 1550 B.C.).

In the Carpathians and Balkans we have yet a third group of metal-working peoples, characterized by their gray-ware pottery and socketed cast axes. This group again does not have glass, although they were spread over the Lower Danube-Anatolian-Iranian axis, while glass is very uncommon in Mesopotamia.

The technology of these three different European groups that appear to migrate eastwards during the second millennium B.C. presents an interesting field for further study, and the writer would suggest that to the conventional Bronze Age should be added the ages of silver and of glass.

References

1. J. E. Dayton, "The Problem of Tin in an Ancient World," *World Archaeology* 3, (1977): 47-70.

2. I. L. Barnes, W. R. Shields, T. J. Murphy, and R. H. Brill; "Isotopic Analysis of Laurion Lead Ores," *Archaeological Chemistry*, Advances in Chemistry Series (Washington, D.C.: American Chemistry Society, 1979).

3. J. E. Dayton, *Minerals, Metals, Glazing, and Man* (London: 1978).

4. J. E. Dayton, "Cobalt, Silver, and Nickel in Late Bronze Age Glazes, Pigments, and Bronzes, and the Identification of Silver Sources for the Aegean and Near East by Lead Isotope and Trace Element Analysis," in *Scientific Studies in Ancient Ceramics* (British Museum Occasional Paper no. 19), M.J. Hughes, ed. (London: 1981).

5. J. E. Dayton, "Sardinia, the Sherden and Bronze Age Trade Routes," *Ann. Ist.Or. Napoli*, 1983 (in press).

6. M. Guido, *Sardinia* (London: 1963).

7. M. L. Ceruti, "Ceramica Micenea in Sardegna," *Riv. Scienza Preist* 34, (1979): 243-253.

8. L. Vagnetti, "Mycenaean Imports in Central Italy," Appendix 2 in E. Peruzzi *Mycenaeans in Early Latium* (Incunabula Graeco LXXV), (Rome, 1980).

9. R. L. Stanton, *Ore Petrology,* (New York, 1972).

10. V. Zwicker, P. Virdis, and M. Ceruti; "Investigations on Copper Ore, Prehistoric Copper Slag, and Copper Ingots from Sardinia," in *Scientific Studies in Early Mining and Extractive Metallurgy* (British Museum Occasional Paper no. 20), P. T. Craddock, ed. (London: 1980).

11. J. M. Coles and A. F. Harding, *The Bronze Age in Europe* (London: 1979).

12. L. Barfield, *Northern Italy Before Rome* (London: 1971).

13. G. Mylonas, *O Taphikos Kyklos B toi Mykinoi* (Athens: 1973).

14. P. T. Craddock, "The Composition of the Copper Alloys Used by the Greek, Etruscan, and Roman Civilizations: 1, The Greeks Before the Archaic Period," *Journal of Archeological Science,* 3, (1976): 93-113.

15. H. Schliemann, *Mycenae.* (London, 1878).

16. W. Dörpfeld, *Troja und Ilion,* (Athens, 1902).

17. H. Hencken, *The Iron Age Cemetery of Magdalenska Gora in Slovenia.* (Mecklenburg Collection, Part II). Bulletin 32, American School of Prehistoric Research, Peabody Museum, Harvard University, 1978.

18. P. S. Wells, *The Emergence of an Iron Age Economy.* (Mecklenburg Collection, Part III). Bulletin 33, American School of Prehistoric Research, Peabody Museum, Harvard University, 1981.

19. T. E. Haevernick, "Zu den Glasperlen in Slowenien," *Situla* 14- 15, (1974a): 61-5.

20. T. E. Haevernick, "Die Glasfunde aus den Gräbern vom Dürrnberg," *Der Dürrnberg bei Hallein II,* F. Moosleitner, et al, (Munich: 1974b).

JONATHON E. ERICSON and HIROSHI SHIRAHATA

Lead Isotope Analysis of Ancient Copper and Base Metal Ore Deposits in Western India

Hoards of beautiful bronze Astadhutu (eight-metal alloy) or Panchaloha (five-metal alloy) images have been found in India, especially in Gujarat, Bihar, Madhya Pradesh and Tamil Nadu.[1,2] Unfortunately, many of these images have not been recovered through proper excavation and have subsequently changed hands many times. As a consequence, their geographical, cultural, and temporal provenance have been lost.[1] This study was designed to acquire lead isotope data from ore deposits in west India to improve the quality of geographic attribution of bronzes produced in this region.

The initial project in 1980-81 involved a field reconnaissance of the ancient copper mines and smelting sites in the Aravalli Hills in the states of Gujarat and Rajasthan in western India.[3] There were two research objectives: (1) to locate and survey the ancient sites in order to observe variations in mining, ore processing, and smelting technologies; and (2) to collect samples of base metal ores for lead isotope analysis and to obtain samples of slags, furnace remains, pottery, charcoal, and potential fluxing agents for future analysis.[4] This paper will focus on our lead isotope work.

Ancient Metals from the Aravalli Hills

The Aravalli Hills in Gujarat and Rajasthan were a dominant source area for copper and base metals in ancient India (fig. 1). Other source areas include Bihar, Kashmir, Uttar Pradesh, and Tamil Nadu.[5-8] From pre-Harrapan times through the medieval period, the ancient copper mining and smelting technology in the Aravalli Hills produced copper and other metals used to produce a variety of artifacts such as tools, bangles, cookware, and idols.[9] Pliny speaks of copper as one of the chief products of India. Copper, iron, and lead are described as arriving at the port of Carmanian on their way to the Persian Gulf. These observations were confirmed by the writer of *Periplus of the Erythrean Sea,* who mentions copper as an export item from the Port of Barygaza (modern Broach on the Gulf of Cambay) to the Persian Gulf.[10] This evidence suggests that copper from the Aravalli Hills was being exported to the Persian Gulf and west Asia by sea from early times.[9] The Aravallis were by no means the sole source of metals transported across the Persian Gulf. Preliminary data suggest that the ancient mining sites in Oman may have supplied smelted copper to Mesopotamia from the third millennium B.C.[11] Future research may reveal the nature of trade and competition between these two important source areas.

Geological Setting

The Aravalli Hills were formed by several episodes of folding and faulting of Precambrian and Mesozoic rocks. Sychanthavong and Desai[12] propose that the processes represent pre-Cambrian plate tectonics of the proto-Indian continent.[14] Though heavily weathered, the Aravalli Hills rise an average of 670 meters above the Thar Desert. The Hills tend to be shear blocks or isolated peaks above the desert floor. To the west some 300km, across the Thar Desert, lies the Indus River and its fertile valley.

The Aravalli Range contains moderate amounts of copper that, though not as abundant as in Eastern India, is heavily worked today. The copper and other base metals occur discontinuously throughout the 1600km-long shear zones that trend northeast to southwest. The copper ores are highly localized having been emplaced by hydrothermal activity along these zones.

The base metal mineralization, shown in figure 2, is divided into two zones of seven shears.[12] The eastern zone comprises the Kho-Dariba copper-lead-zinc belt, also of Udaipur district. The western zone includes the Ambaji-Deri lead-zinc-copper belt of Alwar district, the Rajpura Dariba lead-copper-zinc belt of Udaipur district, and the Zawar lead-zinc-copper belt of Banaskantha and Sirohi district, respectively, the Rajgargh-Khankaria (Ajmer Sandra) lead-copper-zinc belt of Ajmer district and the Khetri copper-lead-zinc belt of Sikar district.

There is an interesting change in the base metal composition of the ore deposits from the northeast to the southwest. The northeast deposits are highly enriched in copper with lead and zinc occurring at trace levels, e.g., at Khetri. In comparison, the southwestern deposits are rich in lead and zinc with copper the least plentiful element. Here, copper ores occur as superficial oxide ores that have been preferentially leached out of the parent lead and zinc ores. In the intermediate deposits the lead exceeds copper, which exceeds zinc. The regional gradient in base metal composition, within the Aravallis, most likely had profound effects on the development of ancient Indian copper technology.

Sample Selection

Ancient copper deposits and smelting sites were located in the field by drawing on the geological literature.[5-7] The following mining and smelting sites, surveyed during 1980, provided samples for lead isotope analysis.

Ambaji and Related Sites

Ambaji (24°20′, 72°53′) is a copper-lead-zinc mining area. There were abundant surface mines indicating the extraction of copper oxide ores. Several deeper shafts were observed. This mining of shallow surface pits is considered to be a very early technology.

The smelting site associated with Ambaji is Kumbhariya (24°19′ 72°52′) is an ancient copper slag site. A series of seven terraces of slag was observed within a square kilometer area. Kiln remains and slags dominate the surface. There appear to be two components to the site: an early one possibly dating back to Harrapan times and a later medieval component associated with building of the Jain Temple complex. Deri is another mining locale located approximately 15km northwest of Ambaji. The ore is located in the same shear zone.

Zawar

Zawar (24°22′, 73°40′) is a predominantly lead-zinc mining area. Here, we observed native silver in one of the hanging walls of an ancient mine.

Pipalwas and Related Sites

Pipalwas (42km northeast of Udaipur) was the location of mines and a smelting site. Apparently, copper oxide ores were extracted from the surface along the 1.2km shear zone that forms a ridge. Deeper shafts with ventilated galleries were also observed. The mining technology was very similar to that observed for Ambaji.

Pipalwas smelting site (1km to the east of the mines) was quite pristine in that evidence of kiln remains and other artifacts remained on the surface. The kiln technology was more similar to Ambaji than to Kho Dariba. The slag on the site was 70 by 2200 meters in area and up to 3 meters thick.

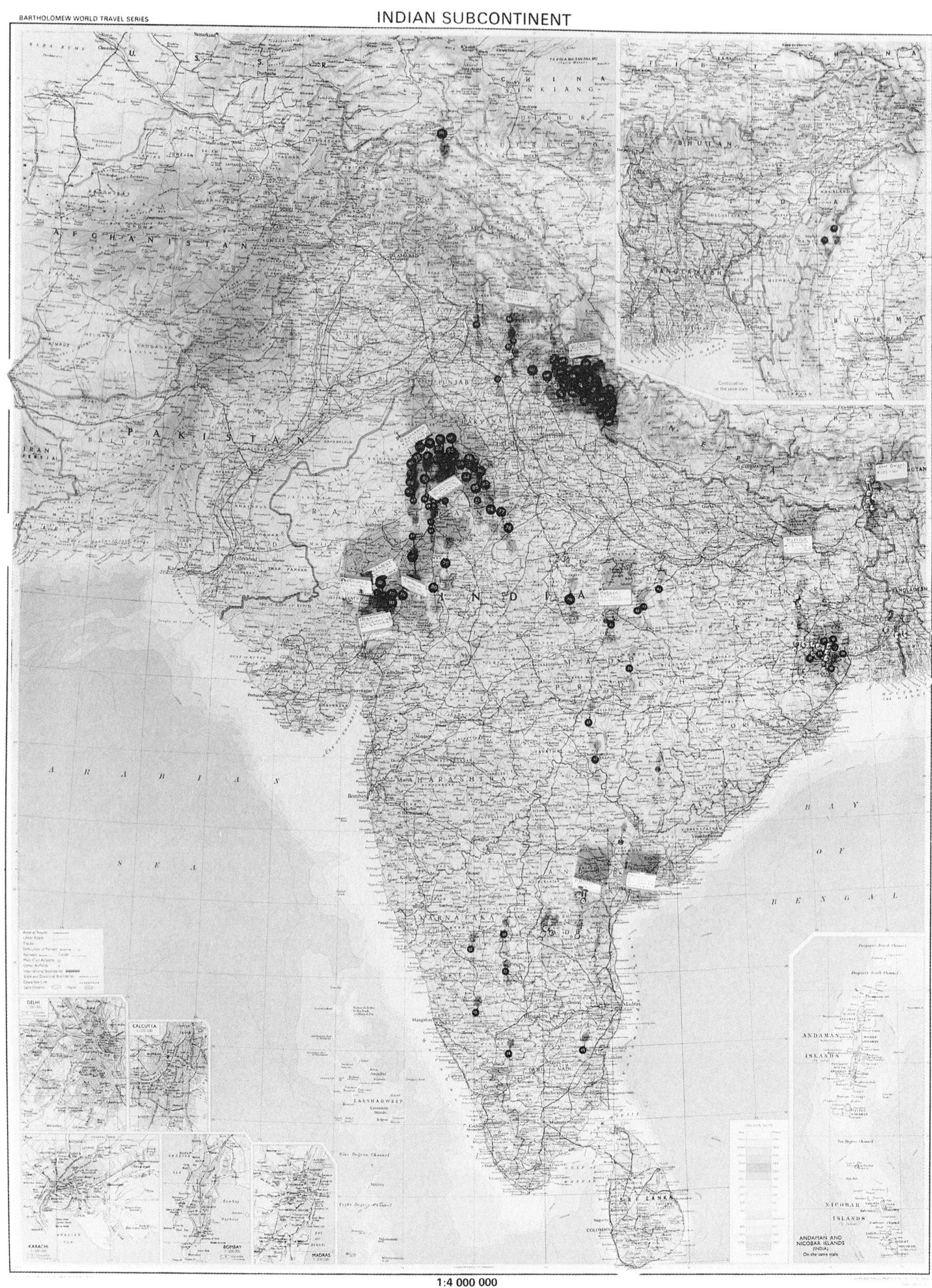

Fig. 1. Map of the subcontinent of India and the distribution of copper and base
metal deposits in the geological literature.[5-7]

Rajpura and Rajpura Dariba

Rajpura (24°58′, 74°36′) is the location of present-day lead-zinc mines. Copper occurs but only in a very selected area. We observed ancient surface mining along the ridge about 5km to the north of the smelting site. Here, miners removed copper oxide ores. We visited the interior of the ancient workings at great depth along the shear zone. We and the geologists were amazed by the tremendous effort and the sophisticated technology of the ancient people. The interior of the ancient mine has wooden beams that have been radiocarbon dated to 2,000 years ago. The interior of these mines was very unpleasant and unsafe due to the soft graphite schist and everseeping water from a huge strip mine approximately one half km in diameter and 120m deep.

Rajpura Dariba (24°57′, 74°08′) was the second largest slag site. Slag creates mounds 1-3 meters high spread discontinuously over 0.5 square km. The slag and the kiln remains point to a copper smelting technology, even in the presence of the lead-zinc-rich ore bodies. Further analysis is warranted at these sites.

Khankaria

Khankaria (26°28′, 74°38′) near Aimer is a copper-lead-zinc deposit. Here we observed shallow surface pits dug to mine the copper oxide ores. We expect this activity to be early. We did not locate the associated slag site.

Darioba Nala and Kho Dariba

Darioba Nala (27°13′, 76°24′) is a copper mining area. Copper occurs as sulfide ore in the softer phyllite within the shear zone, set in quartzite. Mining here was easier than at Khetri. A series of fifteen deep shafts and interconnecting galleries and stages was used to extract the copper ore. The deepest shaft is located 80m below the surface. There are many other mining and smelting sites along this shear zone extending northwest to Bairat (27°26′, 76°11′).

The associated smelting site is at Kho Dariba (27°12′, 76°24′) a copper smelting site. Here the slags form a perimeter to the north of the deserted town of old Kho Dariba. Abundant kiln remains were observed and samples were collected. We observed a slight change in the kiln technology relative to Ambaji.

Khetri and Singhana

Khetri (27°58′, 76°46′) is a copper mining area. The copper ore occurs as copper-iron sulfide in the shear zone of very hard quartzite rock. The ore is exceptionally pure, producing electrical grade copper with simple smelting techniques. Here, ancient mining was probably active during the medieval period. There are many deep shafts opening into large galleries, ventilated by a series of small shafts. We observed charcoal smelting indicated by ancient fire treating of hardrock walls in the bottom Chandmari mine.

Singhana (28°19′, 75°53′), the associated smelting site, is 5km north of the mines. There are two 300m diameter slag heaps thirty and fifty-two meters above the surface. There appear to be two major strata in the slag. Kiln fragments and other important materials were lacking from the surface, due to the daily disturbance by the townspeople and pigs.

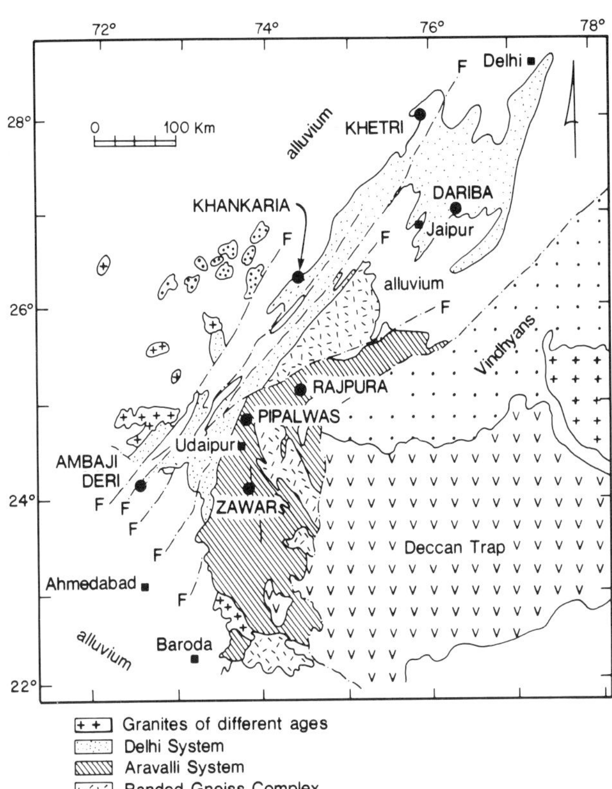

Fig. 2. Map of regional geology of Aravalli Hills and distribution of surveyed copper and base metal ore sites. (Modified after Sychanthavong and Desai) [12: Fig. 9].

Characterization of Metal Alloys

There are inherent limitations to elemental characterization of copper and bronze artifacts. Werner used spectral analysis of the major and minor elements to classify the alloys of Indian bronzes.[13] He noted that high levels of zinc characterized the older (ca. A.D. 700-1200) bronzes of Gujarat and Kashmir. The attempts to characterize bronzes based on trace elemental composition has had varying degrees of success.[14,15] The observed variations in trace element chemistry most likely reflect elemental contributions introduced during mining, concentrating the ore, fluxing, fueling, and smelting. The geochemical structure of the ore deposit, the alloying and metalworking, use, recycling of metals, and corrosion would be contributory factors in changing the composition.[16] For comparison, lead isotopes have successfully characterized both the ore deposits and associated artifacts.[17-19]

Lead isotope analysis was selected for this study to overcome the inherent limitations of elemental characterization. Lead isotope characterization has an important advantage over elemental characterization, since the isotopic ratios tend to remain constant throughout the chemical history of the object.[16] The mixing of the ore lead with leads of different isotopic composition can alter the isotopic composition.[19] Fortunately, Sanskrit texts from the early medieval period mention sanctions against remelting religious statuary into new statues.[20] These sanctions were operative throughout India for both Hindu and Buddhist statues. This evidence suggests that the problem of reuse of metal will be greatly reduced in Indian religious statues.

Lead Isotopic Analysis

Two types of geological samples were analyzed. The ore samples were those with abundant copper, lead, or zinc minerals present. The host rocks were those rocks in the shear zones that contained the ore bodies.

Samples were analyzed by mass spectrometric isotopic dilution. The sample preparation was conducted in the ultraclean laboratories of the Muroran Institute of Technology. Lead was extracted using a standard ion exchange technique, described below.

The sample materials to be analyzed were hand-picked from lumps, put into an FEP beaker, washed in dilute HNO_3 and deionized water. The sample in the beaker was dissolved in aqua regia, and evaporated to dryness. The residue was twice redissolved with HNO_3 and then evaporated to dryness. An aliquot of sample solution was transferred into an FEP separatory funnel.

In analysis of the galena ore sample, lead from the sample solution was separated and purified by extracting with dithizone solution in chloroform in the presence of a buffer of ammonium citrate and potassium cyanide (pH 9). The chloroform layer was washed twice with slightly ammoniated water and the lead was back-extracted into dilute HNO_3. The final solution was drained into an FEP beaker and evaporated to dryness. A portion of the sample residue was loaded with a mixture of silica gel and phosphoric acid onto an out-gassed rhenium filament for isotopic ratio measurements.

In the copper-zinc-lead ore analysis the sample solution in an FEP separatory funnel was first extracted with a dithizone solution at pH 2 to separate the copper and bismuth present, as in the lead extraction. If the sample contained iron, the solution was treated with a reducing agent before the copper and bismuth extraction.

For whole rock leachate, the sample rock chips were pulverized in a stainless steel mortar and pestle to less than 100 mesh. The rock powder was soaked in deionized water in an FEP beaker and dried at room temperature. The dry sample was further ground in an agate mortar and pestle. A split of rock powder was dissolved in a mixture of HF, HNO_3, $HClO_4$ in a PTFE bomb and the resultant solution was evaporated to dryness. The acid dissolution was repeated twice to convert the salt residue to nitrates and perchlorates. The dry residue was dissolved in dilute HNO_3 solution. Separation and purification of lead from the solution was achieved by dithizone-chloroform extraction.

All reagents used were further purified from an ultrapure grade of commercial reagent. Teflon and other ware used were soaked in HNO_3 baths for weeks. Throughout the sample preparation and analysis, laboratory articles were touched and handled with polyethylene gloves in a Class 100 clean laboratory. Total lead blanks were estimated at less than 2ng which is insignificant given the sample weight of 2mg separated. Lead isotope ratios of samples were analyzed using a solid source thermal ionization mass spectrometer with automatic digital read-out and computation. All isotopic ratios obtained

here have been normalized to their absolute values by intercomparison with the NBS SRM981, Common Lead Standard. The standard lead was run before and after two sample analyses. Most "within run" precision for the mass spectrometric analysis of the sample lead was 0.1% relative standard deviation at 95% confidence level. The results are reported in table 1.

Discussion

The data presented in table 1 and figure 3 indicate that copper artifacts from the above copper and base metal sources will be able to be characterized by lead isotope analysis in the future.

The intrasource variation indicates a high degree of reliability of the data and, for certain sources, the degree of internal variation. The ore and host rocks from Ambaji, Kumbhariya, and Deri form a tight group. The ore and host rocks from Zawar are quite similar. The results of each set of duplicate analysis of samples from Pipalwas (14-9) and Rajpura Dariba (16-18) have excellent internal consistency. There are isotopic differences between the ore and host rock from Rajpura Dariba and Khankaria. The results from Khetri (4-6) and nearby Ghatiwali (7-1) are reported in table 1, although not plotted in figure 3. Both samples are quite radiogenic with contribution of lead from uranium mineralization. Due to the apparent variability of the data and the lack of inclusion of an ore sample from the main Khetri mine, additional analyses are warranted for this source. These have been planned and will be reported elsewhere.[8]

The intersource variation indicates that there is a structural and geochemical relationship among the southwestern sources including Ambaji-Deri, Zawar, Khankaria, Pipalwas, and Rajpura Dariba. Characterization of copper artifacts from these sources will be possible in terms of probability of membership. Although geographically separated, Khankaria and Rajpura Dariba cannot be separated due to an overlap of the isotopic data. The ore data from the southwest group is generally consistent with the host rock and the the age of the rock.[12] Dariba Nola and preliminary data from Khetri-Ghatiwali plot quite differently from the southwestern cluster of sources. Most likely three distinctive groups will be shown to exist in this important metalliferous region. Of the 3458 lead isotopic analyses available worldwide, only seven have been published for the Indian subcontinent.[21] The data resulting from this research supplements that literature.

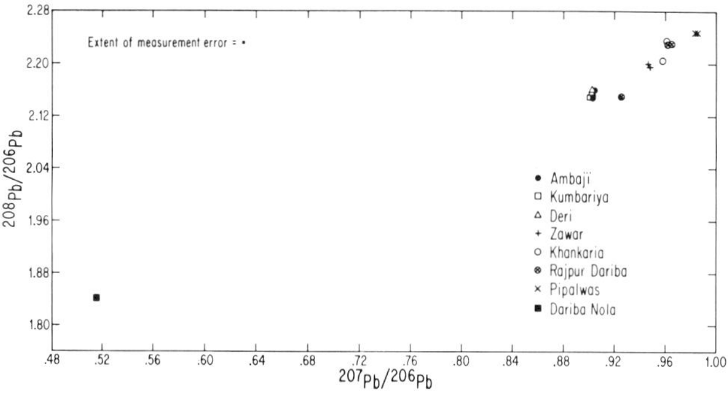

Fig. 3. Isotopic ratios of lead ^{208}Pb-^{206}Pb and ^{207}Pb-^{206} Pb of copper, base metal, and host rocks from the ore deposits in this study.

Table 1. Lead Isotope Ratios in West Indian Ores and Host Rocks

Site	sample	206/204	207/204	208/204	208/206	207/206
Ambaji, ore:						
chalcopyrite with PbS, ZnS	1-1	17.315 ± 0.016*	15.649 ± 0.010*	37.387 ± 0.016*	2.1596 ± 0.0015*	0.90405 ± 0.00046*
Ambaji, host:						
malachite in chlorite schist	1-31	17.127 ± 0.0657	15.411 ± 0.0649	36.810 ± 0.1604	2.1494 ± 0.00310	0.89936 ± 0.00134
Ambaji, host:						
malachite in quartzite, brecciated	1-33	17.337 ± 0.0249	15.648 ± 0.0233	37.329 ± 0.0604	2.1528 ± 0.00206	0.090247 ± 0.00078
Kumbhariya, host:						
talc-chlorite schist	2-10	17.358 ± 0.0136	15.656 ± 0.0008	37.344 ± 0.0172	2.1509 ± 0.00127	0.90165 ± 0.00069
Deri, ore:						
galena	3-5	17.280 ± 0.008	15.608 ± 0.007	37.279 ± 0.036	2.1573 ± 0.0016	0.90343 ± 0.0020
Deri, host:						
ferrugenous quartzite with biotite, possibly pegmatite	3-3	17.292 ± 0.0302	15.614 ± 0.0338	37.336 ± 0.0832	2.1593 ± 0.00131	0.90263 ± 0.00075
Zawar, ore:						
galena	12-5	16.583 ± 0.005	15.698 ± 0.007	36.535 ± 0.046	2.2031 ± 0.0020	0.94671 ± 0.00019
Zawar, host:						
galena and chalcopyrite in quartzite	12-4	16.593 ± 0.0189	15.730 ± 0.0154	36.472 ± 0.0373	2.1977 ± 0.00089	0.94778 ± 0.00039
Pipalwas, galena in chlorite schist, whole sample leachate	14-9	15.682 ± 0.011	15.440 ± 0.0006	35.203 ± 0.025	2.2445 ± 0.0020	0.98443 ± 0.00054
Pipalwas, galena in chlorite schist, whole sample leachate	14-9	15.688 ± 0.016	15.447 ± 0.016	35.233 ± 0.048	2.2457 ± 0.0009	0.98449 ± 0.00043
Rajpura Dariba, ore: galena	16-18	16.061 ± 0.014	15.477 ± 0.040	35.815 ± 0.089	2.2345 ± 0.0021	0.96543 ± 0.00020
Rajpura Dariba, ore: galena	16-18	16.056 ± 0.021	15.485 ± 0.019	35.892 ± 0.039	2.2297 ± 0.0034	0.96441 ± 0.00061
Rajpura Dariba, host: graphitic schist or phylite	16-10	16.800 ± 0.0149	15.571 ± 0.0205	36.237 ± 0.0045	2.1580 ± 0.00281	0.92708 ± 0.00281
Khankaria, malachite stains	11-9	16.122 ± 0.008	15.524 ± 0.007	35.925 ± 0.057	2.2284 ± 0.0032	0.96301 ± 0.00040
Khankaria, host: actinolite tremolite schist; serpentine rock	11-6	16.269 ± 0.0209	15.581 ± 0.0065	35.887 ± 0.0564	2.2050 ± 0.00063	0.95757 ± 0.00031
Dariba nola, galena and pyrite ore	10-2	33.39 ± 0.12	17.04 ± 0.06	61.44 ± 0.26	1.844 ± 0.001	0.5167 ± 0.0002
Khetri, host: quartzite, whole rock acid leachate	4-6	21.311 ± 0.0574	15.923 ± 0.0504	43.496 ± 0.1099	2.0407 ± 0.00316	0.74702 ± 0.00048
Ghatiwali, galena and pyrite in muscovite schist ore, sulfide leachate	7-1	45.91 ± 0.10	18.24 ± 0.05	50.09 ± 0.18	1.091 ± 0.002	0.3976 ± 0.001

*Standard deviation (20).

Acknowledgments

The research was sponsored by M.S. University of Baroda through University Grants Commission and Fulbright-Hays Commission, International Exchange of Scholars. We are grateful to Helena Wing for data on figure 1 and discussions with C.C. Patterson, California Institute of Technology. Special thanks to Professor K. T. M. Hegde, Department of Archaeology and Dr. S. P. Sychanthavong, Department of Geology, M.S. University of Baroda for their discussions in the filed and fieldnotes.

References

1. K. T. M. Hegde, "Sources of Ancient Tin in India." in A. D. Franklin, J. S. Olin and T. A. Wertime, eds., *Search for Ancient Tin*, (Washington, D.C.: Smithsonian Institution, 1978) pp. 39-42.

2. W. A. Oddy and W. Zwalf, *Aspects of Tibetan Metallurgy*, British Museum Occasional Papers 15 (1981):20.

3. J. E. Ericson, K. T. M. Hegde, and S. P. Sychanthavong, Ancient Copper Mining and Smelting Industries of Western India, International Symposium on Archaeometry and Prospection, Brookhaven National Laboratory, New York, 17-22 May 1981.

4. J. E. Ericson and C. Reedy, "Context and Analysis of Copper Slags from Western India," *Science in Archaeology,*" (Philadelphia: Archaeological Institute of America, 1982).

5. J. A. Dunn, "Copper," Geological Survey of India, Bulletin 23 (1965).

6. *Geological Survey of India,* Miscellaneous Publication 13 (1968).

7. R. Chakrabortty, "Bibliography of Early Indian Mining and Metallurgy, *Indian Minerals, Geological Survey of India* 27 (1973): 139-154.

8. C. Reedy, Production Technology of Medieval Western Himalayan Metal Statues. Unpublished PhD. dissertation, Program in Archaeology. UCLA 1984/5.

9. D. P. Agrawal, C. Margabandu, and N. C. Shear, "Ancient Copper Workings: Some New ^{14}C Dates," *Indian Journal of History of Science,* 11 (1976):132-136.

10. H. W. Schoff, ed., *The Periplus of the Eythrean Sea,* (1974) p. 151.

11. G. Weisberber, "Evidence of Ancient Mining Sites in Oman: A Preliminary Report," *J. Oman Studies 4* (1978): 15-28.

12. S. P. Sychanthavong and S. D. Desai, "Proto-Plate Tectonics Controlling the Pre-Cambrian Deformations and Metallogenic Epochs of Northwestern Peninsular India," *Minerals Science Engineering,* 9 (1977):218-236.

13. O. Werner, *Spektralanalytische und Metallurgische Untersuchungen An Indischen Bronzen* (Leiden: E. J. Brill, 1972).

14. K. T. M. Hegde, Technical Studies in Chalcolithic Period Copper Metallurgy in India, Ph.D. dissertation, M.S. University of Baroda (1965).

15. D. P. Agrawal, "Problems of Protohistoric Copper Artifacts and their Ore Correlations," in U. V. Singh, ed., *Archaeological Congress and Seminar: 1972, B. N. Chakravarty, University Kurushetra* (1976) pp. 273-277.

16. I. L. Barnes, J. W. Gramlich, M. G. Diaz, and R. H. Brill, "The Possible Change of Lead Isotope Ratios in the Manufacture of Pigments: A Fractionation Experiment," in G. Carter, ed., *Archaeological Chemistry* II (Washington, D. C.: American Chemical Society, 1977) pp. 273-277.

17. R. H. Brill and J. M. Wampler, "Isotope Studies of Ancient Lead," *Am. J. Archaeol.* 71 (1967): 63-77.

18. I. L. Barnes, W. R. Shields, T. J. Murphy, and R. H. Brill, *Isotope Analysis of Laurion Lead Ores,* Adv. Chem. Series 138 (1974) 1.

19. R. M. Farquhar and I. R. Fletcher, "Lead Isotope Identification of Sources of Galena from Some Prehistoric Indian Sites in Ontario," *Canada Science* 207 (1980) 640-643.

20. C. Reedy, "Metallurgy in Himalayan Culture," *J. Asian Culture* 7 (1983): 72-95.

21. B. R. Doe and R. Rohrbough, "Lead Isotope Data Bank: 3458 Samples and Analyses Cited," *U. S. Geological Survey, Open,File Report,* 79-661 (1977).

GIRAUD V. FOSTER

Xeroradiography: Non-Invasive Examination of Ceramic Artifacts

Xeroradiography is a non-destructive radiological method applicable to the study of archaeological artifacts.[1-3] By detecting minor differences in the radiodensity of solid objects, it has the potential to demonstrate details not seen in conventional radiographs or by direct inspection.

In this paper we report its applications to the study of ceramic artifacts. Investigations have been carried out to determine how vessels are made, figures constructed, repairs detected, inclusions recognized, and mold-produced features identified. Xeroradiographs taken of excavated specimens have been compared to models made under controlled conditions to assure that interpretation of radiological images are correct.

Methods

Xeroradiography

The principles of the method have been reviewed by others.[4] Xeroradiography is similar to conventional radiography with the important difference that instead of film, a plate of amorphous selenium sulfide is employed to record the image produced by x rays. Before use, the plate is charged in an electrical field. The object to be studied is interposed between the radiation source and the plate. Following exposure to x rays, a latent image remains on the selenium sulfide where the charge has been partially dissipated. The image is developed with blue particles of opposite charge, called "toner," and transferred to paper. This produces a blue on white or white on blue picture, depending on whether the charge on the plate was initially positive or negative. During development of the partially depolarized selenium sulfide plate, the number of charged colored toner particles is depleted at the edges of the latent image on the side carrying a similar charge. These colored particles are attracted to the areas of the latent image that have maintained a charge opposite to that of the toner (fig. 1). In this way artificially sharp edges are produced that give the impression of high contrast and make possible measurements that are not feasible in conventional radiographs. The image edge is darker and a halo about the image is caused by the toner being depleted from the periphery.

In the following studies, a 72-inch focal distance was used to minimize distortion. Artifacts were placed on or close to the cassette carrying the semi-conductor and positioned using radiolucent sponge wedges. To improve chances of observing joins, objects were xeroradiographed from several angles.

Clay

Potters' standard clay 153 was used in all studies. Objects were made to demonstrate wheel construction, coil construction, construction of slabs, end-to-end joins, handles, punch-through joins, and joins made in sculptured pieces. In all, several hundred pieces were produced in order to illustrate a spectrum of variations of each of the above techniques. Objects were fired at 850° for four hours in a Norman Test Kiln Model T-H, monitored by an Eagle Pyrometer, and allowed to cool in the furnace to room temperature.

Applications

Recognition of Method of Construction of Vessels

Hand-formed, coil-made, slab-made, and wheel-thrown pots cannot

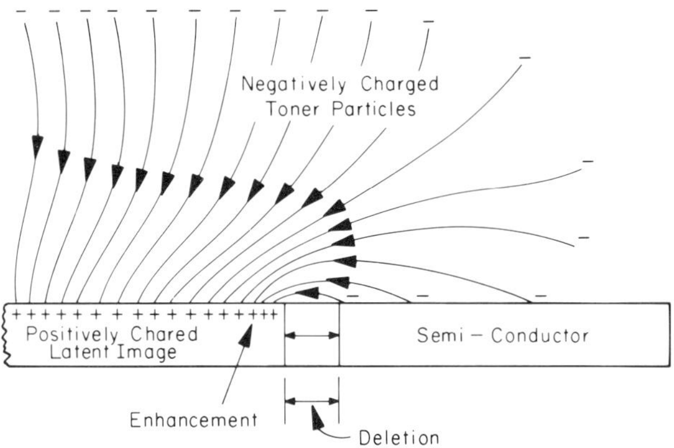

Fig. 1. Schema of toner particle movement during development of xeroradiographs. The charge on a surface of amorphous selenium sulfide, which acts as a semi-conductor, is partially dissipated by ionized radiation, leaving behind a latent image. When exposed to colored toner particles of opposite charge, these are attracted to areas on the latent image on the semi-conductor in proportion to the residual charge. Because of this attraction, the interface of different radiodensities in the object studied are enhanced by adjacent depletion and addition of toner as illustrated.

always be distinguished from one another even with the aid of xeroradiography. Xeroradiography, however, frequently helps to recognize wheel-made pottery when parallel markings, produced by scraping or trimming instruments, are seen. Air inclusions, if present, are invariably parallel to the base (fig. 2). In contrast, coil-formed pots often contain air spaces within and between coils and the heights of coils are not much larger than their widths as seen in cross-section at the edges of the pots. Pots constructed of slabs differ in that the heights of the coils are much greater than their widths. Furthermore, the angle of coil rotation is significantly steeper in slab-constructed pots than in those made from coils. Hand-formed pots, often called pinch pots, usually have walls of variable density that give a "moth-eaten" appearance in xeroradiographs; air inclusions, when present, are randomly oriented (fig. 3).

Not all vessels made by potters in antiquity show these diagnostic signs. The features are easily obscured by use of scraping tools, wet sponges, or compression during final remodeling. Orientation of small air inclusions and high-density particles in the clay body, however, frequently permit distinction between hand-made pinch pots and those made by other methods.

Detection of Joins

End-to-end joins are potentially detectable as straight lines. Joins made by approximating beveled surfaces or by sewing edges together appear as such. When clay is added for support, the additions are usually self-evident in xeroradiographs, although demonstration of such joins frequently requires several exposures at several angles. Joins are particularly difficult to see in thick-walled pots, pots made with acute angles, and pots whose joins have been extensively modeled by the potter.

Fig. 2. Mayan bowl, Classical period. Distribution of air inclusions parallel to the base of the pot are compatible with coil construction. Intercoil spaces were probably obliterated during final reworking, which accounts for the partial upward displacement of some air inclusions. No large radiodense particles are present in the clay suggesting that the clay may have been prepared by extensive lavigation.

Fig. 4. Small blackware pitcher, Greek, fourth century B.C. Lower portion of the handle is attached to body of the vessel with sprigging in the design of a circle.

Fig. 3. Egyptian beaker, red slip with reduced black rim, late Predynastic Period or early First Dynasty. Paucity of radiodense particles is compatible with relatively well-washed clay. The diffuse distribution of air inclusions together with the heavy and irregular base indicate that the vessel was made by hand in the pinch pot manner.

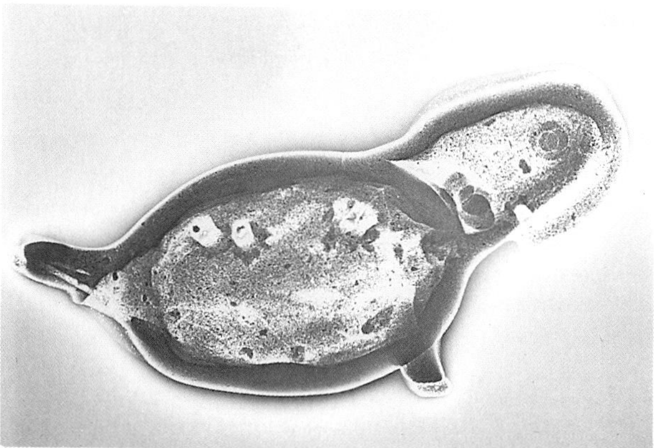

Fig. 5. Mayan ocarina, late Classical period. The trailing on the inner edge of the musical stops supports the conclusion that the holes were made with a solid rod punch. The absence of core buttons indicates that the central cavity was not sealed until after the buttons had been removed.

Side-to side joins, as used to affix handles to the bodies of pots, are made in several ways. Most commonly, a slip is applied to affix two flat surfaces. However, sprigging is also occasionally employed. This entails the regular or irregular etching of one surface to improve stabilization of the join (fig. 4). Unique designs may prove useful as a means of recognizing individual potters within a given culture, place, and period.

Punched-through joins are frequently used to attach handles, feet, and spouts to vessels. Holes are usually made from the outside by methods that are often recognizable. When a solid cylinder is used as a punch, streaming is observed on the interior surface of the hole

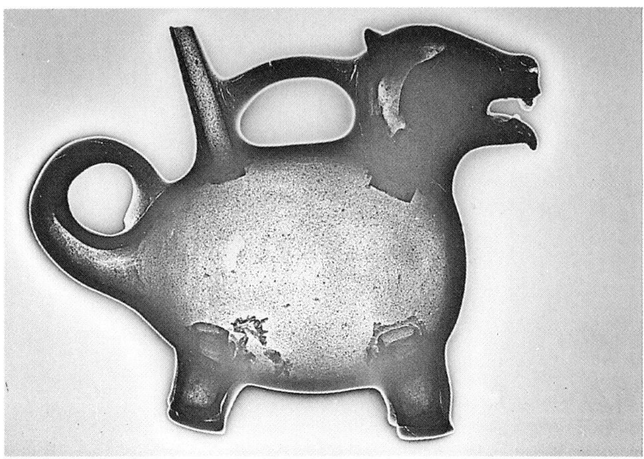

Fig. 6. Jaguar huacos, burnished red slip, Teotino culture (Peru). Irregular shavings lying on the floor of the central cavity are compatible with the holes for insertion of the head and feet having been carved with a sharp blade when the body was in the leather-hard state. The fine linear air inclusions at joins of the tip of the animal's tail to the body and the handle to the spout indicate that wet clay was used to make the seals that support these joins.

and a core with irregular edges may be found within the piece, if it is hollow and completed before the additions were made (fig. 5). Holes produced with a sharp-edged hollow cylinder produce holes without streaming and the cores, if present, are smooth-edged. Irregular inclusions that appear as shavings indicate that holes were carved with a blade when the clay was in the leather-hard state (fig. 6). The tubular elements are also usually added when leather-hard and joined to the parent piece with wet clay. Thin radiolucent lines invariably indicate these unions as well as modern repairs made in the same manner.

Complicated pieces of pottery are most often made in sections. Spheres are construced from two hemispheres that are joined. Necks, lips, and bases are separately made and separately affixed. The component parts, expecially in coarseware, are readily recognizable (fig. 7). However, joins in fineware may be difficult to detect.

Evaluation of Construction of Sculptured Artifacts

Hand-sculptured figurines, particularly those made with sandy clays, frequently contain air inclusions that are good indicators of methods of construction. Air, when entrapped in clay that is rolled into coils, elongates to a ribbon that may be recognizable in xeroradiographs. The direction of air inclusions and the orientation of temper particles suggest construction methods. Iron Age Cypriot votive figurines have been studied by the author in this way.[3] Most of these were horse and rider figurines or four-horse chariot groups. In all cases the horses were found to be manufactured with an economy of effort. Working from a cone-shaped piece of clay, the wider end was squeezed to form a Υ. One arm of the Υ was used to produce the neck and head of the horse. The other arm and tail of the Υ were pulled downwards and split to form the fore and hind limbs respec-

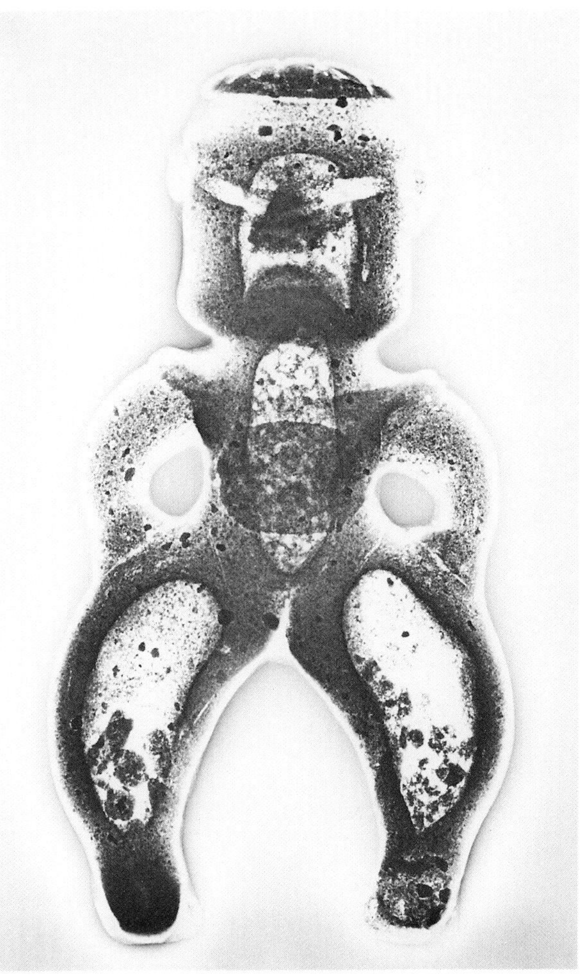

Fig. 7. Rattle, Guanacaste-N-coya (Costa Rica). The fine linear radiolucencies where the arms rest on the hips and where the thighs join the trunk are indicative that the artifact was assembled in pieces. The contents of the rattle remain unknown but their irregularity indicates that they are not washed pebbles. Part of the right foot is missing. The presence of relatively large radiodense particles in the clay is consistent with clay that has not been prepared by sedimentation techniques.

tively. Clay was gathered by scraping the groin of the horse to model a tail for the horse.

The position and shape of air inclusions can prove helpful not only in indicating how clay is bent but also where additions have been made. Air inclusions lying deep in added parts or in the parent body do not overlap. Conversely, if deep air inclusions extend across an area, it is likely that joins are not present there.

Characterization of Temper

Sand, shell, organic matter, and grog, in addition to air inclusions, are readily detectable by xeroradiography. Temper produced by crushing natural objects is invariably angular and readily distinguishable from alluvial sands, the particles of which have rounded edges.

Grog can sometimes be recognized in xeroradiographs. The particles are angular and may have radiodensities that differ from that of the clay. Clay that has not been vigorously wedged or slaked will have air pockets on one or more sides of at least some of the grog particles.

The texture of the clay body itself can be characterized. In our laboratory negative emulsion films of xeroradiographs are scanned by a microdensitometer with data acquisition and image enhancement capabilities. This permits the percentages of particles of different sizes and radiodensity in one sherd to be estimated and compared statistically to those of another.

Assessment of Repairs

X-radiography facilitates detection of repairs. It is not difficult to recognize when plaster of Paris has been used, since it is radiolucent relative to most ceramic materials. Repairs in which sherds from other pieces have been used to replace missing pieces are also easily recognized. However, repairs mimicking the clay body itself, inclusive of temper, may present difficulties. Where modern repairs have been made with wet clay, a radiolucent line in xeroradiographs invariably indicates the junction of the addition.

Measurement of Molded Artifacts

Many objects were manufactured in antiquity either wholly or partially in molds. This is particularly true of terracotta votive figures. Sometimes dozens of different molds were used to make a particular part such as the body or face. Identification of molds is facilitated by taking xeroradiographs of impressions of the original pieces made in Polyform. Polyform is a non-oily plastic material that does not shrink. It can be used in the field and transported to the laboratory where xeroradiographs can be taken. The features recorded are distinct due to edge enhancement and toner depletion producing sharp borders that lend themselves to making measurements. Measurements can be made from xeroradiographs with greater precision than from photographs of the objects or the objects themselves. Measurements facilitate the detection of castings from secondary and tertiary molds. Every time a mold is made of an object and fired, shrinkage occurs. When a cast is made from the new mold and fired, it too shrinks, such that the final product from a secondary mold may be as much as 30% smaller than the original, and that from a tertiary mold a third again as small.

Identification of objects made from secondary and tertiary molds may be of importance in establishing chronology in a sanctuary where votive figurines were periodically buried in pits. For example if two votive pits are present and one contains figures cast from smaller but otherwise identical molds than another, it can be presumed that the former pit is probably the more recent of the two.

Summary

From these studies it is concluded that xeroradiography is useful in determining how ceramic artifacts are constructed or restored. Visualization of temper and air inclusions may suggest how clay was handled, and xeroradiographs taken of impressions of ceramic objects can facilitate the making of measurements with a precision not possible either from photographs of the objects or the objects themselves. The findings support the view that xeroradiography has the potential to help identify and classify ceramic artifacts by schools and possibly by individual potters.

References

1. S. Heinemann, "Xeroradiography: A New Archaeological Tool," *American Antiquity,* no 41 (1976): 106-111.

2. R. E. Alexander and R. H. Johnson, "Xeroradiography of Ancient Objects: A New Imaging Modality," *Archaeological Ceramics,* ed. J. S. Olin and A.D. Franklin (Washington: Smithsonian Press, 1982), pp. 45-61.

3. G. V. Foster, "Xeroradiography: A Study of the Kourion Votive Figures," *MASCA Journal* 3: in press (December, 1983 issue).

4. E. E. Christensen, T. S. Curry III, and J. E. Dowdey, "Xeroradiography," in *An Introduction to the Physics of Diagnostic Radiology,* 2nd ed. (Philadelphia: Lea and Febiger, 1978) pp. 308-328.

H. F. JAESCHKE

Oriental Lacquer Analysis: Thin-Section and Electron Microprobe

In the past, the study of the structure of lacquered objects has been largely limited to visual examination[1] of the surface or damaged areas and the use of x-radiography. A few sections have been examined by reflected light.[2]

Thin sections viewed by transmitted light reveal the fine detail of the structure of the object, the materials used in its fabrication, and the methods of manufacture. They can also disclose repairs, alterations, and areas of deterioration, providing further data on the history and condition of the object.

Thin sections are also well suited to examination by electron microprobe, providing qualitative and even quantitative analyses of separate layers[3] or even of individual particles embedded in the lacquer matrix, thus eliminating some of the doubts and errors that can occur through the use of less discrete samples.

Sampling

Although this technique requires only minute samples (samples less than 1mm^3 have proved adequate in some cases) the nature of the objects themselves restricts the choice of sampling sites. To disrupt an intact lacquer surface on an object could cause increased damage out of all proportion to the size of sample taken. For the present, sampling should be limited to areas that have already been damaged.

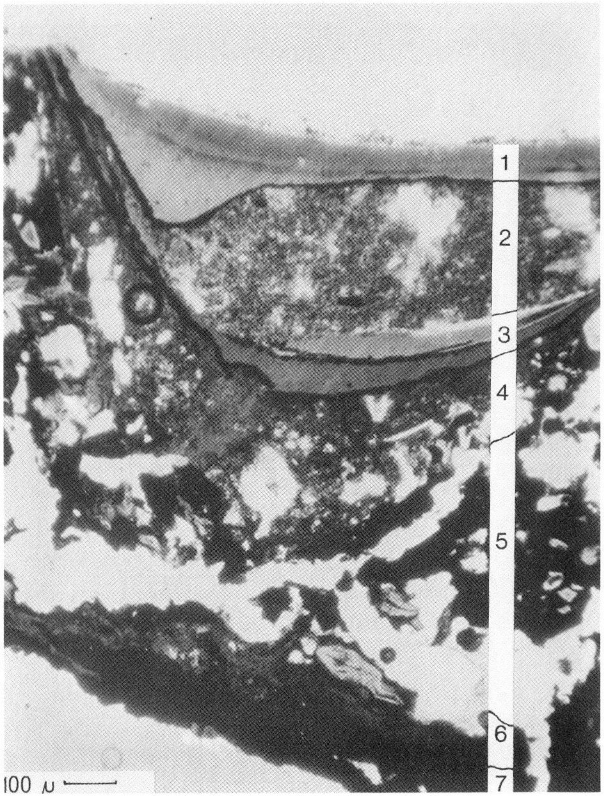

Fig. 1. Thin section: Japanese eighteenth-century black lacquer.

Many of the samples used in this study came from small flakes that had become detached from the object, usually chipped from an edge or from an area weakened by physical damage. It may, however, be necessary to detach a sample from a damaged edge or break. A clean, sharp scalpel blade or a hollow-core dental drill (1-2mm in diameter) may be used. Drills should be hand-operated, as electrical drills, even at low speeds, will overheat, scorching the lacquer.

Sample Preparation

The sample is embedded in polyester or epoxy resin. One face of the specimen is exposed, using a fine jeweler's saw. This face is smoothed using increasingly fine grades of wet and dry abrasive paper. Fine abrasives are then used to polish the face, usually either 6µm, 1µm, and .25µm diamond paste or aqueous suspensions of 5µm and .1µm alumina. The choice of abrasive system depends on the particular vagaries of the sample, the coarseness of its constituents, and their susceptibility to the solvents involved. The abrasives are generally applied to a soft polishing cloth fastened to a slowly revolving disc. Great care must be taken when changing grades of abrasive to ensure that the sample is thoroughly rinsed and that no cross-contaminiation of the polishing cloths occurs. The specimen should be examined under the microscope as the polishing progresses so that the correct materials can be chosen.

The specimen is then cut free from the resin block, using a fine jeweler's saw. It is attached to a glass microscope slide, polished face down, using a transparent adhesive. Clear epoxy and polyester adhesives have proved to be the most suitable. The exposed face is then ground down and polished until the correct thickness is obtained. This can only be judged by viewing under the microscope.

Examination

The studies summarized here were conducted using an Olympus BH microscope with 35mm camera attachment. The sections were examined at different magnifications and whole views or special features recorded on Kodachrome 50 ASA tungsten light film. Color photography is essential for recording the images—black and white film, even with filters, cannot capture the degrees of transparency or the subtlety of the color differences, especially where the lacquer appears orange or red.

A few specimens are included to illustrate results obtainable. Figure 1 is a sample of black lacquer from a metal neckguard, probably from the late eighteenth century, attached to a Japanese helmet of slightly earlier date. The base layer (6) was originally applied directly to the prepared metal (now represented by a layer of iron corrosion products (7)). A foundation layer of lacquer mixed with coarsely ground clay and plant fibres (5) was applied over this, followed by a second foundation layer of lacquer and finer clay (4). On top of the foundation layers can be seen the remains of two layers of transparent lacquer, the original surface (3). These are black lacquer, tinted by a chemical reaction with iron salts rather than with an opaque pigment. Hence the layers appear pale yellow in thin-section. Both of these layers lie directly over a thin opaque band. It is not known exactly what this is, but it could be the remains of the polishing compounds (usually charcoal or bone-ash) used to rub down each layer

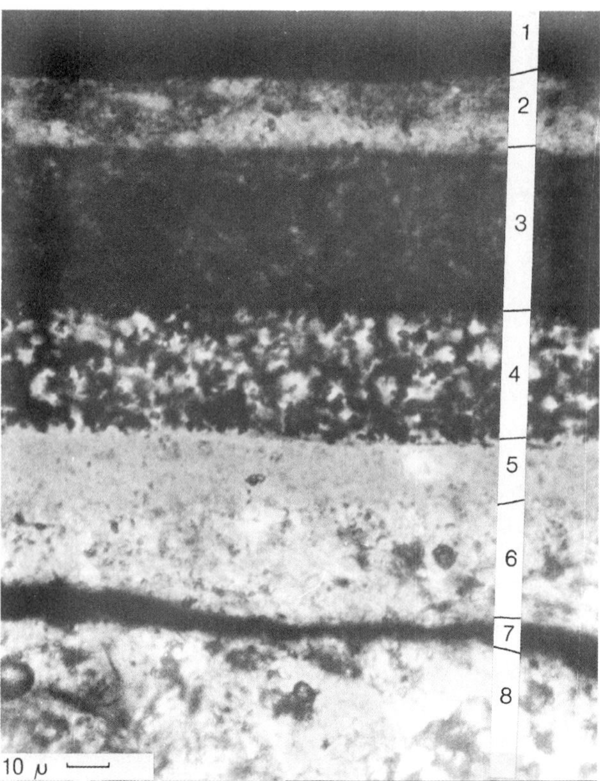

Fig. 2. Thin section: Japanese eighteenth-century red lacquer.

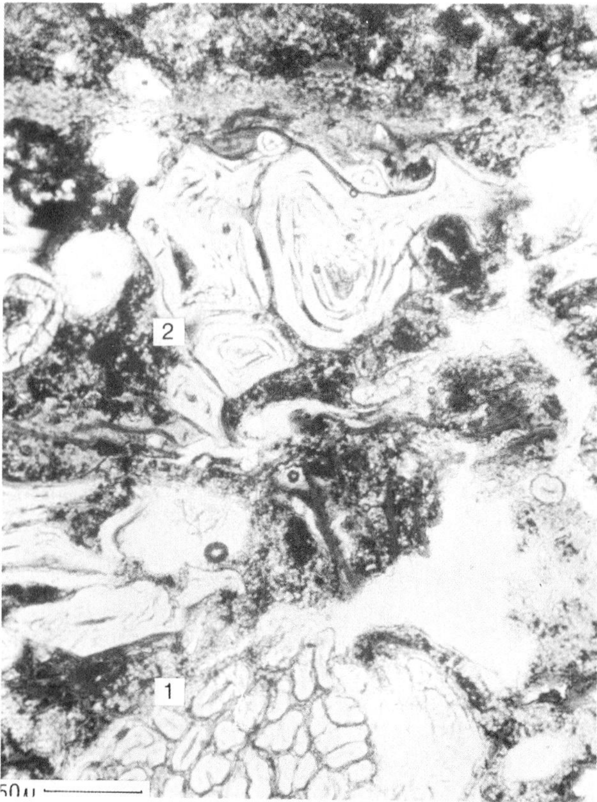

Fig. 3. Ramie and other plant fibers: Japanese eighteenth-century black lacquer.

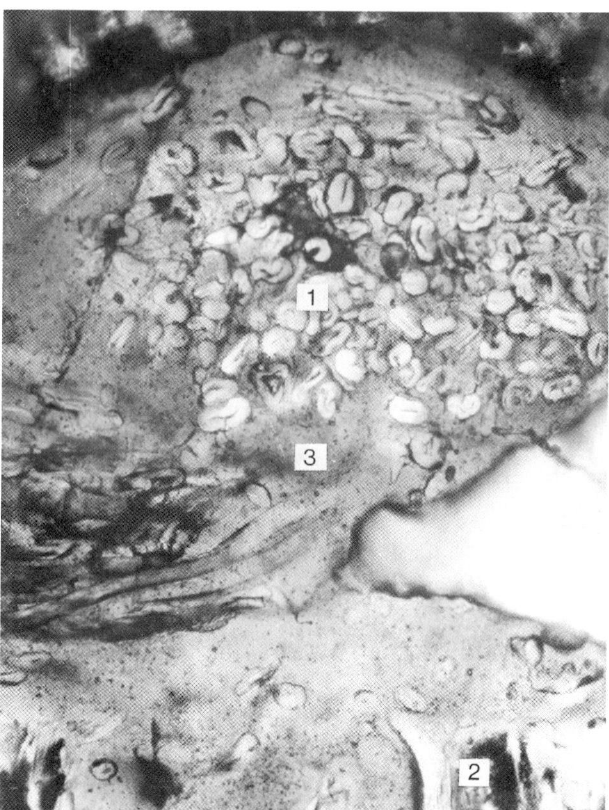

Fig. 4. Cotton fabric on wooden base: Chinese twentieth-century red and black lacquer bowl.

before the application of the next. The finer layers at the surface are usually cleaned thoroughly after polishing, but it is possible that less care was taken with this piece, especially since the compounds would not be readily visible against the black lacquer. The original surface has subsequently broken down, either through wear or through deliberate grinding to reshape the surface. A fresh foundation layer of fine clay and lacquer (2) has been applied, followed by two more layers of transparent black-tinted lacquer (1). The black opaque layer is visible beneath the first of these, but does not appear between the first and second. Note that these two surface layers adhere closely to each other, unlike the two layers of (3).

Compare that section with figure 2, a sample of red lacquer from the same neckguard. Here the foundation layer (8) is much finer and has been sealed with a very thin opaque layer (7) before the application of transparent layers (6, 5, 2) and layers pigmented with different proportions of translucent subangular red grains, later identified as cinnabar (4, 3, 1).

Although the study of the entire cross section of different lacquers yields a wealth of information, the study of individual particles and inclusions can also be of value. Fibers are often used in the base layers, both as textiles and as unwoven fibres and can often be identified in thin-section. Figure 3 shows the clay and lacquer foundation layer of a sample of eighteenth-century Japanese lacquer, containing a bundle of ramie fibers (1) and several larger, loose plant fibres (2). Compare this with figure 4 from a twentieth-century Chinese bowl. Here cotton fabric (1) has been pasted onto the wooden base (2) with a transparent layer of lacquer (3). Note the presence of fibers in cross-section and longitudinal section.

Thin sections can be used to study the particular techniques of dif-

 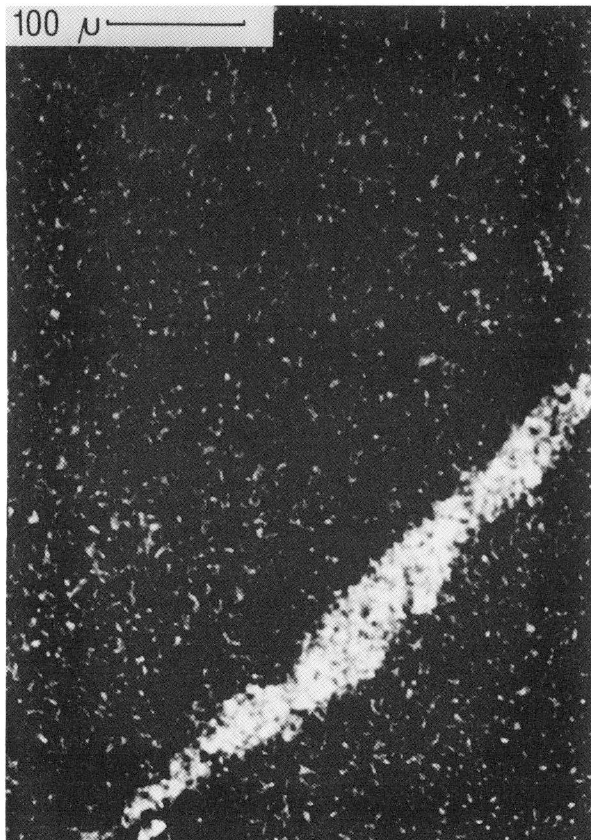

Fig. 5. Electron microprobe scans (left: Hg, right: Si), Japanese nineteenth-century red sake bottle.

ferent periods or individual craftsmen. The foundation layers, for example, often reveal a great deal about the value accorded to a piece and the care used in its manufacture. The size and shape of decorative particles of pigments or metals, for example, can be recorded, and their method of placement studied. When the thin section is examined under the electron microprobe, even more information can be obtained.

Electron Microprobe

The electron microprobe used in this work was a Cambridge Microscan 5, capable of identifying the presence of any element with an atomic number greater than that of boron (5). The beam can be focused down to a point less than .1 µm in diameter (though when used on lacquers it forms a crater several microns in diameter), or set to scan a line or an area of the specimen. This is extremely useful when plotting the distribution of a particular element or group of elements.

The thin section was first examined using the fine focused probe to identify the elements present and to indicate their relative proportion. Each layer in the sample was analyzed at several different places and the results were noted. Then individual particles were analyzed. Finally scans were made of major elements present in the sections. Figure 5 shows two scans of the same area of a sample from a nineteenth-century Japanese red sake bottle plotting the distribution of mercury (left, present as the red pigment, cinnabar) and silicon (right, present in the clays used in the foundation layer).

Figure 6 shows two scans of the same area of fine metal dust in the surface of a brown Japanese writing box, (ca. 1746). Analysis of the surface or a pulverized sample would have revealed the presence of iron, copper, and zinc, but the use of the microprobe reveals that the iron is present only in the lacquer matrix and that the metal dust is a mixture of pure copper and a copper-zinc alloy.

Future Developments

This work represents an initial foray into what looks to be a very promising field of study. To be of maximum use to curators, conservators, and other interested parties, this kind of data should be gathered into a comparative corpus embracing as many areas and ages of lacquer production as possible.

Thin-section analysis is only one of the many valuable methods of studying lacquerware, but it provides so much detailed information that it deserves further consideration. Two particular aspects that may prove of great interest or value are the correlation of analytical results with those obtained by the use of powdered or extracted samples, or by non-destructive analysis of the whole object; and the study of the areas of deterioration and damage to aid in the prescription and evaluation of various methods of cleaning and conservation.

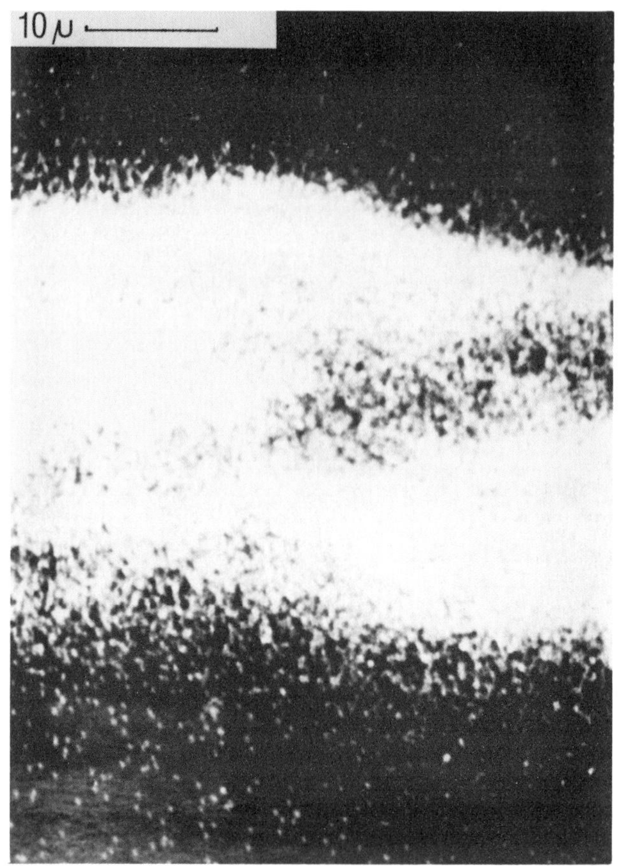
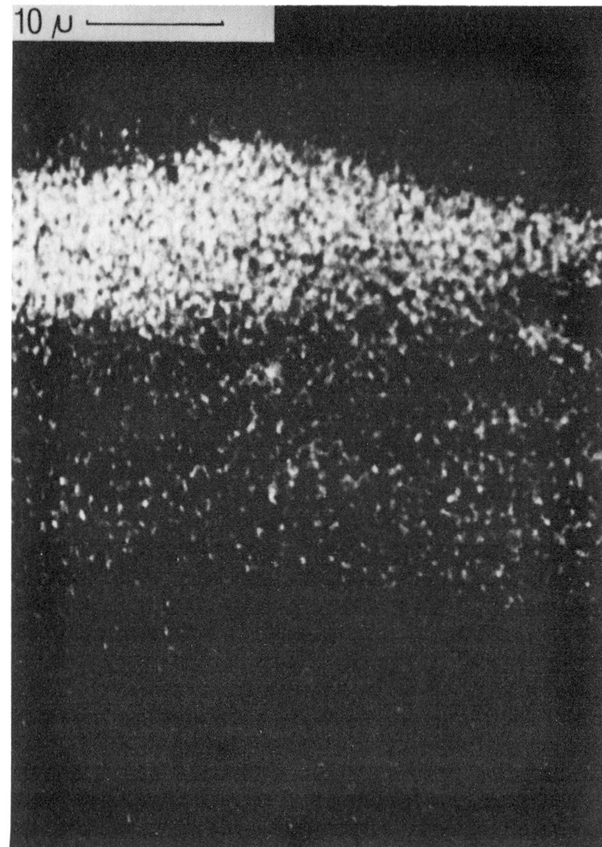

Fig. 6. Electron microprobe scans (left: Cu, right: Zn), Japanese writing box, ca. 1746.

Acknowledgments

I would like to thank the Institute of Archaeology, London, U.K., Department of Conservation and Materials Science, for providing both the topic and a base. Also Peter Dorrell and Stuart Laidlaw of the Photographic Department for their generous help and advice; Eastgate House Museum, Rochester, U.K., especially Michael Moda, Keeper, for providing samples from the Japanese armor collection; the Victoria and Albert Museum, London, U.K., especially members of the departments of Oriental Arts and Conservation, for providing samples from the collection; University College, London, U.K., Geology Department, for the use of their electron microprobe, especially Dr. R. M. Preston for granting permission and Ian C. Young and Keith W. Stephens, for their unfailing help; National Research Institutes for Cultural Properties, Tokyo and Nara, Japan: especially Dr. H. Mabuchi and Masaaki Sawada for generous help and advice; Restoration Workshop, Toshogu Shrine, Nikko, Japan: especially Yoshihara Hokusai and Sato Noritake for providing samples and information; and finally, The Winston Churchill Memorial Trust Fund for financing the trip to Japan.

References

1. T. Nakasato, various articles, especially "Technical Data of Lacquer Work in the Heian Period," (ix) *Science for Conservation* 19 (1980): 91-105.

2. Sir H. Garner, "Technical Studies of Oriental Lacquer," *Studies in Conservation* 8 (3) (1963): 84-95.

3. A. P. Hornblower, "Some Examples of Work with the Electron Probe Microanalyser," *Archaeometery* 61 (1963): 37-42.

K. A. JAKES and L. R. SIBLEY

Chemical and Physical Analysis of Textile Fabric Pseudomorphs

Textile fabric products function in a close association with the people who manufacture, use, and discard them. Their study as archaeological artifacts has been hindered as few survive the long-term degradative influences of the environments in archaeological contexts. Surviving occasionally as part of the corrosion products surrounding metal objects, however, are mineralized formations termed textile fabric pseudomorphs that display the physical shapes of textile fibers, yarns, and fabrics. They are formed in a petrification process in which mineral compounds replace the organic structures of fibers while retaining their physical identities. Very little is known about their composition, the nature of the replacement process involved, the factors that encourage their formation, or their evidence of textile products. Where textiles do not survive and textile fabric pseudomorphs are found, the latter may be analyzed as if they were fragile textiles in order to reconstruct the cultural use of textile products. In the case of metal artifacts with unknown provenance, determination of the composition of the pseudomorphs may indicate such conditions present in the archaeological contexts as acidity, oxidation potential, and elemental content, thus contributing essential information about the artifacts.

The pseudomorph phenomenon is not restricted to one particular location, type of fiber, type of metal, or period of history. Rather its incidence is far more frequent geographically and culturally than had been anticipated in the past. Such a frequency implies the availability of textile evidence hitherto unknown. The impact of this type of information upon the reconstruction of such cultural processes as trade and exchange networks, of prestige hierarchies, and of technological developments is being assessed. It could well be substantial. Biek was the first to use the term "pseudomorph" to characterize a mineralized textile fragment that he found on the surface of a brass ring from the medieval Chertsey Abbey.[1] Sylwan described entities that may be pseudomorphic as "preserved fragments of silk belonging to the patina" of a Shang bronze axe and *chi*,[2] and Vollmer reported the finding of pseudomorphs on seventy-four Shang period bronzes in the Royal Ontario Museum.[3] Carroll identified flax pseudomorphs attached to a bronze Etruscan bowl found at Veii.[4] Jakes and Sibley described the textile fabric pseudomorphs displayed on a Shang period bronze halberd and proposed a model for their formation.[5] A more detailed review of the literature on the topic of textile fabric pseudomorphs can be found in an earlier article by Sibley and Jakes.[6]

Before being analyzed as examples of textile fabric structures, pseudomorphs must first be described as mineralogical structures. This is essential for the determination of the environments conducive to the occurrence of the phenomenon, the verification that mineral replacements do in fact maintain the fiber shape, and the determination of the organic materials previously present that affected the replacement process. The purpose of this investigation was to accomplish the chemical and physical characterization of pseudomorphs on a select number of artifacts.

No textiles have been found from the Shang period, but recovery of numerous bronze artifacts from that period as well as the environmental conditions of their burial makes the Shang corroded metals a logical choice for the investigation of the phenomenon of textile fabric pseudomorphism. In this work, three bronze weapons dated to the Anyang phase of the Shang period (ca. 1300 B.C.) were ob-

tained on loan from the Museum of Anthropology, University of Missouri, Columbia. Two of the weapons are halberds, labeled UMC 1 and 2; the third weapon is a spear point labeled UMC 3. These objects were examined for evidence of pseudomorphism. They were photomicrographed, and small samples were removed for elemental and compositional analyses.

Experimental Procedures

Photographs of the three bronze weapons were taken with a Nikon F camera with a micro Nikon lens using Kodachrome type A film. Each object was placed on a sheet of glass on which was drawn a grid of 1cm by 1cm. The glass, suspended beneath the camera and over a gray matboard, was angularly illuminated in a 2:1 lighting ratio in order to eliminate background shadows and to highlight the surface texture of the object. To achieve true color, the photographs were developed by the Cibachrome process in lifesize and twice lifesize prints.

A Wild M400 photomacroscope with zoom settings yielding magnifications of up to 20.4X, a Wild MPS55 Electrical Control Unit, fiber optics illumination, and Panatomic X and Kodachrome type A film were used in taking the photomicrographs. The objects were supported on a stage that was marked in a 1cm by 1cm grid and moved by means of microscope positioning screws so that specific locations on each object could be found repeatedly.

Samples of pseudomorphs from halberd 1 and the spear point were cut by scraping carefully along the edge of the sample with a blade until it became loosened from the corrosion material around it. Although these samples were detached from the other material in which they were embedded, some material on their surfaces could not be removed, and therefore some contamination was expected in the mineralogical analysis. Small samples that were intended for x-ray diffraction were ground into a powder and collected into a small wax bead or spindle mount. X-ray diffraction powder patterns were developed using a chromium source and a Phillips Debye-Sherrer camera with the length of exposure time for the samples ranging from five to twenty hours. The film was measured in accordance with standard procedures,[7] and the resulting diffraction spacing values were compared to standard values of known materials.[8]

Samples for scanning electron microscopy were cut and mounted in such a way that both an upper surface and a fresh-cut cross-section were visible. Samples from halberd 1 were coated with palladium and gold; samples from the spear point were carbon coated. Micrographs were taken at levels of magnification from 20X to 20,000X using a Cambridge S-4 Stereoscan Electron Microscope. Energy dispersive x-ray analysis was performed using an EDAX International model 707B connected to the SEM unit. EDAX counts were taken in both scanning and point focus modes.

Results

Halberd 1 is 23.8 by 5.8cm, halberd 2 is 25.6 by 7.5cm, and the spear point is 22 by 5.5cm. Each is encrusted with green, blue green, blue, black, and red orange corrosion. On one side of each object there is a pattern that is distinctly different from the remainder of the corrosion. The absence of materials of organic composition on all three weapons was confirmed by neutron radiography performed at the University of Missouri, Columbia.[9]

Halberd 1

Close scrutiny of halberd 1 reveals the presence of pseudomorphic structures in the corrosion. The pattern and regularity of the structures as well as their long, narrow shape distinguish them from the randomly located nodular growths characteristic of the corrosion products of bronze. Upon magnification, the intricacy of the patterns of the pseudomorphs becomes more apparent. An interlacing of filamentlike and yarnlike elements in a fabriclike structure is shown in figure 1. Upon higher magnification of the same area, heavy yarn pseudomorphs and thin filament pseudomorphs are visible (fig. 2). In some areas on the blade, the pseudomorphs are separate and distinct enough to show that each filament is actually two filaments used as one. This is characteristic of silk fiber alone and is one of the observations used in determining that the pseudomorphs on halberd 1 are those of silk. A more detailed discussion of the textile information derived from these pseudomorphs is given in reference 5.

Scanning electron microscopy of the green and black pseudomorphs showed that the interior of each, as observed in cross-section, was solidly filled. The green pseudomorphs were formed from platelike crystals; the black pseudomorphs were amorphous. As shown by energy dispersive x-ray analysis, both types of pseudomorphs contain copper and aluminum. X-ray diffraction powder patterns of the green pseudomorphs indicated that they contained malachite. Those of the black pseudomorphs could not be attributed to a cuprous crystalline material, and it was concluded that they contained tenorite, an amorphous, cuprous corrosion product of bronze.[10] The presence of separate black and green pseudomorphs in the same location on the halberd has led to some speculation concerning differential replacement of the silk fibers. No conclusive statements can be made at this point; research continues in the study of the replacement process. Data derived from the chemical and physical analyses of halberd 1 are summarized in table 1.

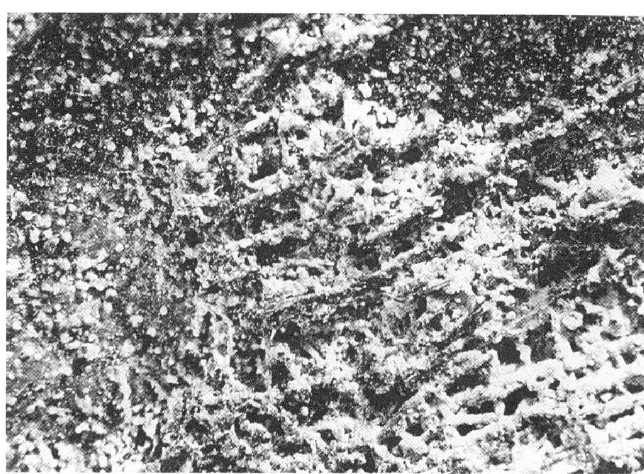

Fig. 1. Textile fabric pseudomorphs, 21.2X magnification.

Fig. 2. Dark heavy yarn pseudomorphs interlaced with pale, thin (green) filament pseudomorphs, 54X magnification.

Table 1
Chemical Analyses of Halberd UMC 1

Neutron Radiography	No organic compounds		
Micrography	Green pseudomorphs	Black pseudomorphs	Sandy-colored interstitial grains
Scanning Electron Microscopy	Microcrystalline	Amorphous	Amorphous
Energy Dispersive X-Ray Analysis	Cu, Al	Cu, Al	Si, Al, K, Ca, Fe
X-Ray Diffraction	Malachite Some Aluminum Hydroxide or Oxide	Tenorite Some Aluminum Oxide or Hydroxide	

The Spear Point

The pseudomorphs contained in the corrosion on the spear point are visible without magnification. In some areas, one can see a change in direction of the weave, as if the fabric had been crumpled or twisted at these points. Magnification of the pseudomorphs reveals the intricacy of the fabric pattern. The red-colored pseudomorphs are made from multiple element units in an unbalanced plain weave construction with the visual effect of a ribbed fabric. Beneath the red-colored pseudomorph, there is a layer of green-colored pseudomorphs of similar pattern. Figure 3 displays a patch of red-colored pseudomorph that has broken away to reveal the green pseudomorphs beneath it. The similarity of these pseudomorphs to a nineteenth-century ribbed silk fabric is striking (fig. 4).

Study of a cross-section of a small sample of the red-colored layer revealed that a majority of that layer was muddy green in color with a thin red coating on both top and bottom surfaces. Scanning electron microscopy of this sample showed no crystalline character in any of the layers. Elements contained in both layers, as determined by EDAX, were copper, aluminum, silicon, sulfur, potassium, calcium, and iron. Sulfur appeared to be localized, as indicated by its presence in some point focused EDAX analyses and its absence in others. The soil from the Yellow River area in China is loessial, containing silicon, aluminum, potassium, and calcium and probably is the source of these elements.[11] Sample contamination with soil and other surface particles was expected; apparently, soil elements also participated in the replacement throughout the whole pseudomorph layer.

Interpretation of x-ray diffraction patterns of mixtures is difficult, but among the bands displayed in the pattern produced by this red-colored sample were those bands attributable to cuprite.

When viewed with the scanning electron microscope, a sample of the green pseudomorph lower layer can be seen to comprise plate-like crystals. The elemental composition determined by EDAX is copper, aluminum, silicon, sulfur, and chlorine. These facts, combined with the clear green coloration of the structures, indicate that they are formed from malachite. Elements other than copper could be attributable to possible soil contamination; pale brown encrustations are apparent in the interstices between the fiber pseudomorphs.

Fig. 4. Nineteenth-century ribbed silk fabric.

Fig. 3. Surface layer of (red) fabric pseudomorphs over a lower layer of (green) fabric pseudomorphs, 54X magnification.

Table 2
Chemical Analyses of Spear Point UMC 3

Neutron Radiography	No organic compounds	
Micrography	Red-coated muddy green pseudomorphs	Green pseudomorphs
Scanning Electron Microscopy	Amorphous	Crystalline
Energy Dispersive X-Ray Analysis	Cu, Al, Si, S, K, Ca, Fe S is localized	Cu, Al, Si, S, Cl
X-Ray Diffraction	Complex pattern due to number of compounds present; probably cuprite plus others	

As can be seen in figure 3, these pseudomorphs are three-dimensional rod shapes similar to fiber filaments. The red pseudomorphs looked flatter, as if the red material were a coating that filled the interstices between the filament pseudomorphs enclosed in it. An explanation for this phenomenon cannot be given until further study of the pseudomorph formation process is conducted. Data derived from the chemical and physical analyses of the spear point are summarized in table 2.

Halberd 2

Of the three bronzes studied, halberd 2 appears to display the most readily discernible pseudomorphs because of the geometric pattern of colors and shapes that are apparent even at some distance. Upon close examination, however, the pattern proves to be pattern alone and is not due to three-dimensional mineralized structures of pseudomorphs. With increasing magnification, the pattern can be seen to be comprised of patches of color with no regular structure. Perhaps a mineralized fabric was present in the corrosion on the halberd and broke away, leaving these traces. Or, perhaps the fabric that was adjacent to the bronze influenced the formation of the corrosion products while it, itself, degraded. Further study of the process of pseudomorph formation will clarify these questions and determine to what extent the patterns such as this one can be studied as evidence for textiles.

Conclusion

Three bronze weapons, dated ca. 1300 B.C., were examined for pseudomorphic structures of textile fabrics. Two of the weapons, halberd 1 and the spear point, exhibited textile fabric pseudomorphs.

Scanning electron microscopy, energy dispersive x-ray analysis, and x-ray diffraction were used to determine the composition of these pseudomorphs. Halberd 1 contained pseudomorphs of malachite and tenorite; the spear point contained pseudomorphs of cuprite and malachite. Fiberlike, yarnlike, and fabriclike structures characteristic of silk were evident as were structures typical of an unbalanced plain weave. Though a pattern was evident on the surface of halberd 2, it was shown to be devoid of textile fabric pseudomorphs. Whether fabric pseudomorphs were ever formed on the halberd is not known at this time.

The basic mineralogical information determined in this work is now being applied to an examination of the pseudomorph formation process including the consideration of the environments conducive to their formation and the determination of information concerning their previous organic content. The evidence of textile structures is also being studied further.

The phenomenon of textile fabric pseudomorphism is useful as a means of reconstructing ancient cultures. The quantity of information that could be determined from the three objects without provenance suggests that much more could be learned from the study of pseudomorphs displayed on other metal objects about which site information is known.

Acknowledgments

The authors would like to acknowledge the support of the dean of the College of Home Economics for portions of this work. The generous assistance of Dr. J. H. Howard, Department of Geology, is greatly appreciated.

References

1. L. Biek, *Archaeology and the Microscope* (London: Butterworths, 1963).

2. V. Sylwan, "Silk from the Yin Dynasty," *Museum of Far Eastern Antiquities Bulletin,* 9 (1937):119.

3. J. Vollmer, "Textile Pseudomorphs on Chinese Bronzes," in *Proceedings, Irene Emery Roundtable on Museum Textiles,* (Washington, D.C.: The Textile Museum, 1974).

4. D. L. Carroll, "Etruscan Textile in Newark," *American Journal of Archaeology,* 77 (1973): 334.

5. K. A. Jakes and L. R. Sibley, "An Investigation of the Phenomenon of Textile Fabric Pseudomorphism," in *Archaeological Chemistry III* (Washington, D.C.: American Chemical Society, 1983).

6. L. R. Sibley and K. A. Jakes, "Textile Fabric Pseudomorphs: A Fossilized form of Textile Evidence," *Clothing and Textiles Research Journal* (1982): 24.

7. B. D. Cullity, *Elements of X-Ray Diffraction* (Reading, Mass.: Addison Wesley, 1978).

8. "Joint Committee on Powder Diffraction Standards," *Powder Diffraction File Reference* (Philadelphia: Joint Committee on Powder Diffraction Standards, 1967).

9. L. R. Sibley, L. Korslund, and R. M. Rowlett, "Silk Pseudomorphs on Shang Dynasty Bronze Artifacts: A Preliminary Investigation," in *Combined Proceedings, Association of College Professors of Textiles and Clothing* (Salt Lake: Brigham Young University Press, 1978).

10. R. J. Gettens, "Patina: Noble and Vile," in S. Doeringer, ed., *Art and Technology, A Symposium on Classical Bronzes* (Cambridge: MIT Press, 1970).

11. W. F. Collins, "The Corrosion of Early Chinese Bronzes," *Journal of the Institute of Metals,* 45 (1931):23.

R. J. KOESTLER, A. E. CHAROLA and G. E. WHEELER

Scanning Electron Microscopy in Conservation: The Abydos Reliefs

The Abydos Reliefs, the bas-relief sculptures from the small temple of Ramesses I at Abydos in the collection of the Metropolitan Museum, New York, have had an eventful history. The temple, built around 1310 B.C., was buried in shifting sand and soil and was subjected to periodic wetting and drying by the Nile's flooding waters.

A village was later built above the area, and the temple was fortuitously found by the villagers during the digging of a water well in A.D. 1910. The excavation that followed was carried out by the villagers themselves with little concern for proper archaeological procedures. A detailed description of the temple, its reliefs, and their history up to the time they were donated to the Metropolitan Museum, is given by Winlock.[1,2]

The Museum received the reliefs in 1911-12. Since then their story has not been much happier. The problems associated with their rapid deterioration due to high salt content and unsuccessful previous treatments were described in more detail in previous reports.[3,4] The reliefs were removed from exhibition in 1966.

Instrumentation

In the present study the scanning electron microscope (SEM) was used to examine the condition of the limestone, to assess the degree of penetration of the consolidant and the effectiveness of the cleaning method, and to compare the microstructure of the treated and untreated samples.

The use of SEMs in the conservation of works of art has not been fully exploited. A wide understanding of the operation of the instrument, its strengths, and its weaknesses is essential for its maximum utilization.

The SEM is basically a "reflecting" microscope (fig. 1) employing a fine beam of electrons instead of light photons. Electrons are emitted from a heated filament and are driven by a high voltage variable from one to fifty kV, through electromagnetic lenses that focus the electrons to a beam of the order of 2nm in width. The final electromagnetic lens carries coils employed to drive the focused beam in a scanning raster of x-y form. The specimen is normally attached to a special holder, called a stub, which provides a stable conducting platform for the specimen. The stub is placed in a stage that has mechanical or electrical feedthrough controls that allow for movement of the specimen while it is under vacuum.

The signal from the interaction of the electron beam with the specimen surface results from the generation of secondary electrons that reach an electron multiplier detector that generates a visible signal on a cathode ray tube (CRT). The displayed signal follows the point by point tracing of the scanning raster by the electron beam.

The number of secondary electrons collected at each point—a function of the surface topography, specimen tilt, and other factors—determines the point by point contrast on the visual CRT. Thus an image of the specimen is generated on the CRT. The image may also be recorded on film or videotape, or stored in a computer for further processing. More detailed information on the operation of the SEM may be obtained from a variety of sources, such as Goldstein et al.[5] or Heinrich.[6]

Other useful signals are x rays, cathodoluminescence, backscattered electrons, auger electrons, transmitted electrons, and specimen current. To collect each signal requires different detectors and different specimen preparation procedures.

Specimen Preparation

Specimens to be viewed in the SEM generally require some special preparation depending on the type of signal desired and the nature of the specimen: whether it is a conducting material, whether it contains moisture, whether it can withstand a vacuum.

For the samples examined in this study it was found that fractured surfaces did not provide enough information, so another method for specimen preparation was developed following Lewin's suggestion.[7] In this procedure the specimen was sawed and the surface was polished and then lightly etched with 1M HCl. All samples were sputter coated with 10nm of gold.

Examination of the Abydos Reliefs Stone

Before any treatments were attempted, the cause of the deterioration and the condition of the stone itself had to be established. The deterioration could be attributed to the high concentration of such soluble salts as halite, soda-niter, and magnesium chlorides, sulfates and nitrates, present in the stone. The highly hygroscopic magnesium salts—which have a tendency to deliquesce—draw in sufficient quantities of water to dissolve other soluble salts. When the relative humidity changes, these salts, in turn, can recrystallize, producing the mechanical disruption of the stone matrix.[8]

The paraffin wax and tung oil treatments that the reliefs received during the years 1911-13 produced a thin, leatherlike layer of "consolidated" stone, which was broken up and lifted off the rest of the stone when salts crystallized under it. Figure 2 shows a general view of the surface of the stone with the "skin" peeling off. Details of the edge of the skin, showing halite and gypsum crystals, are shown at higher magnification in figures 3, 4 and 5.

Laboratory Testing of the Consolidant

Possible consolidants were first tested on a marly limestone from Dendera similar in texture and composition to the Abydos stone. Paraffin wax and tung oil were applied to this limestone following the procedures of 1911-13. The treated stone was then artificially aged to produce conditions similar to those of the stone from the reliefs.[8] After testing of some consolidants, methyl trimethoxy silane (Dow Corning Z-6070 or T-4- 0149) was chosen.

The first specimens were prepared from fractured surfaces of the treated stone. Figure 6 shows the microstructure of the Egyptian limestone treated with paraffin wax and tung oil and after being consolidated with the silane. The consolidant is not readily visible in this picture. To find a better means of observing the consolidant and assessing the degree of consolidation the etching technique previously described was used. Figure 7 shows the appearance of the consolidant matrix that is left behind after the etching has removed the exposed calcareous part. Notice that this matrix is homogeneous indicating the solubilization of the wax and oil mixture by the silane.

Testing of the Cleaning Procedure

Since the reliefs had been treated with paraffin wax and tung oil, a procedure that, as noted by Pliny the Elder,[9] considerably darkens the light-colored surface of stone, cleaning was necessary. Usually this is undertaken before any consolidation is carried out. In the special case of the Abydos Reliefs the surface was so delicate that it

Fig. 1. Diagrammatic representation of a scanning electron microscope.
(a) filament; (b) anode; (c) condenser lens; (d) scanning coils; (e) final lens;
(f) specimen holder and specimen; (g) detector.

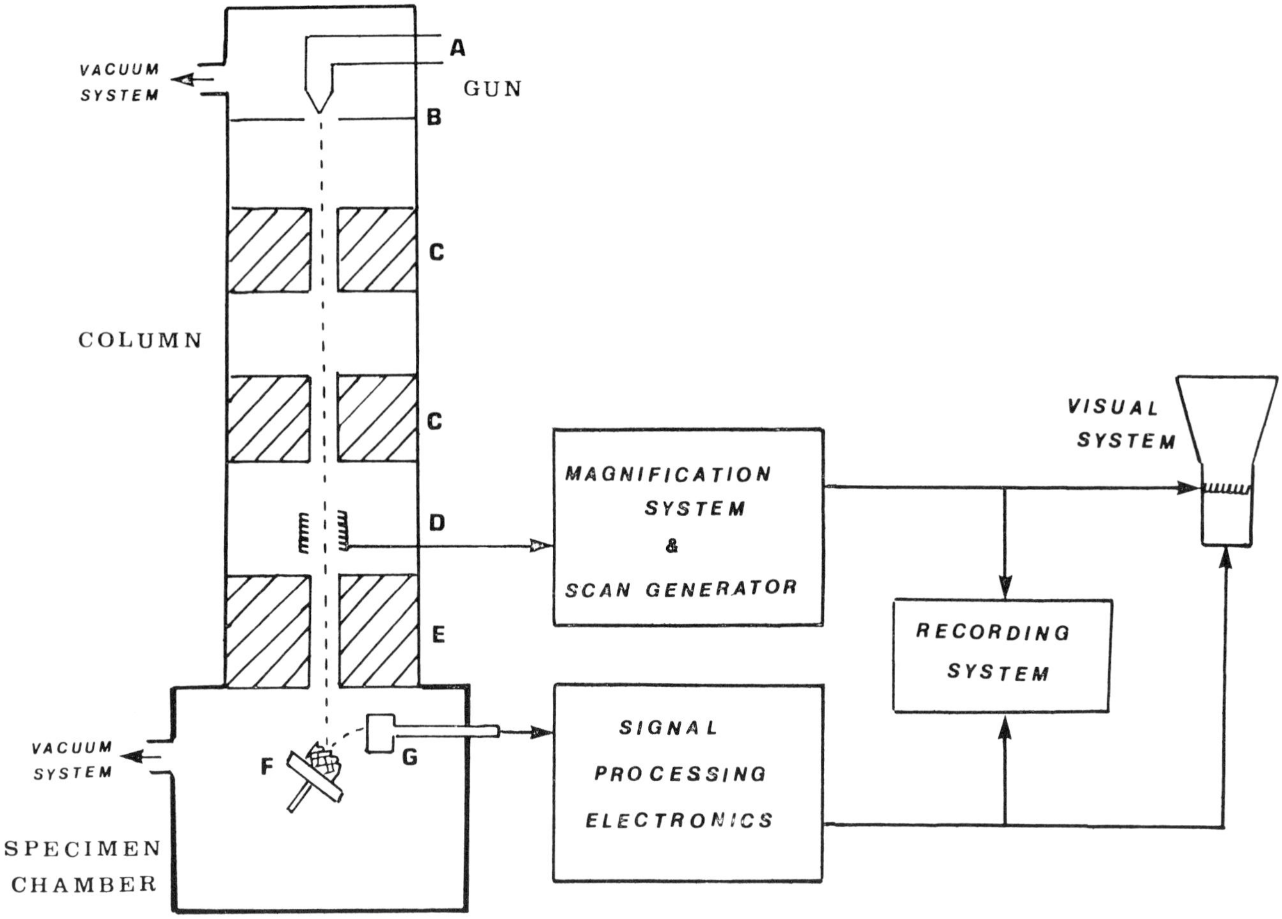

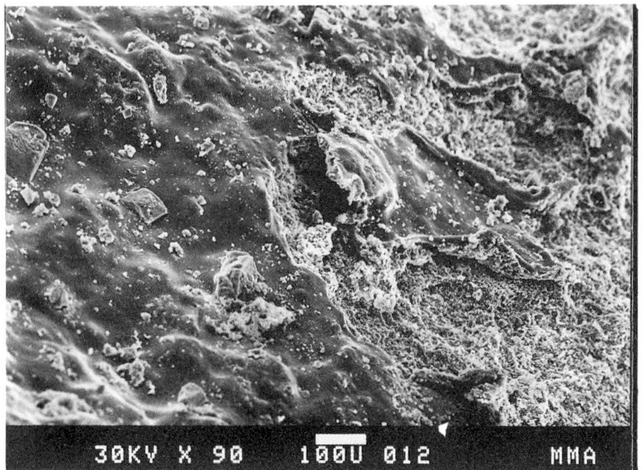

Fig. 2. Leatherlike layer of "consolidated" stone produced by paraffin wax and tung oil treatments. Layer lifts off from the rest of the stone as salts crystallize under it.

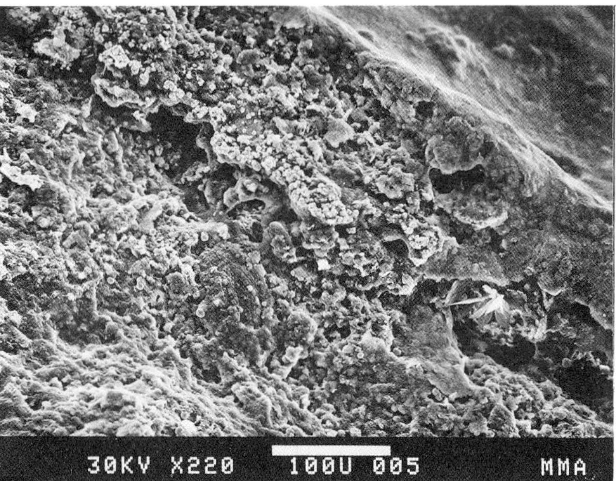

Fig. 3. Edge of leatherlike layer showing the separation from the underlying stone.

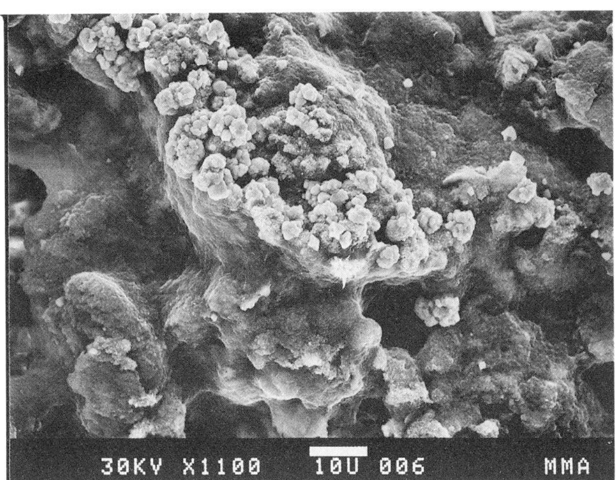

Fig. 4. Detail of edge at higher magnification showing crystals that are responsible for lifting of layer.

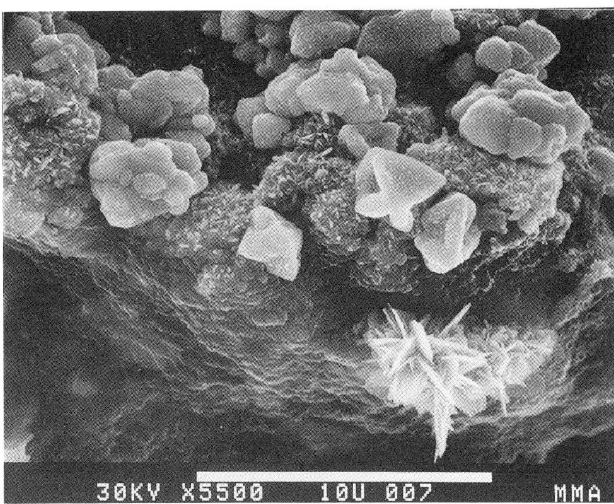

Fig. 5. Detail of edge showing masses of halite crystals and rosettes of gypsum crystals.

Fig. 6. Microstructure of Egyptian limestone treated with paraffin wax and tung oil and consolidated with methyl trimethoxy silane. The consolidant is not readily visible.

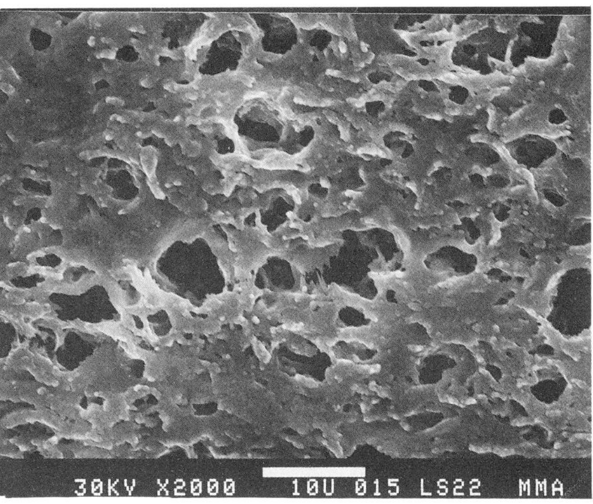

Fig. 7. Consolidant matrix left behind after specimen was treated by the polishing and etching technique. White rounded protrusions are produced by the solidification of the wax-oil mixture.

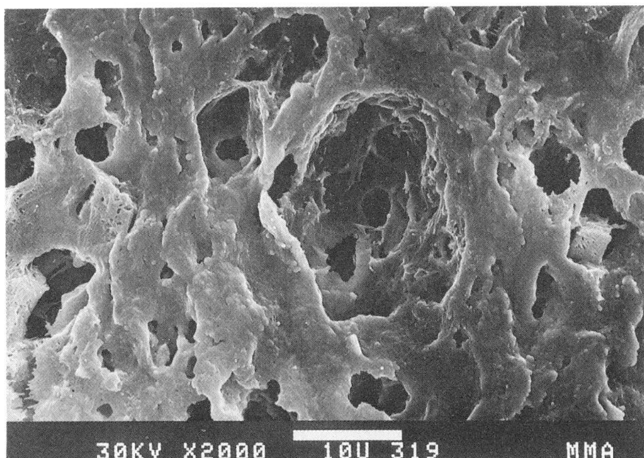

Fig. 8. Appearance of the consolidant matrix after application of one poultice (PERC in attapulgite). Notice the diminution of the number of white protrusions.

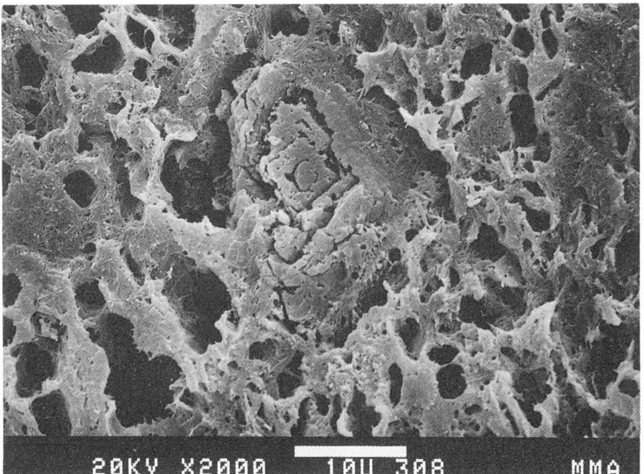

Fig. 9. Appearance of the consolidant matrix after application of two successive poultices. Practically no wax-oil mixture is left behind.

was considered necessary to consolidate first and clean after the curing of the consolidant. A more detailed description of the decision-making process is given elsewhere.[10]

Laboratory tests were carried out to ascertain the feasibility of cleaning after consolidation had taken place. Samples for these tests were also examined under SEM. Figure 8 shows the appearance of the Egyptian limestone sample after the application of one poultice (perchloroethylene in attapulgite). Figure 9 shows the appearance after the application of a second poultice. Both specimens were prepared using the polishing and etching technique.

Testing on the Reliefs

Samples from the sides of the reliefs were used for laboratory testing. Figure 10 is a fractured surface of one of these specimens. Figure 11 shows another fractured surface after consolidation with silane. Again the consolidant is not too evident. Figures 12 and 13 are micrographs of samples prepared by the polishing and etching technique. The matrix of the consolidant is clearly visible. It can also be noticed that the silane forms a homogeneous matrix with the wax-oil mixture from the previous treatments.

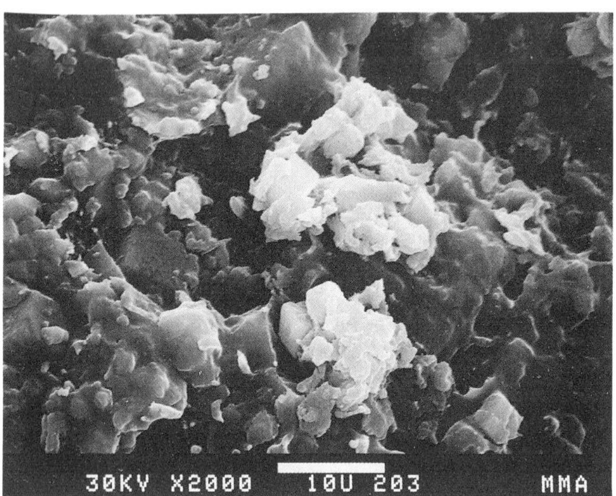

Fig. 10. Fractured surface of a flake from the Abydos Reliefs.

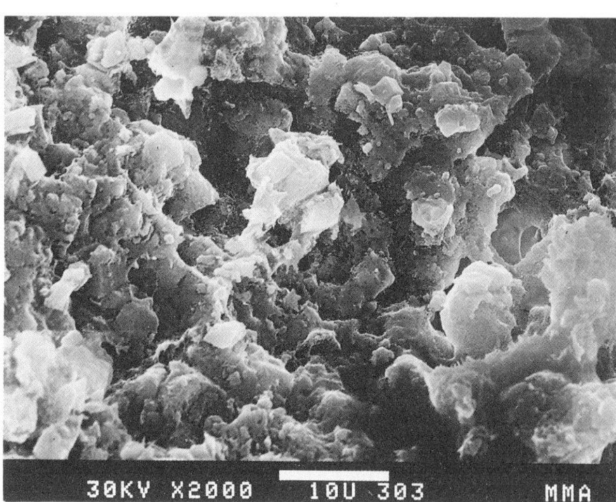

Fig. 11. Fractured surface of another flake from the Abydos Reliefs after consolidation with methyl trimethoxy silane. The consolidant is not clearly visible.

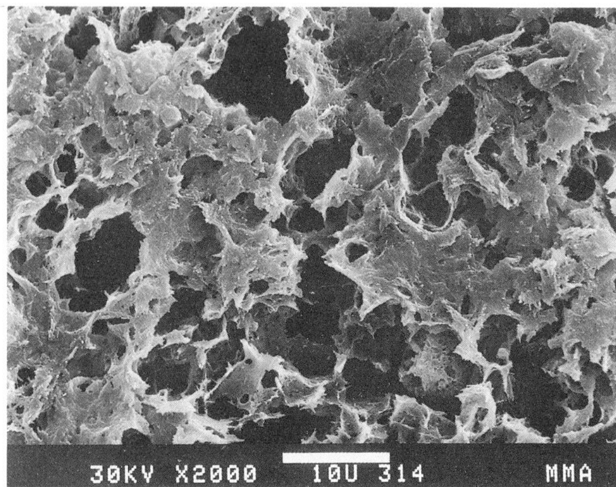

Fig. 12. Polished and etched specimen of the Abydos Reliefs. The hydrophobic nature of the wax-oil mixture in the stone shapes the residue.

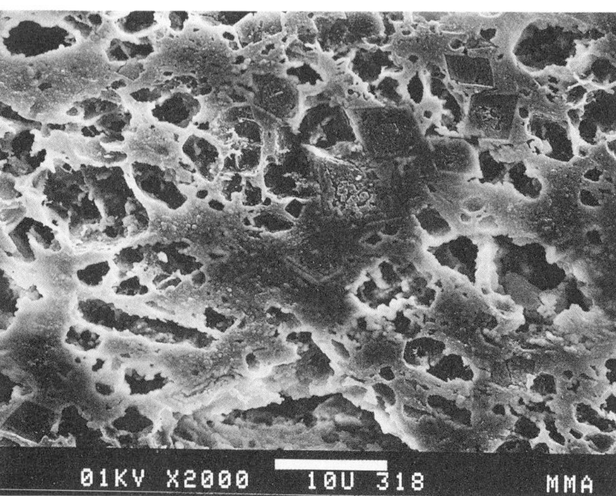

Fig. 13. Polished and etched specimen of the Abydos Reliefs after consolidation with silane. The consolidant forms a homogeneous matrix with the existing wax-oil mixture.

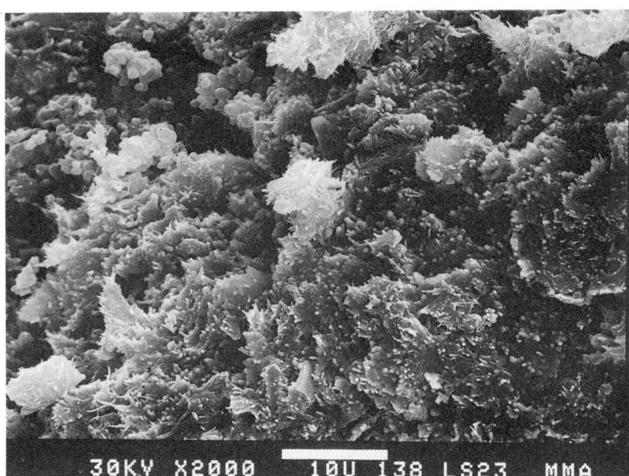

Fig. 14. Fractured surface of Egyptian limestone after treatment with wax-oil mixture, silane, and one poultice application. The micrograph is from an area near the edge of the specimen which is richer in oil. The white rounded protrusions and needle-like formations are due to the wax-oil mixture that solidified after being partially mobilized by the solvent (PERC).

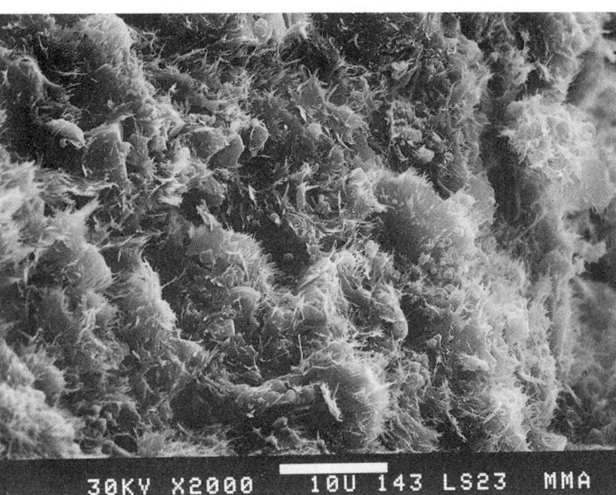

Fig. 15. Micrograph of the same sample but towards the interior of the specimen. The thin white needles are mainly paraffin wax that crystallized after the evaporation of the solvent that mobilized it.

Discussion

Through scanning electron microscopic examination of the microstructure of the samples it is possible to visualize the consolidating action of a particular compound and to deduce the mechanisms involved during cleaning procedures. The microstructure of fractured surfaces does not by itself provide all the information and can in some cases be misleading (compare figs. 10 and 11). The use of the polishing and etching technique permits a better assessment of the consolidating action of the silane (figs. 9 and 13).

The consolidant forms a mesh that holds the particles of the stone together. Methyl trimethoxy silane also forms a uniform "solution" with the wax and oil in the reliefs (figs. 6 and 13). This feature was important in the choice of the consolidant because the poor condition of the reliefs made it necessary to consolidate the stone without removing the previous treatments. That cleaning after consolidation can take place is shown in the series of micrographs (figs. 7-9) in which the decreased concentration of wax in the matrix of the consolidant is clearly evident.

The mechanism through which the wax is removed during poulticing is easier to deduce from the examination of fractured surfaces (figs. 14 and 15) than from polished and etched surfaces. Figure 14 corresponds to an area toward the edge of the specimen that is richer in tung oil. This component did not penetrate as readily into the stone as did the silane or even the molten paraffin. Figure 15 corresponds to an area towards the center of the sample. The fibrous-looking crystals are the paraffin wax that crystallized after the evaporation of the solvent (PERC) that mobilized it. The difference in morphology in the crystals from these two micrographs is due to the different composition of the wax-oil mixture. It would appear that the tung oil is also mobilized along with the wax. The poulticing technique is therefore a series of wax dissolution-crystallization steps, with each step drawing the wax and any oil that can be mobilized further out from the stone.

To achieve a thorough understanding of stone, or any sample, by SEM examination, special specimen preparation techniques may have to be developed. No one technique can provide all the necessary information. The different specimen preparation techniques selected should be used to provide complementary information and to ensure that accurate interpretaion of the images is derived.

Acknowledgments

The authors wish to thank R. Sheryll for the photographic reproduction of the SEM micrographs and help in SEM sample preparation.

References

1. H. E. Winlock, *Bas Reliefs from the Temple of Ramesses I at Abydos*, Papers Vol. I Part I. (New York: The Metropolitan Museum of Art, 1921).

2. H. E. Winlock, *The Temple of Ramesses I at Abydos*, Papers No. 5 (New York: The Metropolitan Museum of Art, 1937).

3. L. Aussenberg, Preliminary Report on the Abydos Reliefs (unpublished). On file at the Egyptian Department, MMA, N.Y., 1974.

4. G. M. Helms, "Conservation of Egyptian Limestone: The Abydos Reliefs," *3rd Annual Conference of Art Conservation Training Programme*, Queen's University, Kingston, Ontario, Canada (1977) p. 40.

5. J. I. Goldstein and H. Yakowitz, *Practical Scanning Electron Microscopy, Electron and Ion Microprobe Analysis* (New York: Plenum Press, 1977).

6. K. F. J. Heinrich, *Electron Beam X-ray Microanalysis* (New York: Van Nostrand Rheinhold Co., 1981).

7. S. Z. Lewin, personal communication.

8. A. E. Charola, G. E. Wheeler and R. J. Koestler, "Treatment of the Abydos Reliefs: Preliminary Investigations," *Proceedings of the 4th International Congress on Deterioration and Preservation of Stone Objects*, Louisville, KY (1982) pp. 77-88.

9. Pliny, *Natural History* Book 36.30.

10. G. E. Wheeler, J. K. Dinsmore, L. J. Ransick, A. E. Charola and R. J. Koestler, "Treatment of the Abydos Reliefs: Consolidation and Cleaning," *Studies in Conservation*, 29 (1984), pp. 42-49.

GEORGE MAVROV

Technical Investigation on the Conservation of an Eighth-Century Rock Relief

The *Madara Horseman* is a monument in northeast Bulgaria. In 1980, it was listed by UNESCO as an object of international importance. The relief, including the inscriptions, is seven meters wide and six meters high and is cut out of the lower part of a rock wall that is 100 meters high. The horseman, accompanied by a dog, is depicted having stabbed a lion with his lance; the lion is between the forelegs of the horse (figs. 1, 2).

The monument and inscription are together considered to be the first official document of the Bulgarian state. Dating from the early eighth century A.D., it commemorates a military victory over Byzantium. The relief is in a grave condition. Many details are worn and much of the inscription is illegible because of progressive erosion. The rock wall is a calcareous limestone of varying composition. At the level of the relief, it is 50:50 quartz:calcite, has a porosity of 20-30% and a compression strength of 219 kg/cm². Three deep cracks run across the relief dividing it into two irregular prisms, the larger of which is estimated to weigh 80-100 tons. Technical problems make the conservation of the horseman very difficult. Nonetheless, Bulgarian and foreign experts are unanimous in calling for immediate treatment.

Technical Investigations

These experiments were carried out in order to select the most appropriate method and material for the consolidation. The actual consolidation will be the task of other experts. The following areas had to be considered: (1) The investigation of the physical, mechanical, and structural changes in the consolidant during polymerization and artificial aging; (2) the technical analysis of the Madara rock; (3) the investigation of the physical, mechanical, and structural changes in the consolidated stone during artificial aging.

Consolidants

The selection of the consolidant was governed by three considerations. First, the rock had to be waterproof, not only on the surface but also at depth. Second, the relief is in a sunny position so no color change can be permitted. Third, internal stresses at the interface between consolidated and unconsolidated stone had to be avoided.

Three different silicone materials were selected for these experiments: (1) Drisil 773 (Dow Corning): 60% polysiloxane in white spirit (here referred to as DS 773); (2) X 54-802 (Rhone Poulenc): methyl-methoxysilane, depositing ca. 30% dry substance; and (3) Sandsteinverfestiger OH (Wacker-Chemie): tetramethoxysilane, giving amorphous SiO_2 as an end product (here referred to as SVOH).

Investigation of the Consolidants

The parameters investigated were hardness, hydrophobity, susceptibility to an atmosphere of SO_2 and to Na_2SO_4 solution, and the mechanical changes during frost-thaw cycles.

Table 1 lists the viscosity of the resins used in the experiments: DS 773 was used in a number of different dilutions with white spirit, as indicated in the table.

Table 1

The viscosity of the resins used in the experiments.

Viscosity, Pa.s.10³

Resin °C	SVOH	X 54-802	DS 773 (% in white spirit)					
			10	20	30	40	50	60
20	1.877	2.389	1.222	2.142	3.503	5.101	12.261	23.58
35	1.554	1.870	1.022	1.503	2.723	4.750	8.855	16.12

Table 2

Contact angles (o) of drops of water on the film surface.

Resin		σ, Degrees		
		Drop Volume, μ l		
		50	40	30
	60	98	98	100
DS 773, % in	30	105	107	108
white spirit	10	100	114	123
X 54-802		95	98	100

Fig. 1. General view of the *Madara Horseman* and part of the rock massif.

Fig. 2. Aerial view of the Madara Plateau indicating the position of the relief.

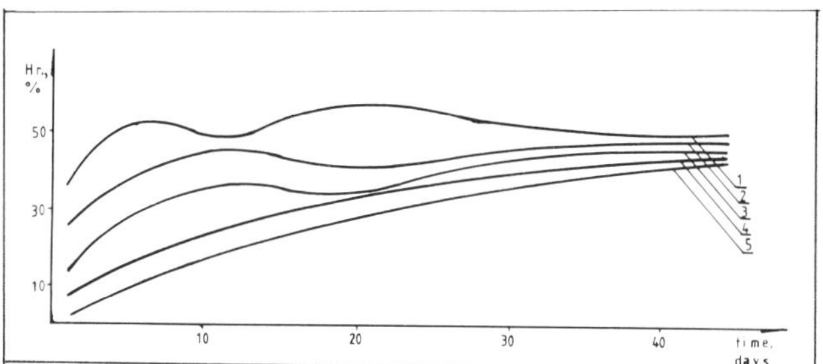

Fig. 3. The change in hardness (Hr) during polymerization of DS 773 at varying film thicknesses.

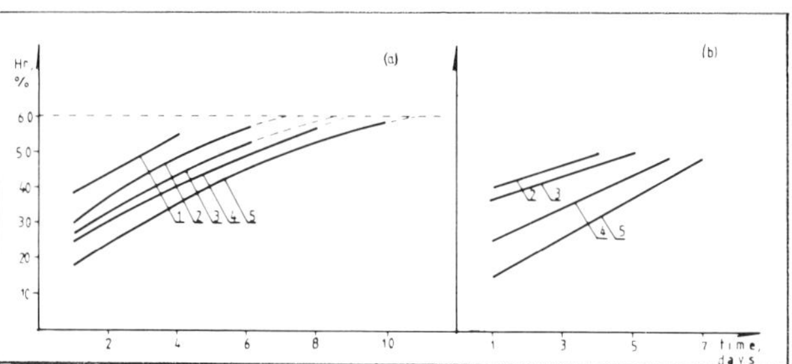

Fig. 4. The change in hardness (Hr) during polymerization of SVOH (a) and X 54-802. (b) at varying film thicknesses.

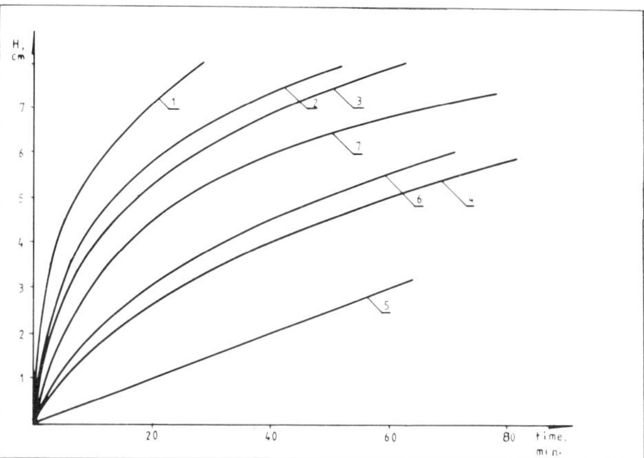

Fig. 5. The rate of uptake of water, solvent, and each consolidant (1) water, (2) white spirit, (3) 10% DS 773, (4) 30% DS 773, (5) 60% DS 773, (6) X 54-802, and (7) SVOH.

Hardness

The hardness of the materials during polymerization was measured using a standard pendulum hardness meter (figs. 3 and 4). The different thicknesses of the film (in the liquid state) are 0.1-0.5mm, referred to respectively as 1-5 on the graphs. Unfortunately, the X 54-802 films cracked after 6-8 days and those of SVOH after 9-14 days. DS 773 hardens more slowly than the other materials and reaches 45% after 50 days under laboratory conditions.

Hydrophobity

SVOH was not included in this experiment.

Hydrophobity was evaluated by measuring the contact angle σ of a water drop (of a uniform volume) on the film surface (see table 2). A second angle, α, was also measured: this was the angle at which a drop of water of a certain volume began to run over the film surface. At volumes less than 10 μl, the drops are retained even when the surface is vertical, giving an assessment of the water-repelling characteristics of the film.

Investigation of the Madara Rock

The rate of uptake of water, solvents, and consolidants gives an indication of the expected time and depth of impregnation for each consolidant. Figure 5 illustrates that 60% DS 773 (5) is unsuitable for

impregnation treatments whereas 10% DS 773 (3) and SVOH (7) are more satisfactory.

The Strength of the Consolidated Stone

Both pressure and bending strengths were measured. The results are indicated in tables 3 and 4 and in figure 6. The range of the results is due to the inhomogeneity of the stone.

Migration of Consolidants During Impregnation

Blocks of stone 7 x 3.3 x 3.3 cm³ were prepared for this experiment. Impregnation was by capillary action. After polymerization, slices 2mm thick were cut from the blocks every 0.5cm along the length. These slices were treated with HCl for two hours. Figures 7 and 8 show them before and after acid treatment. The poor performances of 50% and 60% DS 773 is thought to derive from their high viscosities and associated slow migration.

Susceptibility to an SO_2 Atmosphere, in Na_2SO_4 Solution and to Freeze-Thaw Cycles

The films are affected by SO_2 but are stable to Na_2SO_4 solution. The results of the freeze-thaw part of the investigation determined that the silicone films suffered from many surface defects including openings, crackings and tearings.

Table 3

Compression strength of the stone after impregnation with each resin. The increase over unconsolidated stone (219 kg/cm^2) is also indicated.

Resin		Compression Strength kg/cm^2						
		1	2	3	4	5	Mean	Increase %
	10	235	225	210	195	210	215	—
DS	20	275	245	240	235	225	241	9.5
773	30	290	285	275	260	245	271	23.18
% in	40	308	295	290	280	255	285	29.5
white	50	308	295	260	255	255	276	25.7
spirit	60	295	280	270	253	270	255	22.7
SVOH		276	280	290	260	245	270	22.7
X 54-802		275	270	265	250	230	258	17.2

Table 4

Bending of the stone after impregnation with each resin. The increase over unconsolidated stone (259 kg/cm^2) is also indicated.

Resin		Bending Strength, kg/cm^2						
		1	2	3	4	5	Mean	Increase %
DS	10	436.8	373.8	292.9	439.5	384.8	385.4	41.0
773	20	478.5	528.0	463.0	551.0	590.6	502.2	83.8
% in	30	581.1	636.4	639.0	402.0	510.7	555.3	103.4
white	40	604.0	567.0	661.4	605.0	619.0	611.2	123.8
spirit	50	687.8	589.0	246.0	411.0	607.6	508.3	86.0
	60	410.7	592.7	622.8	480.3	625.0	547.2	100.0
X 54-802		462.0	495.0	558.0	553.0	580.0	529.0	93.8
SVOH		470.0	490.0	538.0	437.0	505.0	488.0	78.7

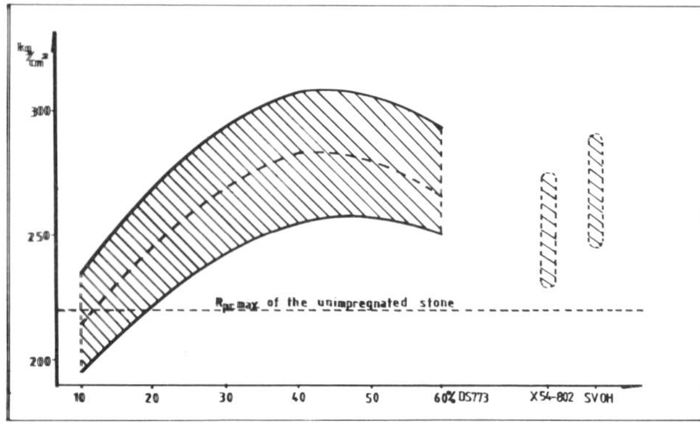

Fig. 6. The relation between compression strength kg/cm^2 and the consolidant type and its concentration.

Artificial Aging in SO$_2$ Atmosphere, Na$_2$SO$_4$ Solution, and Freeze-Thaw Cycles

These tests are to assess the duration and effectiveness of the treatment. Scanning electron microscopy was used to follow the structural changes in the stone and resin. Of note are the following observations: dilute DS 773 solutions form fragmentary films on the stone surface. X 54-802, SVOH and DS 773 (over 20%) form torn, cracked films and polymer bridges in the pores. See figure 9 (X 54-802) and figure 10 (SVOH). With repeated treatments, the pores are gradually filled. See figure 11 (DS 773).

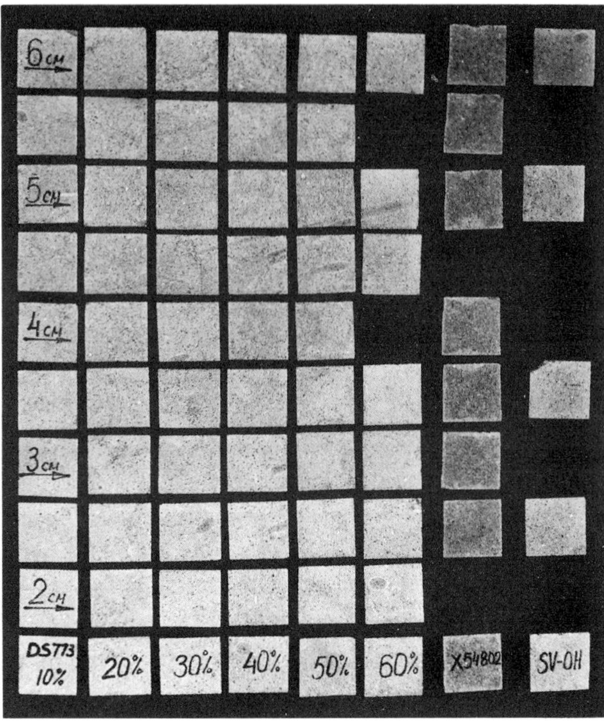

Fig. 7. Before treatment with dilute HCl.

Fig. 9. X 54-802 bridges in a stone pore X2500.

Where there are hygroscopic salts in the pores, the silicones cover them with a polymer film (fig. 12). This is not, however, a reliable or permanent protection against salt action and in time, the silicone film will be torn by the action of the salt. Migrating salts, crystallizing at depth, will block the pores when the stone surface is protected by a polymer film (fig. 13).

Conclusions

The experiments suggest that the most suitable consolidants would be 25-35% DS 773 and X 54-802. These also provide the necessary strengthening and protection of the Madara rock.

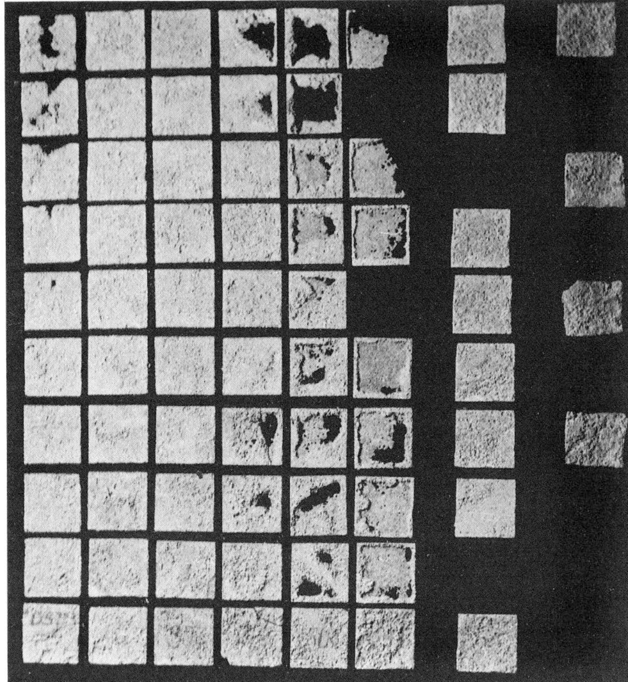

Fig. 8. After treatment with dilute HCl.

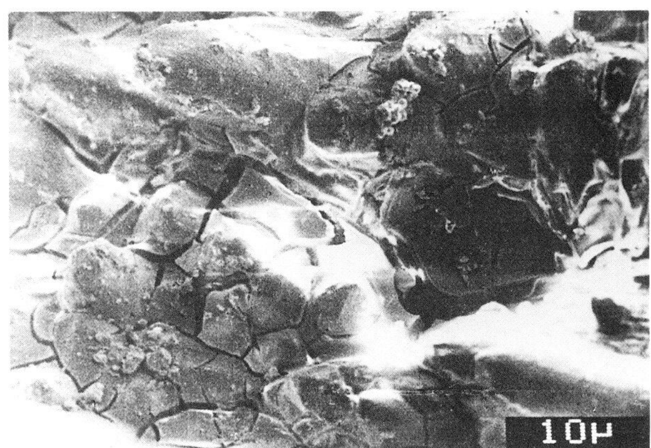

Fig. 10. SVOH cracked formations X1000.

Fig. 11. DS 773 (60%) filling surface pores X100.

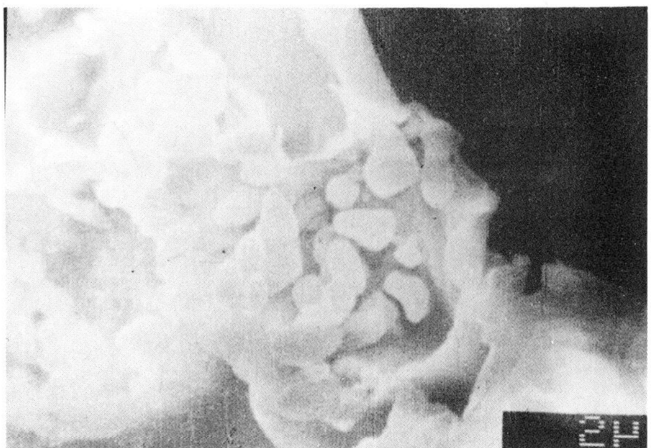

Fig. 12. Hygroscopic salts blocked by DS 773 X5000.

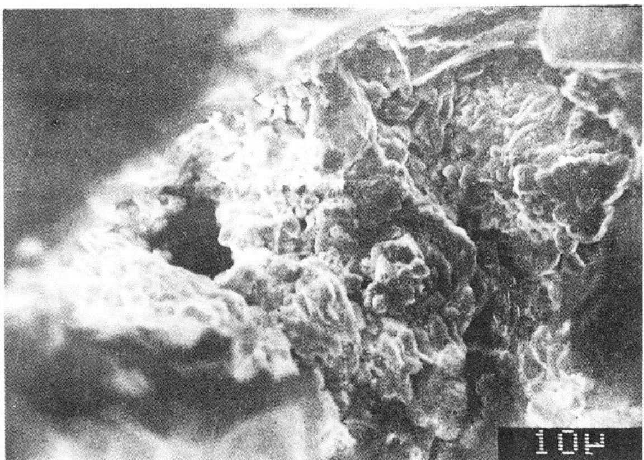

Fig. 13. Closing of inner pores because of hydrophobic surface barrier X1000.

ELIZABETH A. MOFFATT, NEIL T. ADAIR, and GREGORY S. YOUNG

The Occurrence of Oxalates on Three Chinese Wall Paintings

Oxalate salts have been identified on three large Chinese wall paintings belonging to the Royal Ontario Museum in Toronto. Two of these paintings, each of which measures approximately 3m by 10m, are Taoist from southern Shansi province and are thought to be Yuan Dynasty (A.D. 1279-1368). The paintings, *The Lord of the Northern Dipper* and *The Lord of the Southern Dipper,* together entitled *Homage to the First Principle,* are mounted on the east and west walls respectively of the Bishop White Gallery. A larger painting, the Buddhist *Maitreya Paradise,* which is also Yuan Dynasty from southern Shansi, is mounted on the north wall.

The Buddhist painting was brought to the Museum from China in 1928. The two Taoist paintings were purchased in 1937 from a New York dealer in Far Eastern art. The paintings were executed on temple walls built up of progressively finer layers of clay. The *Maitreya Paradise* was mounted by George Stout in 1933 and the Taoist paintings were mounted in the late 1930s by Museum staff. Further details on the paintings including iconography, painting technique, and treatment history can be found in a publication by Phillimore.[1]

The two Taoist paintings recently underwent extensive conservation treatment. In support of this, in 1982, the Analytical Research Services Laboratory of the Canadian Conservation Institute was asked to undertake the analysis of the materials from *The Lord of the Southern Dipper.* The original pigments (hematite, goethite, vermilion, red lead, azurite, malachite, and kaolinite), grounds (gypsum, kaolinite, illite, and quartz) and materials from previous treatments (notably polyvinyl acetate consolidant) were identified.

During the analysis, calcium oxalate hydrates were found in most of the samples of original pigment, in the clay support, and in three of the eight twentieth-century overpaints that were analyzed. Copper oxalate hydrate was identified as a major component of many of the blue and green pigment samples. Subsequent analysis of blue and green pigments from *The Lord of The Northern Dipper* and the *Maitreya Paradise* also revealed the presence of calcium and copper oxalates. Microscopic examination of the blue and green pigments indicated that the copper oxalate was an alteration product of the original azurite and malachite.

The occurrence of oxalates is unusual and, to our knowledge, unreported in studies on other wall paintings of the same period. Three potential sources of oxalates have been proposed. These include the presence of oxalates as original constituents, their introduction during a conservation treatment, or their production from microbiological activity. To date we have been unable to conclusively explain the origins of the oxalate formation. In the event that similar oxalate occurrences may be found by other laboratories, we felt it important to report our observations.

Method of Analysis

Under a stereomicroscope, minute samples of the pigments, clays, and overpaints were removed from various panels of the paintings using a sharp scalpel. A total of fifty-six samples from *The Lord of the Southern Dipper,* seven from *The Lord of the Northern Dipper* and six from the *Maitreya Paradise* were obtained. All samples were analyzed by infrared spectrophotometry, x-ray diffraction and x-ray fluorescence spectrometry. Certain samples were also examined by light microscopy and scanning electron microscopy.

For infrared spectrophotometric analysis, a portion of each sample was mounted in a diamond micro-sample cell,[2] which was positioned in a Perkin-Elmer 4X reflecting beam condenser. The spectrum was recorded using either a Perkin-Elmer model 283 dispersive infrared spectrophotometer or a Nicolet 5DX spectrophotometer. The sample was then removed from the diamond cell and mounted on a glass fiber with silicone grease adhesive for insertion into the sample holder of a Gandolfi x-ray diffraction camera. The diffraction pattern was recorded using $CuK\alpha$ x-ray radiation and CEA 25 x-ray film after which the sample on the glass fiber was mounted in a Norelco wavelength dispersive x-ray fluorescence macroprobe for elemental analysis.

Three green pigment samples were mounted in a viscous oil medium (Carmount, n = 1.66) for examination with a Leitz Orthoplan microscope using polarizing optics. Other samples were embedded in a low viscosity epoxy resin (Spurr's medium), ground, and polished to reveal a cross-section of the pigment layer. The microscope, with incident illumination, was used for recording 35mm color photomicrographs of the cross-sections. Following this examination, a radio frequency excited oxygen plasma microincinerator was used to "ash" away the epoxy medium from the surface of the cross-sections. Ashing proceeded for ten minutes, but to reduce heating it was carried out one minute at a time with one to two minutes in between for specimen cooling. This technique allowed the pigment particles and voids in the pigment layers to be examined without being obscured by the embedding medium. The sections were sputter coated with gold-palladium in preparation for observation with an Hitachi HHS-2R scanning electron microscope (SEM) at an operating voltage of 20kV.

Results

Calcium Oxalates

Calcium oxalate hydrates were identified in the majority of samples from the three wall paintings. On *The Lord of the Southern Dipper,* calcium oxalates were present in at least one sample of each color and were not restricted to any particular areas of the painting. More than three-quarters of the thirty-nine samples of original pigments contained calcium oxalates. They were also detected in the clay support and in three of eight twentieth-century overpaints but not in the twentieth-century clay fills. All seven samples from *The Lord of the Northern Dipper* and five of the six from *Maitreya Paradise* were found to contain calcium oxalates. The calcium oxalates were present in a number of hydration states, primarily weddellite ($CaC_2O_4.2H_2O$) and whewellite ($CaC_2O_4.H_2O$), with significant amounts of other hydrated forms up to $CaC_2O_4.3H_2O$, as well as some anhydrous material. The compounds in this series are colorless. The infrared spectrum of a synthetic calcium oxalate standard is shown in figure 1.

Copper Oxalates

Copper oxalate hydrate was identified in all nine copper-based original green samples, in two of six original azurite samples and in two overpaints (a blue and a green) from *The Lord of the Southern Dipper.* Malachite was detected in three green samples that were examined by polarized light microscopy, but the concentrations of this

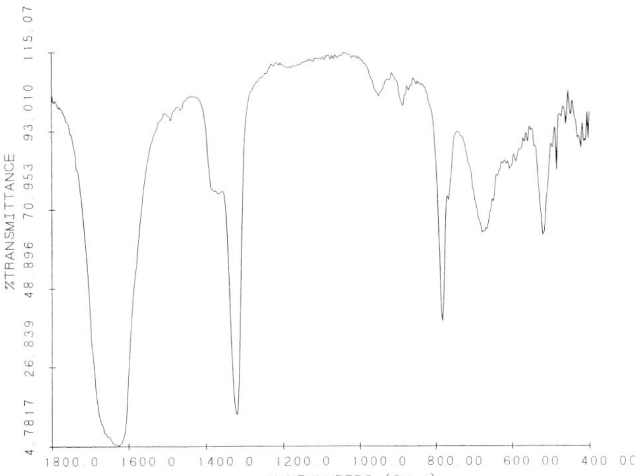

Fig. 1. Infrared spectrum of calcium oxalate. Major absorptions occur at 1650 and 1320cm⁻¹.

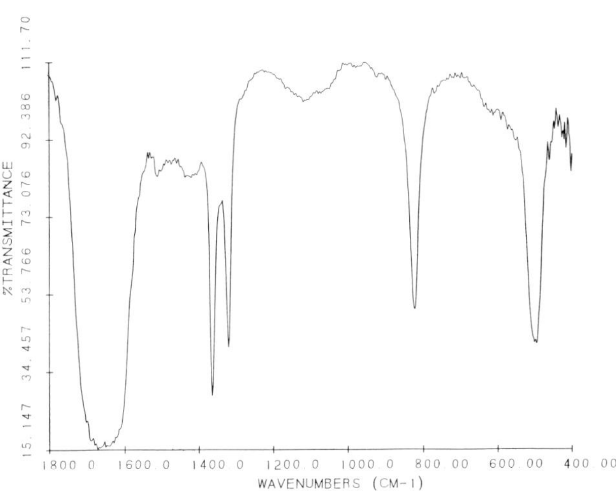

Fig. 2. Infrared spectrum of copper oxalate. Significant absorptions are found at 1650, 1365, and 1320cm⁻¹.

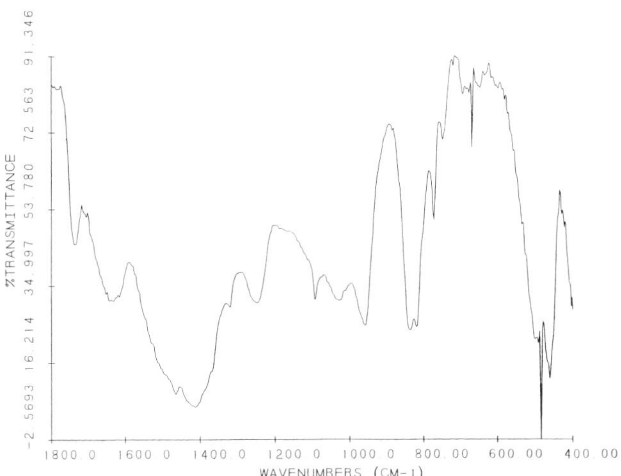

Fig. 3. Infrared spectrum of Blue I, an oxalate-free sample from *The Lord of the Southern Dipper*. An intense absorption due to azurite is centered at 1410cm⁻¹.

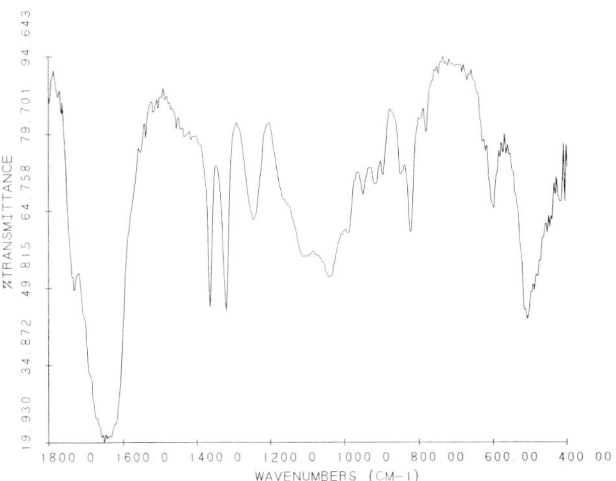

Fig. 4. Infrared spectrum of Blue II, an oxalate-containing sample from *The Lord of the Northern Dipper*. Note the absorptions due to oxalates at 1650, 1365, and 1320cm⁻¹, as well as the absence of the azurite absorption at 1410cm⁻¹.

pigment were below the detection limits of the XRD and IR techniques employed. Six of seven malachite samples and five of six azurite samples from *The Lord of the Northern Dipper* and the *Maitreya Paradise* were also found to contain copper oxalate. The infrared spectrum of a synthetic copper oxalate standard is shown in figure 2.

Preliminary tests have indicated that both azurite and malachite can be converted to copper oxalate by treatment with concentrated aqueous oxalic acid. The copper oxalate hydrate formed by the reaction of malachite and oxalic acid was a finely divided turquoise powder.

Microscopy

Light microscopical and scanning electron microscopical studies of selected samples from the three paintings were undertaken to document any morphological details that might indicate the source of the oxalates. The morphology of two typical blue samples, one without and one with copper oxalates, designated Blue I and Blue II respectively, will be described.

The oxalate-free Blue I was composed of azurite, quartz, and polyvinyl acetate. Its infrared spectrum is shown in figure 3. Blue I consisted of transparent, blue, angular, separated crystals with a size range of up to 40μm in any given direction. Its morphology is shown in figure 5.

Blue II was a complex mixture of calcium and copper oxalates, azurite, quartz, gypsum, kaolinite, chlorite-montmorillonite and polyvinyl acetate. The infrared spectrum of Blue II is shown in figure 4. A fragment of Blue II was composed of translucent, turquoise nodules of approximately 1μm average diameter in a cemented matrix. Approximately one-quarter of the pigment layer was taken up by interstitial space and a few anglular crystalline particles were embedded within the layer. The morphology is shown in figure 6.

Two networks of nodular filaments were found within the interstices of the pigment layer in this fragment of Blue II. These filaments are illustrated in the electron micrograph, figure 7. The filaments, which have an average diameter of 1μm, interweave, branch, and adhere to the sample matrix.

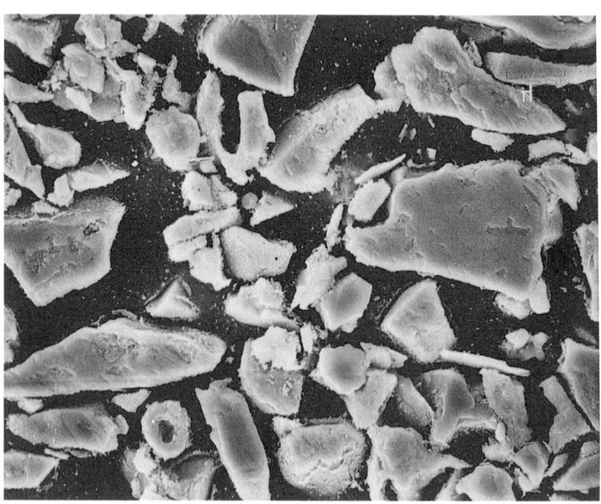

Fig. 5. Electron micrograph of Blue I showing transparent, angular, separated crystals with a size range up to 40μm in any given direction.

Discussion

The mechanism of the incorporation of calcium and copper oxalates into the pigment layers of the paintings has not been determined. The widespread distribution of oxalates included in the original pigments, clay support, and overpaints suggests an overall, non-specific process of oxalate introduction. Calcium oxalate could either have been deposited directly on the paintings or it could have been formed *in situ* by the reaction of the oxalic acid with a calcium-containing material in the paintings. The copper oxalate present in most of the blue and green pigment samples is apparently an alteration product of the original azurite and malachite. Microscopic examina-

tion of representative samples revealed a difference in morphology between the oxalate-free and the oxalate-containing samples. The increase in morphological heterogeneity and the precipitated matrix appearance of the samples that contain copper oxalate are consistent with the conversion of the pigment and the subsequent precipitation of copper oxalate. The morphology and the fact that copper oxalate was present in a wide range of admixtures with the azurite and malachite indicate that an alteration of the original pigments has occurred.

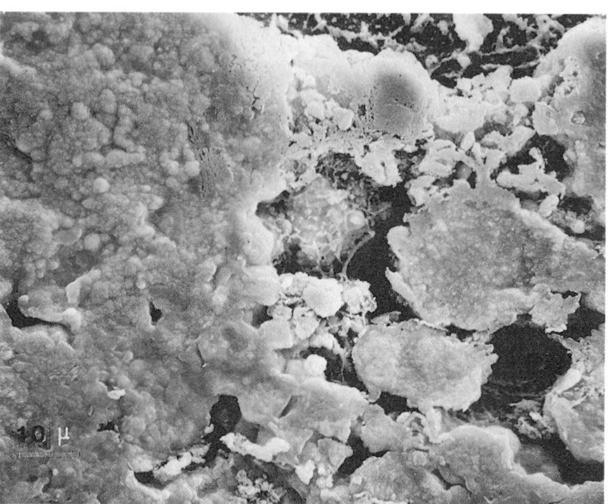

Fig. 6. Electron micrograph of Blue II showing translucent nodules of 1μm average diameter in a cemented matrix.

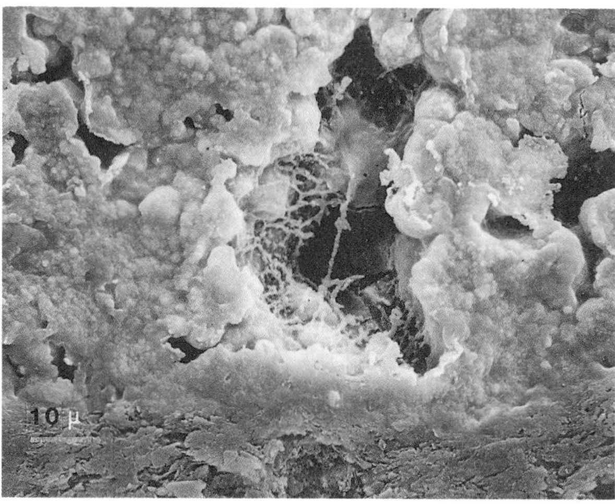

Fig. 7. Electron micrograph of Blue II. Filaments are visible in the interstitial spaces of the pigment layer of one fragment of Blue II.

Azurite and malachite are considered to be stable under ordinary circumstances, but are known to be affected by acids, warm alkalis, and heat.[3,4] There are reports in the literature of the alteration of these pigments under other conditions. Evidence has been found that azurite in prolonged contact with sodium chloride on a wall painting was converted to paratacamite, basic copper chloride.[5] Another related phenomenon is "malachite staining" on oriental works of art. Acid released on the hydrolysis of alum in Dosa size reacts with malachite to form soluble copper salts that stain surrounding materials.[6] Considering the susceptibility of azurite and malachite to acids, even in dilute solution, it is not unexpected that the pigments would be affected by oxalic acid. It has not been determined whether the partial conversion of the pigments to copper oxalate would have an effect on the appearance of the paintings.

Three possibilities for the source of the oxalates have been proposed: the presence of oxalates as original constituents, the introduction of oxalates during a conservation treatment, and the production of oxalates *in situ* by microorganisms.

Oxalates as Original Constituents

No reference has been found to the intentional use of oxalates as original constituents on Chinese wall paintings. However, since there are naturally occurring sources of oxalates, including plant materials and urine,[7] oxalates could have been incorporated into the paintings when they were executed. For example, oxalates might have been used in the preparation of clay, pigments, or media of the paintings. However, the identification of oxalates in three of the twentieth-century overpaints tends to discount this possibility.

Treatment History

The application of oxalic acid or an oxalate salt during a conservation treatment would explain the widespread occurrence of oxalates on the paintings. The treatment information compiled by the museum has been examined in detail for a reference that would suggest the use of oxalates.

Oxalic acid or an oxalate salt would be an unusual treatment material for such wall paintings but it is possible that the oxalate was introduced unintentionally, perhaps as a minor constituent of a proprietary product. For example, oxalic acid was an ingredient in a nineteenth-century recipe for the preparation of animal glue.[8] An oxalate treatment would explain the observed phenomenon but there is no documentation of its use on these paintings or tradition for its general use on wall paintings. However, the gaps in the treatment records do not permit the possibility to be dismissed. The presence of oxalates in some of the twentieth-century overpaints from *The Lord of the Southern Dipper* would be consistent with an oxalate treatment of the surface of the painting.

Microbiological Activity

The formation of oxalates by microorganisms such as lichens and fungi is well documented.[7,9-11] Oxalates can account for up to 45% of the dry weight of the mycelium in some species of fungi.[7] Calcium oxalates have been identified on a variety of stone surfaces including monuments, sculpture, and buildings.[12-14] Lichens were shown to be responsible for the deposition of calcium oxalates on marble columns[14] and both lichens and fungi are known to contribute to the

deterioration of rock-forming minerals.[15,16]

A number of scanning electron microscope studies have investigated the relationship between the fungal hyphae and the oxalates. One study suggests that acicular calcium oxalate crystals form inside fungal hyphae, enlarge, and puncture the walls to form a profusion of crystals on the hyphae.[9] Another report states that oxalic acid is exuded from hyphae and reacts with solubilized calcium in soils to crystallize as calcium oxalate adjacent to the hyphae.[10] In either case, a close association between the crystals and the fungal hyphae is confirmed.

This association would be consistent with the hypothesis that the filaments observed in one fragment of Blue II (fig. 7), are permanent inorganic casts of fungal hyphae. These casts would not readily be destroyed by low-temperature oxygen plasma ashing and thus would remain to be documented by SEM after removal of the embedding medium. The size (average diameter 1 μm) and morphology of these filaments are consistent with those of fungal hyphae[9,10,17] but the rough surface texture is not characteristic of living hyphae.

Although the observation of these filaments is supportive of a microbiological source for the oxalates, it is important to note that the filaments were observed in only one sample. As a result, there is insufficient evidence to conclude that the oxalates originated from microbiological activity. While it has been reported that the oxalic acid produced by microorganisms can convert calcium-containing minerals to calcium oxalate,[16] it is not known if an analagous microbiological process would produce copper oxalate from azurite and malachite as observed in the paintings. The implication of the presence of oxalates in some of the modern overpaints for the theory of microbiological activity as the oxalate source has not been determined.

Conclusion

This study has shown that there is a widespread distribution of calcium oxalates in the original pigment layer, the clay support, and some modern overpaints of *The Lord of The Southern Dipper*. The original azurite and malachite pigments in all three Chinese wall paintings have undergone a conversion, to various extents, to copper oxalate. It is not known if the partial conversion of the pigments to copper oxalate would have affected the appearance of the paintings.

The source of the oxalates was not identified. It is highly unlikely that they were original constitutents of the paintings. Although there was no indication that oxalates were applied during a treatment, this possibility cannot be dismissed. In addition, there was insufficient evidence to conclude that the oxalates were the result of microbiological activity. Consequently, studies on other similar paintings are required to examine this phenomenon in more detail to determine the extent of the occurrence and to elucidate the mechanism. It is hoped that this information assists such a study.

Acknowledgments

The contribution to this study by the members of the Chinese Wall Painting Project of the Royal Ontario Museum, particularly Elizabeth Phillimore, conservator, Special Projects, Elizabeth Walsh and Eric Gordon, is gratefully acknowledged. We also wish to express our appreciation to John Taylor and Marilyn Laver of the Analytical Research Services Laboratory for their assistance in the preparation of this manuscript.

References

1. E. Phillimore, "Initial Report on the Treatment of Two 13th-Century Chinese Wall Paintings in the Royal Ontario Museum, Toronto," preprints of Papers Presented at the Tenth Annual Meeting American Institute for Conservation, Milwaukee, Wisconsin, May 26-30, 1982, pp. 150-159.

2. M. E. Laver and R. S. Williams, "The Use of a Diamond Cell Micro-sampling Device for Infrared Spectrophotometric Analyses of Art and Archaeological Materials," *Journal of the International Institute for Conservation , Canadian Group,* 3(2) (1978): 34-39.

3. R. J. Gettens and E. West FitzHugh, "Azurite and Blue Verditer," *Studies in Conservation,* 11 (1966): 54-61.

4. R. J. Gettens and E. West FitzHugh, "Malachite and Green Verditer," *Studies in Conservation,* 19 (1974): 2-23.

5. R. J. Gettens and G. L. Stout, "A Monument of Byzantine Wall Painting — The Method of Construction," *Studies in Conservation,* 3 (1958): 107-119.

6. Y. Emoto, "Deterioration of a Japanese Wooden Panel Painting — An Approach to Study of Malachite Staining, 'Rokusho-Yake,' " in *Conservation of the Far Eastern Art Objects,* Proceedings of the International Symposium on the Conservation and Restoration of Cultural Property, Tokyo, November 26-29, 1979, (Tokyo: Tokyo National Research Institute of Cultural Properties, 1980), pp. 49-62.

7. A. Hodgkinson, *Oxalic Acid in Biology and Medicine* (London: Academic Press, 1977), pp. 13-16.

8. F. Dawidowsky, *A Practical Treatise on the Raw Materials and Fabrication of Glue* (Philadelphia: Henry Carey Baird & Co., 1884), pp. 58-60.

9. H. J. Arnott, "Calcium Oxalate (Weddellite) Crystals in Forest Litter," *Scanning Electron Microscopy,* 3 (1982): 1141-1149.

10. W. C. Graustein, K. Cromack, and P. Sollins, "Calcium Oxalate: Occurrence in Soils and Effect on Nutrient and Geochemical Cycles," *Science,* 198 (1977): 1252-1254.

11. A. Dietz, "Crystal Production by Micro-Organisms," *Microscope* 29 (1981): 71-76.

12. V.A. Rossetti and M. Tabasso-Laurenzi, "Distribuzione degli Ossalati di Calcio $CaC_2O_4.H_2O$ e $CaC_2O_4.2.25H_2O$ nelle alterazini delle Pietre di Monumenti Esposti all' Aperto," *Problemi di Conservazione,* (1973): 375-386.

13. U. Plahter and L. E. Plahter, "Notes on the Deterioration of Donatello's Marble Figure of St. Mark on the Church of Orsanmichele in Florence," *Studies in Conservation,* 16 (1971): 114-118.

14. O. Salvadori and A. Zitelli, "Monohydrate and Dihydrate Calcium Oxalate in Living Lichen Incrustations Biodeteriorating Marble Columns of the Basilica of Santa Maria Assunta on the Island of Torcello (Venice)," *The Conservation of Stone II,* Preprints of the Contributions to the International Symposium, Bologna, October 27-30, 1981, pp. 379-390.

15. A. Paleni and S. Curri, "Biological Aggression of Works of Art in Venice", in A. H. Walters and E. H. Hueck-Van der Plas, eds., *Biodeterioration of Materials,* 2 (New York: Halsted Press, 1972), pp. 392-400.

16. D. Jones, M. J. Wilson, and W. J. McHardy, "Lichen Weathering of Rock-Forming Minerals: Application of Scanning Electron Microscopy and Microprobe Analysis," *Journal of Microscopy,* 124(1), (October 1981): 95-104.

17. K. Cromack, P. Sollins, W. C. Graustein, K. Speidel, A. W. Todd, G. Sypcher, Y. L. Ching, and R. L. Todd, "Calcium Oxalate Accumulation and Soil Weathering in Mats of the Hypogenous Fungus *Hysterangium crassum*," *Soil and Biochemistry,* 11(5), (1979): 463-468.

WILLIAM MOUREY

A Proposal for an Index Card for Metallic Objects' Analytical Data

Archaeologists and curators are using more and more analytical data in the study of metallic objects. Unfortunately. when the results of these analyses are published. much data is lost. For instance. very often the name of the laboratory is omitted or the method used is not mentioned. In the special case of elemental analysis the percentage error of the results is almost always omitted.

For these reasons. it is proposed to introduce an index card which would be included in the publication. The data could be used repeatedly and referred to when using analytical techniques.

The analytical index card must contain information on the laboratory responsible for the analysis. the analysed object, the sampling procedures, the analytical technique, and the results of the analysis.

One index card for each analysis would be preferable. but for small publications, one index card for several analyses can be used if, of course, the same experimental conditions applied.

Laboratory

This information is the most important because communications might lead to further research on parameters not included in the index card. We think that this information must include the name of the laboratory, its address, the name of the technician who completed the analysis, and the number of the index cards used by the laboratory during the analyses. The last information is important for rapid retrieval.

Analyzed Object

For classification and statistical studies on a series of objects from the same period or of the same origin, and if the analytical technique is the same, it seems necessary to know the object type, its discovery site, deposit location, inventory number, and attribution data. Mention should be made of any conservation or restoration treatment before analysis; if the object has already been treated the results of the analysis can, of course, be quite different from those obtained before treatment.

Sampling

According to the analytical technique employed, it may be necessary to get samples from the object. If sampling is not necessary, the place of the examined part must be clearly indicated. When sampling is necessary, one must mention the exact place of the sampling on the object, the quantity, and reference number of samples removed and their weight.

Analytical Characteristics

Two fundamental options exist in analysis: elemental analysis or structure determinations. Therefore, the index card is in the form of a table divided in two parts according to these options. Other information given to characterize the analysis includes type of analysis (i.e., x-ray fluorescence or mass spectrometry); equipment (i.e. californium source, x-ray tube of n keV); supplier (i.e. Philips, Tracor); date of equipment; percentage of error on majors, minors, and traces, use of standards, source of the standards, and references of standards.

Results

The main information on the techniques having been already given, the results can be presented in percentage for elemental analysis and in components for structure determinations.

ANALYTICAL INDEX CARD

1. LABORATORY:
1.1. Name:
1.2. Address:
1.3. Date:
1.4. Analyst:
1.5. Analysis number:

2. ANALYZED OBJECT:
2.1. Type (i.e. sword):
2.2. Site of discovery:
2.3. Location of deposit:
2.4. Inventory number:
2.5. Dating:
2.6. Has the object been conserved or restored before analysis?   Y   N

3. SAMPLING:
No sampling.   Sampling
3.1. Place of the examined   3.2. Quantity of samples:
part:.   3.3. N° of samples:
                                              3.4. Place of sampling   3.4.1. Edge
                                                                                           3.4.2. Center
                                                                                           3.4.3. Complete
                                              3.5. Weight:

4. ANALYTICAL CHARACTERISTICS:
                                      4.1. Elemental analysis   4.2 Structural analysis
4.3. Type of analysis
4.4.1. Equipment
4.4.2. Supplier
4.4.3. Date of equipment
4.5.1. % error Majors
4.5.2. % error Minors
4.5.3. % error Traces
4.6. Standards   Y   N   Y   N
4.6.1. Supplier
4.6.2. Reference

5. RESULTS:

N.B. Laboratory, analyzed object, sampling and analytical characteristics can easily be computerized on an APPLE II using 36 attributes.

RICHARD NEWMAN

The Stone Sculpture of India: A Petrographic Study of the Materials Used by Indian Sculptors from the Second Century B.C. to the Sixteenth Century A.D.

Stone has been an important medium of sculpture in India since the time of Asoka (third century B.C.). In spite of this, very little information has been published on the types of rocks used by Indian sculptors, and such information is seldom very specific or particularly accurate.

This paper briefly presents the results of a pilot study of samples from 187 Indian sculptures dating from about the second century B.C. to the sixteenth century. Sculptures from nine American museums and private collections were sampled: Fogg Art Museum, Boston Museum of Fine Arts, Brooklyn Museum, Asia Society, Nelson Gallery-Atkins Museum, Norton Simon Museum of Art, Los Angeles County Museum of Art, David Nalin Collection, and Arthur Sackler Collection. A more detailed discussion of the research can be found elsewhere.[1]

The samples from the sculptures were in most instances small chips of rock (about 1cm by 1cm, 0.2-0.5cm thick), that provided thin sections for examination with a petrographic microscope, a scanning electron microscope with attached energy dispersive x-ray fluorescence spectrometer, and an electron beam microprobe.

The purposes of the study were to identify as precisely as possible the varieties of rocks used in the sculptures, to establish a geologically correct nomenclature for the rocks, and to discuss quarry sources for the rocks based on a survey of the literature on the geology of India.

Few of the sculptures in this study come from known sites. For the most part, however, the objects can be assigned on stylistic grounds to specific periods and areas of India. On the basis of these assignments, and assuming that the sources of the rocks were "local," the geological formations that are possible or probable sources can be surmised.

Discussion of the Results

For the purposes of discussion, the sculptures were grouped into several general regions: (1) southern India (Karnataka and Tamil Nadu states, parts of Andhra Pradesh and Maharashtra), 47 sculptures; (2) northeastern India (Bihar, West Bengal, Orissa states) and Bangladesh, 36 sculptures; (3) central India (primarily Rajasthan and Madhya Pradesh states), 84 sculptures; (4) Gandhara region (northern Pakistan, northwestern India, part of Afghanistan), 14 sculptures. Six sculptures were of unknown geographical origin.

A map of the modern Indian states is included in figure 1 for reference. A simplified geologic map of India is given in figure 2.[2]

Southern India

Southern India is underlain for the most part by early Precambrian rocks. Pleistocene alluvium occurs along the coasts. The predominant geologic features in the western part are a series of "schist belts" (fig. 2: C,D-5,6; black shading). The majority of these are volcano-sedimentary sequences (Dharwar Supergroup). Older than the Dharwars are a number of scattered enclaves (Sargur Group), dominated by ultramafic rocks.

The granulites of Tamil Nadu were probably formed deep in the earth's crust, and have been exposed by prolonged erosion. The prominent hill ranges in Tamil Nadu are primarily made up of these rocks. The granulites are divided into two broad groups; those of igneous origin (charnockites and charnockitic rocks) and those of sedimentary origin (khondalite).

The Cuddapah basin in Andhra Pradesh is one of the later Precambrian Indian sedimentary basins. The sediments in this basin, which have been little metamorphosed, are classified into the Cuddapah and Kurnool supergroups (fig. 2: D,E-5,6).

The Andhran empire extended from northern India to Andhra Pradesh. During the later part of this period (first through third century) stupas were built at several sites along the Krishna River (fig. 2: E-5). Sculptures from these sites are carved from limestone, often pale green in color, which comes from the Kurnool Supergroup, extensively exposed in this area.

The Pallavas controlled the eastern part of southern India between about the fifth and ninth centuries. Two of the Pallava sculptures in this study are carved from granulite (charnockitic rocks), a rock also used in the famous reliefs and small buildings at Mamallapuram near Madras (fig. 2: E-6), which were carved directly into outcrops. The other Pallava object is carved from periclase marble, which may come from the area east of the Cuddapah basin, not too far south of the Krishna.

The Cholas were the dominant power in southern India by the ninth century. The major rock of the Cholas is nearly always described as "granite," but most of the Chola sculptures in this study are carved from granulites (charnockitic rocks), rocks that superficially resemble some granites. They are usually light gray with black specks, although some are very dark gray. Many of the prominent Chola temples are located in the Kaveri River valley (fig. 2: D,E-7). If granulite was the principal sculptural material in these temples, the rock may very well have come from the Shevaroy Hills, located to the north of the Kaveri, to the west of Tanjuvar.

Amphibolite-facies quartzo-feldspathic rocks were also used by Chola sculptors, the sources of these rocks being within the largely unstudied gneissic terrain of southern India. Greenstone (metabasalt) was used in some sculptures from the Kanchipuram area to the west of Madras (fig. 2: E-6), the source of which was quite local.

The Hoysalas became dominant in Karnataka in the twelfth century. Most of the Hoysala sculptures of this study are carved from fine-grained gray or gray green magnesian schists, some of which are certainly metaultramafic rocks and others of which may be metamorphosed dolomitic sediments. The metaultramafics probably come from the Sargur Group, several exposures of which are located within thirty miles of the principal Hoysala sites of Belur, Halebid, and Somnathpur (fig. 2: C-6). The possible metasedimentary rocks come from the Dharwar Supergroup.

The nearly perpindicular bluffs of the Tertiary period Deccan lava flows in the Western Ghats (Maharashtra) are the sites of many "cave temples," all of which are located within about 200 miles of Bombay (fig. 2: B-4); these include Bhaja, Karli, Ajanta, and Ellura. One sculpture in this study is carved from Deccan basalt.

Fig. 1. Map of the modern Indian states.

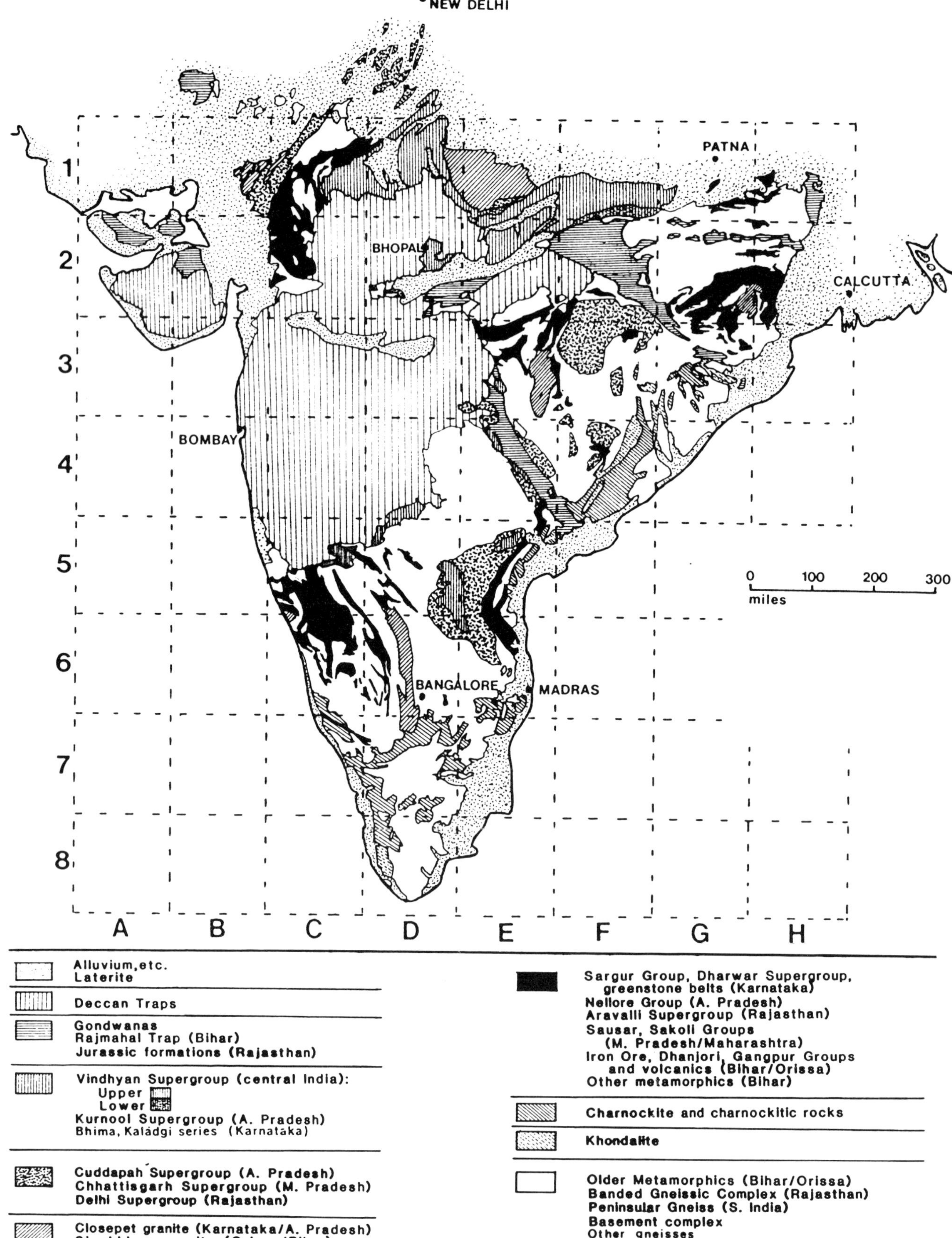

Fig. 2. Simplified geologic map of India, excluding the northeastern states and the Himalayan region. (After Geological Survey of India Misc. Publ. no. 30, 1974; and Geological Survey of India Geological Map of India, 1962.)

Northeastern India

Northern Bihar is covered by alluvium of the Ganga River. Gondwana period sandstones and shales occur in an east-west trending belt that follows the Damodar and Auranga valleys through the center of the state (fig. 2: G,H-2). Central and southern Bihar are underlain primarily by Precambrian gneisses. Xenoliths of metasedimentary rocks, probably Precambrian, are scattered throughout the gneissic terrain (fig. 2: G,H-1,2; black shaded areas). The metasedimentary rocks belong to two distinct groups: a predominantly argillaceous group, of medium metamorphic grade; and, further north, just to the south of the Ganga, a lower-grade group of quartzites and meta-argillites.

The Palas gained control of northeastern India in the eighth century. They were succeeded by the Senas in the twelfth century, who succumbed to the Moslems in the following century. Literally thousands of stone sculptures survive from these centuries. The rocks used by sculptors are phyllites and schists. Sculptural centers seem to have shifted eastward, from northern Bihar to northern West Bengal, around the eleventh century. Sculptures of the ninth through eleventh centuries are carved from pelitic schists, brown gray or greenish gray in color, some of whose mineral assemblages suggest greenschist-amphibolite border facies, while others are probably greenschist facies. Sculptures from the eleventh and twelfth centuries are usually carved from a dark gray carbonaceous phyllite. This rock is still frequently referred to as "black chlorite," a geologically meaningless name first suggested by Bhattasali in 1929.[3] It is often stated that the Rajmahal Hills were its source, but this is not likely since basalt is virtually the only type of rock in these hills (fig.2: H-1). The source may be in a metasedimentary xenolith near Monghyr, just to the south of the Ganga and to the east of Patna (fig. 2: G-1). At least some of the schists used in the earlier Pala-Sena sculptures probably come from the metasedimentary rocks exposed in central Bihar, while others may come from the lower-grade group further north.

About three-fourths of Orissa consists of Precambrian rocks. At the western margin of the coastal alluvial plain on both sides of the Mahanadi River are outcrops of the Gondwanan Athgarh sandstone (fig. 2: G-3). To the south of the river, several small hills made up of this rock were the sites of Jain rock-cut monastic buildings dating as far back as the second century B.C. As early as the seventh century, temples were constructed on the coast at Bhubaneswar. The last major monument in this region is the Surya Deul temple at Konarak, also on the coast. At Konarak, sandstone (probably Athgarh sandstone), khondalite, and metaultramafic rocks were used. Khondalite is extensively exposed at the eastern edge of the gneissic terrain. This rock weathers badly, the garnets breaking down to limonitic blotches on the surface. Most of the Orissan sculptures in this study are carved from an unusual weathered metamorphic rock (gneiss) that resembles weathered khondalite. But this rock is more accurately a "gossan" or cap rock; the yellow orange blotches are skeletal iron oxide "boxworks," replacing sulfide minerals that have broken down.

Central India

Stone carving was introduced into India by contacts with the Hellenic and Iranian West, and stone first became an important medium of sculpture during the Maurya period (322-185 B.C). The Mauryan sculptures were carved from a cream-colored sandstone that presumably came from quarries at Chunar (Uttar Pradesh) on the Yamuna River, located on the northern edge of the Vindha Mountains (fig. 2: F-1). These mountains are made up of largely unmetamorphosed sedimentary rocks of another of the late Precambrian sedimentary basins.

Sandstone was the major structural and sculptural material in the central part of India throughout the period covered in this study, and the source of most of these sandstones was undoubtedly exposure of the Vindhyan Supergroup. Silica-cemented quartz arenites are the most weather-resistant of sandstones, and the majority of the rocks encountered in this study are this type of sandstone.[4]

Among the sites located on or very near outcrops of the Vindhyans are Mathura, Sanchi, Bharhut, Sarnath, Allahabad, Besnagar, and Khajuraho. Sandstone and quartzite are also exposed in other formations in central India, and some of these rocks were also probably used.

Marble and other calcareous metamorphic rocks are exposed in several areas of Rajasthan and Gujarat. Many of the purer marble outcrops belong to the "Raialo series," a problematic group exposed at a few places between the Precambrian Aravalli and Delhi supergroups east of the Aravelli Range in Rajasthan (fig. 2: C-1,2). The marble used in the Taj Mahal and other Mughal period buildings came from Raialo outcrops near Makrana.

Two sculptures in this study come from the Mt. Abu area in the southern Aravelli Range (fig. 2: B-1), the site of several marble Jain temples. Mt. Abu itself consists of granite and the marbles used in these buildings probably came from the Delhi Supergroup, exposed to the southeast down the side of the mountain. The Aravalli Range is predominantly made up of Delhi Supergroup rocks.

Two Gupta period sculptures from the southern part of Rajasthan to the east of the Aravalli Range are carved from metamorphic rocks that come from the Aravalli Supergroup, outcrops of which are extensively exposed in that area (fig. 2: C-1,2; indicated by black shading).

Gandhara Region

Gandhara is the ancient name for the region that included northwestern India, northern Pakistan, and part of Afghanistan. The region was conquered by Alexander the Great in 327 B.C. In the first century A.D., the Kushans, a tribe originating in northwestern China, gained control of the area. Peshawar in northern Pakistan was the northern capital of the Kushan empire.

The rock almost invariably used in Gandharan sculptures of the Kushan period (first through fourth centuries) is a silvery dark gray carbonaceous phyllite. The source of this rock is difficult to ascertain, but probably was the Salkhala Formation, a group of Precambrian rocks that includes graphitic schist, quartz schist, marble, and quartzo-feldspathic gneiss. Marshall thought that the quarries were in the Swat River valley,[5] north of Peshawar, an area whose geology has been only sketchily studied. Studies of areas of northern Pakistan to the east of the Swat River valley have shown that "Salkhala" type rocks—always including graphitic schist—are present, and they may also occur in the Swat valley.

Table 1
Summary of Some of the Rocks Identified in Southern Indian Sculptures

Period	No. of sculptures	General rock type	Mineral assemblage
Western Chalukya,	1	metasiltstone	quartz-illite/sericite-chlorite
10th-11th c.	1	magnesian schist	talc-chlorite-calcite
	1	schist	margarite-muscovite
Hoysala, 12th-13th c.	6	magnesian schist	talc-chlorite chlorite-actinolite ± calcite talc-serpentine chlorite-talc-anthophyllite-carbonate
Pallava, 8th-9th c.; Chola, 11th-14th c.; Vijayanagar	15	granulite	± potassium feldspar ± plagioclase feldspar (oligoclase-andesine) ± quartz ± orthopyroxene ± clinopyroxene
Pallava	1	marble	periclase-forsterite-dolomite
Chola, 9th-11th c.; Vijayanagar, 16th c.	4	greenstone	chlorite-actinolite
Chola	2	quartzo-feldspathic rock	± potassium feldspar ± plagioclase ± quartz-hornblende-biotite
Late Andhra, Amaravati area, 2nd-4th c.	7	limestone	calcite

Table 2
Summary of Some of the Rocks Identified in Northeastern Indian Sculptures

Period	No. of sculptures	General rock type	Mineral assemblage
Gupta, 7th c.	1	schist	andalusite-biotite-quartz-muscovite-sericite
Pala/Sena, 9th-11th c.	15	schist	andalusite-biotite-quartz-muscovite-sericite biotite-quartz-muscovite-sericite chlorite-muscovite/sericite ± quartz
Pala/Sena, 10th-12th c.	8	phyllite	quartz-muscovite-chloritoid
Orissa, 8th-13th c.	5	weathered gneiss	kaolinite-quartz-graphite-iron oxides
	1	magnesian schist	talc-serpentine

Table 3
Summary of Some of the Rocks Identified in Central Indian Sculptures

Period	No. of sculptures	General rock type	Mineral assemblage
Shunga through medieval, 2nd c. B.C.-12th c.	59	quartz or quartzose arenite	
	12	quartz or quartzose wacke	
Gujarat/Rajasthan, ca. 11th c. on	6	marble	calcite calcite-dolomite calcite-dolomite-tremolite calcite-quartz-biotite-muscovite
Gupta, Rajasthan/Gujarat, 5th-6th c.	1	metasiltstone	muscovite-chlorite-quartz
	1	schist	chlorite

References

1. R. Newman, *The Stone Sculpture of India: A Study of the Materials used by Indian Sculptors from ca. 2nd century B.C.-16th Century* (Cambridge, Massachusetts: Center for Conservation and Technical Studies, Fogg Art Museum, 1983).

2. The most recent single text on Indian geology is: M. S. Krishnan, *Geology of India and Burma,* 5th ed. (Madras, India: Higginbothams, 1968). Unfortunately, much of the information in this book is somewhat outdated. The discussions in the text of this paper are based on numerous publications, particularly articles in the *Memoirs* and *Records* of the *Geological Survey of India.* Recent reviews of the geology of the Indian states may be found in *Geol. Survey of India, Misc. Publ. no. 30* (1974).

3. N. K. Bhattasali, *Iconography of Buddhist and Brahmanical Sculptures in the Dacca Museum* Dacca: Rai S. N. Bhadra Bahadur, 1929).

4. "Quartz arenite" is here defined as a sandstone with less than about 10-15% fine-grained matrix, whose detrital grains are more than about 95% quartz.

5. J. Marshall, *Taxila. An Illustrated Account of Archaeological Excavations...*, 3 vol.(Cambridge, England: University Press, 1951).

T. WALL, J. R. BIRD, J. BROWN, M. MAYNARD, and C. KENNARD

An Overpainted Photograph: Investigation by Autoradiography

Swedish born artist Carl Magnus Oscar Friström (1856-1919) is first recorded in Brisbane in 1884 when he exhibited at the Queensland National Association Exhibition. The following year he was in partnership with a photographer, D. H. Hutchinson, and exhibited photographic portraits from life, landscape, and genre, some retouched and colored. He later became known chiefly as a portrait painter in oils and pastels. He was involved in Brisbane art societies and in the intellectual life of Brisbane.

Friström also painted the female figure. One such painting (fig. 1) is held in the Queensland Museum, Ethnology Collection, QE 3613. It is in oils on a panel, 106.7 x 44.0 cm, and signed, "Oscar Friström 1897." A label on the back of the painting reads:

Subject: An Australian water nymph
Name of Artist: Oscar Friström
Address: Oakden Chambers, Queen Street, Brisbane, Queensland.
Price: £30.0.0
Date: 1897

The University of Queensland Department of Fine Arts has obtained a black and white gouache (14 x 34 cm) from the Thompson Collection of Indooroopilly, Brisbane. It is much smaller but practically identical in subject to the museum oil painting. It is mounted on brown board and signed "Oscar Friström 97" with an inscription on the mount in Friström's hand: "Fran die Vau" (From the Swamp) (fig. 2).

Because of Friström's association with photography and his known use of photographs as models for some of his works, it was thought possible that Friström had used a technique of overpainting a photograph. Although the pose in the two works is similar, the facial expressions are different, with the larger painting being natural in appearance and the gouache considerably less so. There are many differences in the treatment of the background in the two pictures.

Neutron activation followed by autoradiography has been used in a number of studies of oil paintings to reveal brush techniques, overpainting, signature changes, pigment types, and other information. The materials used for preparation of the gouache were probably chalk, Indian ink, and water. Apart from possible impurities, these materials do not activate strongly with thermal neutron irradiation whereas silver is readily activated. It was felt that this technique would involve favorable conditions for the detection of a photographic image under the gouache.

Activation

The Moata reactor at Lucas Heights is an Argonaut type reactor that can be operated at power levels up to 100kW. Large shield plugs can be removed from the top shield to provide access to the graphite reflector. During a shutdown period, the gouache was wrapped in a polythene sheet and placed against the reflector. The shielding was then replaced. The reactor was operated for 3 1/4 hours to produce a neutron irradiation averaging 2×10^{14} cm^{-2} at the position of the gouache. A beam hole in the graphite face produced a somewhat greater dose over a 7.5cm diameter circle near the center of the picture. With the reactor again shut down, the picture was removed and the radioactivity allowed to decay for four hours before commencing a series of gamma-ray spectra measurements and autoradiographs.

The gamma-ray spectra were measured with a 120cm^3 lithium-drifted germanium detector contained within a massive lead and bor-

ated paraffin shield. A 7.5cm diameter aperture in the shield allowed gamma rays to be detected from a selected region of the gouache, which was placed against the outside of the shield. The induced activity was too low to permit the use of a smaller aperture for the study of finer details of the activity produced in the picture.

Spectra were measured from three regions of the picture. The darkest (center right) and lightest (top left) regions were selected as well as the bottom right, which showed a dark patch in early autoradiographs. A typical spectrum obtained from the center right position at 210 hours after the irradiation is shown in figure 3. Estimates of the amount of various nuclides present were made from the observed gamma-ray intensities and these are given in table 1 together with details of the radionuclides used for these determinations. Gamma rays from ^{56}Mn (half-life 2.6h) (not included in Table 1) and ^{24}Na (15h), which were dominant in the earliest spectra, are likely to arise from components of the board and paper used in the gouache.

Table 1
Surface Densities of Some Trace Elements in the Gouache (μg.cm^{-2})

ELEMENT:	Na	Au	La	As	Ag	Sc	Zn	Eu	Cs	Co
HALF-LIFE	15h	65h	40h	27h	252d	84d	245d	16y	2.1y	5.3y
REGION										
Bottom Right	360	0.03	4		0.60	80	5			
Center Right	350	0.06	5		0.04	70	1			
Top Left	330	0.02	6		0.03	8	1			
Face Region								80	0.1	0.8	0.7

The concentrations of sodium, scandium, and lanthanum are essentially identical for the lightest and darkest regions, which is consistent with the interpretation that they are components of the backing. Scandium is high in the bottom right area and arsenic is also very high, as indicated by the observation of a strong 560keV gamma ray from ^{76}As (26.5h). Apparently some material containing arsenic had been spilled or splashed on the backing used for production of the gouache. It is of interest that arsenical solutions were also used for cleaning plates in early photographic procedures. Zinc was only observed at later times when the total activity was very weak and spectra measurements were made only on the face region.

After about 50 hours, the 660keV gamma-ray from ^{110m}Ag (252d) became evident and from 100 hours ^{198}Au (65h) was also observed. The silver concentrations correlate reasonably well with the blackness of the three regions of the picture measured and provide support for the hypothesis that Friström used a photograph as an underlay for the preparation of the gouache. However, the low silver concentration from the light area at the top of the gouache suggests that perhaps this area, which contains none of the detail present in the oil painting, was not included in the photographic print.

The gold observed in the intermediate stages of the decay is also significant. Descriptions of early photographic techniques indicate that it was common to use a gold compound, with an albumenized paper, as a toner in the printing process. This was a standard photographic printing procedure in Europe until about 1890 when it was superseded by the use of silver halide gelatin papers.

Separate tests were carried out using a small photograph of simi-

Fig. 1. Friström's *An Australian Water Nymph*, painted in oils on a panel 44 x 107 cm (Queensland Museum, Ethnology Collection, QE3613).

Fig. 2. Friström's black and white gouache *From the Swamp* (14 x 34 cm).

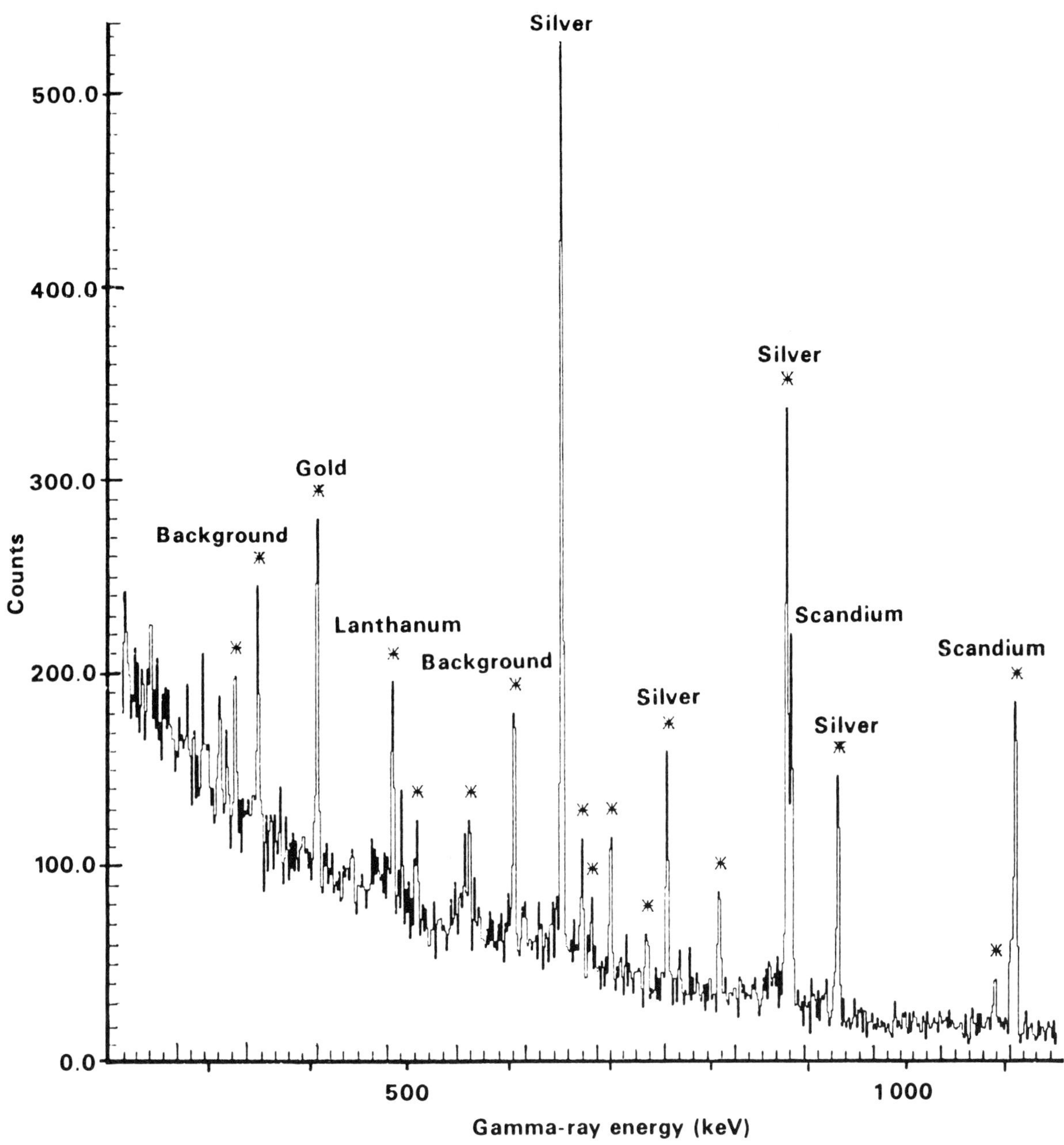

Fig. 3. Gamma-ray spectrum from the center right region of the gouache after activation and subsequent decay for 210 hours. Prominent lines from ^{110m}Ag and ^{198}Au are evidence of the existence of a photograph under the gouache.

lar age to the gouache and using a modern print made from the photograph. Results similar to those described above were obtained from the old print, with silver and gold both being present. The modern print showed only silver.

Autoradiographs

A total of twelve autoradiographs were taken at various times after activation and for different exposure times (table 2). The activated gouache, without frame, was pressed directly against D2, D7, or D10 x-ray film and kept wrapped in a darkroom for the duration of each exposure. The resulting images are chiefly produced by beta-ray absorption in the emulsion, gamma-ray absorption being negligible in such a thin layer. Some contribution may have arisen from pure beta emitters, but the majority of the exposure is expected to be caused by those radioisotopes that have been identified by their gamma-ray emission. The nuclides that are important at the time of each autoradiograph are included in table 2. Positron activity is rare in neutron activation and this is confirmed by the small 511 keV peak observed in the gamma-ray spectra, which can be attributed to background from annihilation events produced by high energy gamma rays.

The early autoradiographs are dominated by a dark patch at lower right and numerous other small spots. These are visible in autoradiograph 3, which is reproduced in figure 4. This was exposed from 51 to 115 hours after the irradiation and was the first to show good detail over most of the picture. A slightly darker region centered to the left of the face is produced by the larger neutron flux density in this region. The explanation of the dark spots as contamination by arsenical solutions has been mentioned above.

Apart from the dark spots, autoradiographs 3 through 8 are very similar in appearance to the gouache. Significant dark features, such as the branches of the bushes at upper right and lower left, and the two tree reflections adjacent to the right arm and the signature, match those in the gouache and differ from the oil painting. We must therefore conclude that one or more of the intermediate half-life nuclides is present in the black pigment used for the dark features of the gouache. This is most likely to have been carbon black and, although carbon is not activated significantly by thermal neutrons, sufficient impurities are undoubtedly present which could produce the observed image. This situation unfortunately obscures any silver image from an underlying photograph.

Whilst some dark features reproduce accurately those in the gouache, the larger dark areas such as those on the body and shadows do not show the same detailed correlation. This may imply that not all the black pigment has the same impurity content—possibly indicating that different mixes were made at different times. The sensitivity of the autoradiograph technique is illustrated by the lines of dots that occur around the outline of the overlapping cardboard frame. These presumably represent the remnant of dust collections after the surface of the gouache was cleaned.

About one year elapsed between the taking of autoradiographs 9 and 12 and there is a marked difference between the early and late autoradiographs. This is clearly seen in figure 5, which is a print of autoradiograph 12. During the relevant period, the gamma-ray emission from ^{46}Sc would decrease by an order of magnitude relative to that from ^{110}mAg. The scandium can therefore be assumed to be the important impurity in the black pigment (Indian ink) and the new fea-

tures seen in the later autoradiographs must be attributed to either silver or zinc, which have sufficiently long half-lives.

Table 2
Details of Autoradiographs and Dominant Activities

No.	1	2	3	4	5	6	7	8	9	10	11	12
Film Type	D7	D7	D2	D7	D7	D7	D7	D7	D7	D7	D10	D7
Elapsed Time (d)	0.5	1.5	3.5	5.5	11.5	17.5	22.5	27.5	52.5	240	300	420
Exposure Time (h)	12	16	64	19	113	114	114	117	402	1000	1000	3000
Dominant Activities	^{24}Na					^{46}Sc						
				^{140}La					^{110}mAg			
										^{65}Zn		

Among the features that reveal striking similarities between autoradiograph 12 and the oil painting, but not the gouache, are marks in the water adjacent to the right leg and left elbow, the shape of the bushes along the waterline at the left, the structure of the bush at the lower left, and the pattern of light and shade on the body and rocks.

Measurements on autoradiograph 12 and the oil painting show that the relative positions of these features as well as the relationship between the figure, waterline, and large rock are so accurately proportioned that a reduced-scale photograph of the oil painting will fit the gouache to within 1mm precision or better. This fact in itself supports the hypothesis thay they are not independent freehand versions of the same subject, but copies prepared with some mechanical assistance. The theory that a photographic image was used for this purpose is supported by the appearance of a new straight border well within the boundary of the gouache image, the importance of silver and gold in the gamma-ray spectra, and the change in general appearance of the figure and rock in the autoradiographs that match the features of the oil painting rather than the gouache.

Conclusions

There is convincing evidence that the gouache is applied over a photograph of the oil painting. However, there are features of both apparent in the late autoradiographs. Some dark brush strokes in the gouache are present in these but many of the dark areas are completely missing (for example, to the right of the photographic border where it passes across the water at the right of the picture). On the other hand, some of the contrasting features in the autoradiograph arise from the white pigment in the gouache which apparently is thick enough to be able to attenuate beta intensities from within the photograph and backing.

There are a number of as yet unanswered questions raised by details of the autoradiographs. The absence of detail at the top and bottom implies absence of a photographic image in these regions but no clear boundary is apparent. The partial reproduction of dark brush strokes implies that several batches of pigment were used, not all of which contained impurities that have been activated. On the other

Fig. 4. Autoradiograph 3, taken five days after irradiation of the gouache. Intense dark spots arise from arsenic contamination. The appearance of the image resembles that of the gouache.

Fig. 5. Autoradiograph 12, taken fourteen months after irradiation. The image now resembles figure 1 in many details but also retains some features from figure 2. A new border at the right probably defines the limit of the underlying photographic image.

hand, the white pigment must have been particularly free of impurities. The presence of gamma rays from ^{65}Zn in the final gamma-ray spectrum cannot indicate the use of a zinc white since this would lead to a blackening of the autoradiograph at the wrong places. It is plausible to assume that the zinc is a component of the backing materials

It is clear that Friström used a photograph of his own oil painting as a starting point for producing a smaller copy in black and white gouache. The obvious similarity between the two works is thus explained and additional information has been gleaned about the materials and methods that he employed. It is not surprising that the immense detail available from autoradiographs and associated measurements of activity raise many new questions that could be the subject of further investigations. The work thus confirms the power of these techniques especially when an experimental reactor facility is available that can be adapted and operated to meet the requirements of individual tasks, including the irradiation of whole paintings.

Acknowledgments

We wish to acknowledge the permission of the Queensland Museum to reproduce the oil painting discussed in this work and their cooperation during the study; also the support of the Australian Institute of Nuclear Science and Engineering, the contribution by P. Gillespie, Materials Division, AAEC (who prepared the autoradiographs) and the photographic work carried out by D. Elkins.

Radiography Applied to the Study of a Portrait of Philip IV in the Museum of Fine Arts, Boston

The recent cleaning of the portrait of Philip IV in the Boston Museum of Fine Arts has offered an opportunity for a renewed study and assessment of the work. The full-length portrait of the young Spanish monarch was purchased by the museum in 1904 as a work by Velazquez and soon became the subject of heated controversy concerning its attribution.[1] At the time, a second version, now in the Altman Collection at the Metropolitan Museum, New York, was known, but unavailable for comparison.[2]

These two paintings together with a bust-length portrait in the Meadows Museum, Dallas, similar in type to the Metropolitan Museum and Museum of Fine Arts pictures, and a somewhat different full-length portrait in the Prado which in turn closely resembles the composition of another full-length version in the Isabella Stewart Gardner Museum, Boston, form a tightly interrelated group of paintings that are of particular interest as Velazquez's earliest representations of King Philip IV.[3]

While the extent of workshop participation in the execution of these paintings has been a point of contention among scholars, the attribution to Velazquez of the Prado, Metropolitan Museum, and Meadows pictures has received widespread support. The Museum of Fine Arts version is almost universally given the lesser status of a workshop copy and has even been considered a later copy by some scholars.[4] Radiography has proven to be an effective method for exploring this attribution problem and has brought forth parallels in quality and technique between the Museum of Fine Arts painting and the related works.[5]

Velazquez visited Madrid for the first time in April, 1622, having made the trip from his native town of Seville at age twenty-three with the hope of receiving an introduction to the court. During this stay he produced the portrait of Luis de Gongora, now in the Museum of Fine Arts collection, but it was not until his return more than a year later that he received the awaited invitation to paint his first portrait of Philip IV, an event that is chronicled in the writings of Velazquez's teacher Francisco Pacheco.[6]

There is no documentary evidence to indicate which if any of the existing portraits can be identified as the original. However, radiographs show that the Prado picture was extensively revised by Velazquez some time after it was originally painted. The underlying version appears well developed, resembling the Museum of Fine Arts and the Metropolitan Museum paintings quite closely in its major compositional features and is regarded by Lopez-Rey as the original portrait cited by Pacheco.[7] Although this is largely a matter of conjecture, these radiographs do indicate that the Metropolitan Museum and Museum of Fine Arts portraits are the only surviving versions of the earliest known compositional type, which certainly places them in a special category.

Recent x radiographs of the Museum of Fine Arts painting have yielded an image of remarkable clarity. The face and hands exhibit a deliberate handling with concern for sculptural modeling reminiscent of Velazquez's Sevillian period, while the treatment of the garments is more fluid, anticipating the rapid brush stroke and terse rendering that characterize his mature style (figs. 1-5).

The unusual clarity of the x radiographic image produced by all of the paintings demonstrates their underlying technical similarity.[8] This is due in part to the low density of the red earth ground that allows the x rays to pass freely without obscuring the pictorial image.[9,10] Absence of significant pentimenti which might otherwise interfere with a clear reading of the radiographs is also a key factor in their legibility and a feature common to all but the Prado version.

Perhaps most surprising, as well as difficult to explain, is the fact that the details and form of the black garment of the Museum of Fine Arts version, which have become thoroughly illegible to the unaided eye owing to abrasion and sinking of the dark colors, are made completely visible in the radiographs. Nuances of light and shadow and effects of volume are enhanced in the radiographs (fig. 2). The diminution over time of this passage may in fact help to explain why the painting has been largely dismissed as a workshop copy, and it renders all the more fortunate this opportunity to gain an insight into the painting's original appearance.

The x radiographs of the Museum of Fine Arts and Meadows pictures indicate that the facial structures are carefully developed with brush strokes that are nearly identical in their variations of paint densities (figs. 1,6). Compare the areas around the eyes, the nose and mouth, the sagging muscles of the cheek and the projecting chin. The boldness and self assurance of the handling, as well as the beauty and economy of the technique are surprisingly consistent. The musculature is articulated in a fashion that parallels the segmentation of planes that is found in the portrait of Luis de Gongora. By contrast, the modeling in the head of the Metropolitan Museum painting (fig. 7) is softer and there is less of an anatomical emphasis.

Comparison of the handling of the garments in the upper body in the Musuem of Fine Arts and the Metropolitan Museum paintings (figs. 2,8,10) shows that the Museum of Fine Arts version is more elaborate; the modeling of the folds, especially in the arm, is more articulated, and the brushwork, while remaining clearly legible, is also more rapid and complex. In general, the superior quality of this passage in the Boston painting argues against the notion that the Museum of Fine Arts version might have been copied from the Metropolitan Museum painting. If anything, the reverse is suggested. As seen in the radiographs the treatment of the garment in the Metropolitan Museum painting seems rather closer to that of the Meadows bust portrait (fig. 6).

In contrast to the clarity of the other versions, radiographs of the Prado painting are complex and difficult to interpret because of the superimposition of two images (fig. 13). The changes, however, do not seem to be the result of a series of spontaneous alterations but appear to represent studied reworking of the entire design. The overlapping of two relatively dense passages makes interpretation in the area of the face especially difficult (fig. 12).[11] However, the position of the hands is changed so that in this area both the initial and later versions can be clearly read (fig. 14). This affords an excellent opportunity to compare the modeling of the hands in the Prado painting with those in the related portraits. Here it may be seen that the treatment of hands and white cuffs in the first Prado version (fig. 14) is similar to that of the Museum of Fine Arts and Metropolitan Museum paintings (figs. 3,4,10). Compare the systems of highlights and shadows in the cuffs, in the note and in the hands. Once again the manner of manipulating paint, articulating the fall of light, and defining volume is seen to be consistent, the product of a single mind.

Compare also the manner in which the gold chains are painted in

Fig. 1. X radiograph detail of head, *Portrait of Philip IV,* Museum of Fine Arts, Boston.

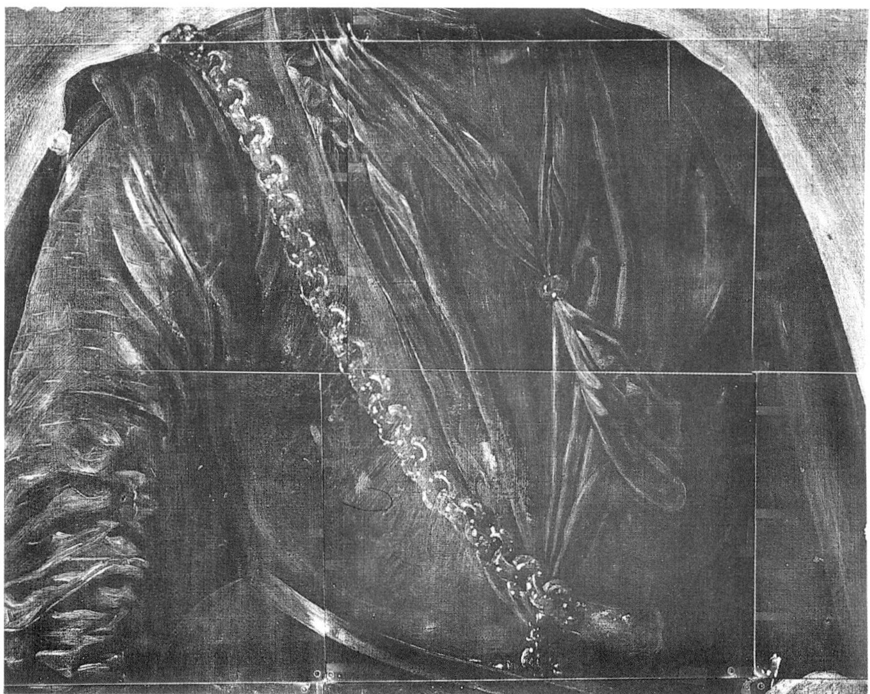

Fig. 2. X radiograph of torso, *Portrait of Philip IV,* Museum of Fine Arts, Boston.

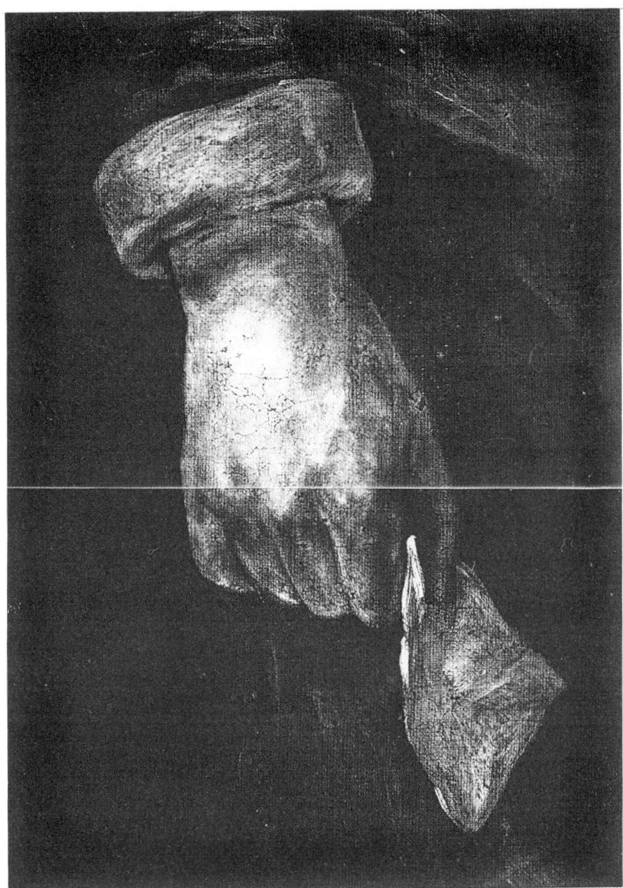

Fig. 3. X radiograph of right hand, *Portrait of Philip IV,* Museum of Fine Arts, Boston.

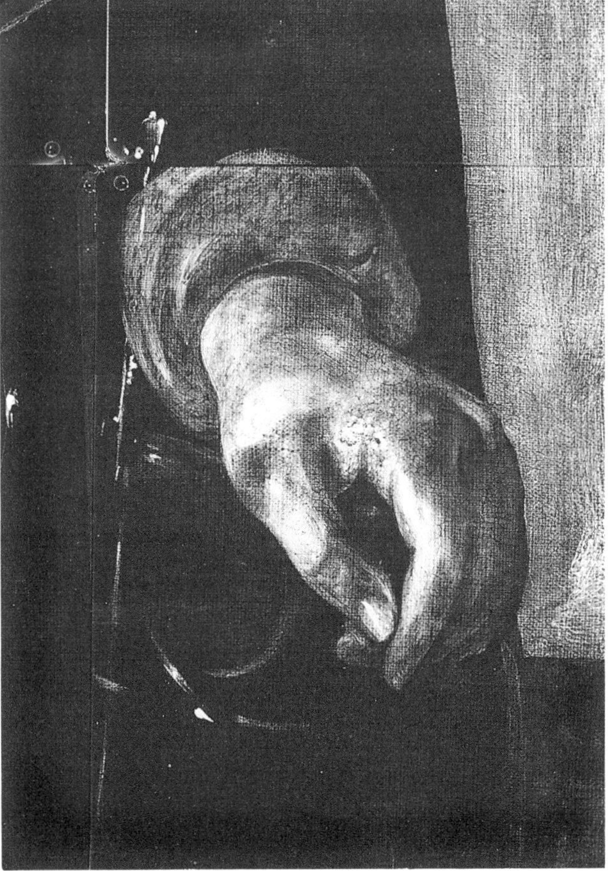

Fig. 4. X radiograph of left hand, *Portrait of Philip IV,* Museum of Fine Arts, Boston.

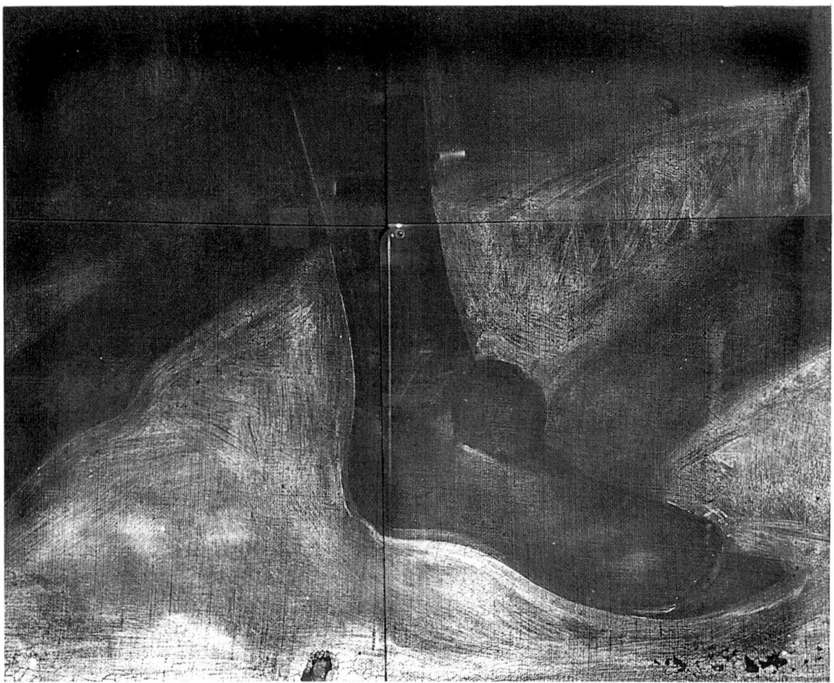

Fig. 5. X radiograph of left foot, *Portrait of Philip IV,* Museum of Fine Arts, Boston.

the Museum of Fine Arts and Metropolitan Museum paintings (figs. 2,8): a striking illusion, created with the utmost economy.

Further comparison of the secondary features of flow and background demonstrate a consistent style in the application of paint in those broadly painted areas. Compare the left shoulders of the Metropolitan Museum and Museum of Fine Arts paintings (figs. 9, 2), where the background is brought up to the contour with a bold, steady sweep of the brush. This is also seen in the Prado painting (fig. 13). There is a characteristic razor crispness of edges and a typical build up of paint at these junctures. Compare the points at which the necks meet with the white collars in all of the portraits (figs. 1,6,7,12,15) where there is an idiosyncratic gap. Similar gaps are seen around the hands (figs. 3, 4, 10, 14, 16).

Compare the treatment of the hair in the Museum of Fine Arts, Meadows Museum, and Metropolitan Museum pictures (figs. 1, 6, 7). There is also a strong similarity to be found between the hair of the Prado and Gardner versions (figs. 12, 15).

In the area of the feet, the stylistic features are consistent throughout the five pictures (figs. 5, 11). The modeling has fullness and clarity, and the outlines are crisp. There are characteristic ridges of paint along the contours. The brushwork in the floor has a distinctive appearance, and the cast shadows are left in reserve so that they appear dark in the radiographs.

To be sure, there are differences among the paintings that are perhaps more apparent in the radiographs than on the surfaces of the paintings themselves. Each has unique qualities, some are softer and others firmer in the character of the modeling, and certain ones exhibit a more painterly manner than the others. This may reflect an evolution in the artist's style and technical method or it may indicate

differences in the artist's concerns stemming from the purpose or the patrons for which the paintings were intended. Finally, one must consider the possibility of participation by studio assistants in the execution of these paintings. However the distinct lack of information about the size and organization of Velazquez's studio during this period complicates speculation on this question.

While recognizing subtle differences between the pictures along with variations in quality within each one, we are in the final analysis confronted with an interconnecting set of stylistic and technical links that unites the paintings and points to the presence of a masterly hand.

In light of evidence that indicates a common technique, style, and quality among the five paintings, it would seem arbitrary as well as inaccurate to portray the Museum of Fine Arts painting as being inferior. On the contrary, it is when viewed as a group that the paintings offer the greatest potential for shedding light on the working methods and preoccupations of Velazquez during this early phase of his career.

Fig. 6. X radiograph of head, *Portrait of Philip IV*, Meadows Museum, Dallas.

Fig. 7. X radiograph of head, *Portrait of Philip IV,* Metropolitan Museum of Art,
New York.

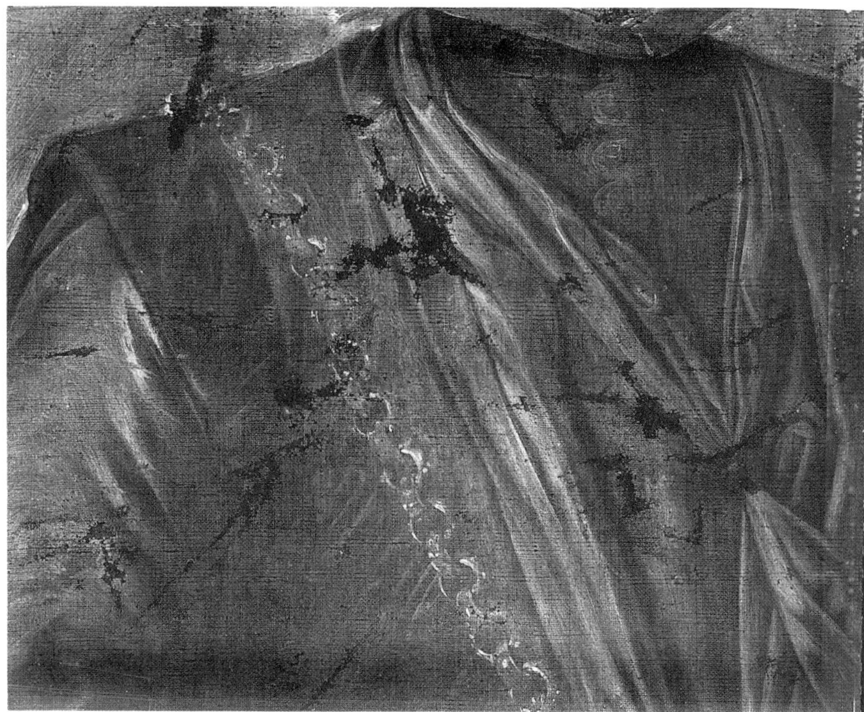

Fig. 8. X radiograph of torso, *Portrait of Philip IV,* Metropolitan Museum of Art,
New York.

Fig. 9. X radiograph of shoulder, *Portrait of Philip IV,* Metropolitan Museum of Art, New York.

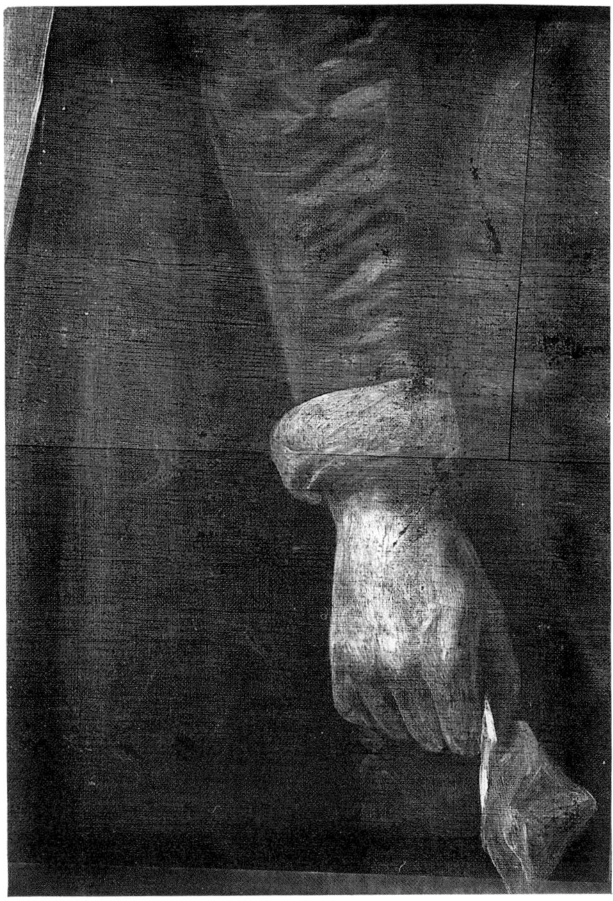

Fig. 10. X radiograph of right hand, *Portrait of Philip IV,* Metropolitan Museum of Art, New York.

Fig. 11. X radiograph of left foot, *Portrait of Philip IV,* Metropolitan Museum of Art, New York.

Fig. 12. X radiograph of head, *Portrait of Philip IV,* Prado, Madrid.

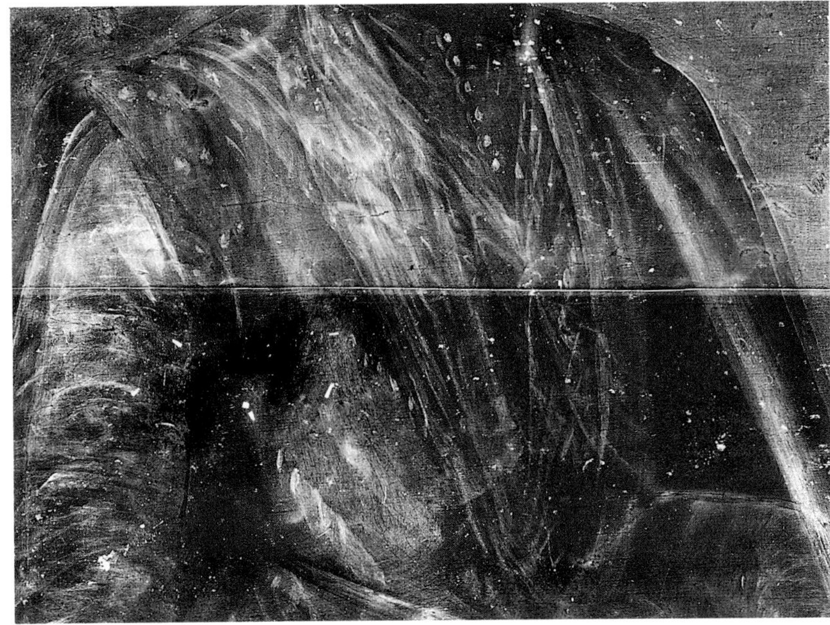

Fig. 13. X radiograph of torso, *Portrait of Philip IV*. Prado. Madrid.

Fig. 14. X radiograph of hands, *Portrait of Philip IV,* Prado, Madrid.

Fig. 15. X radiograph of head, *Portrait of Philip IV,* Isabella Stewart Gardner Museum, Boston.

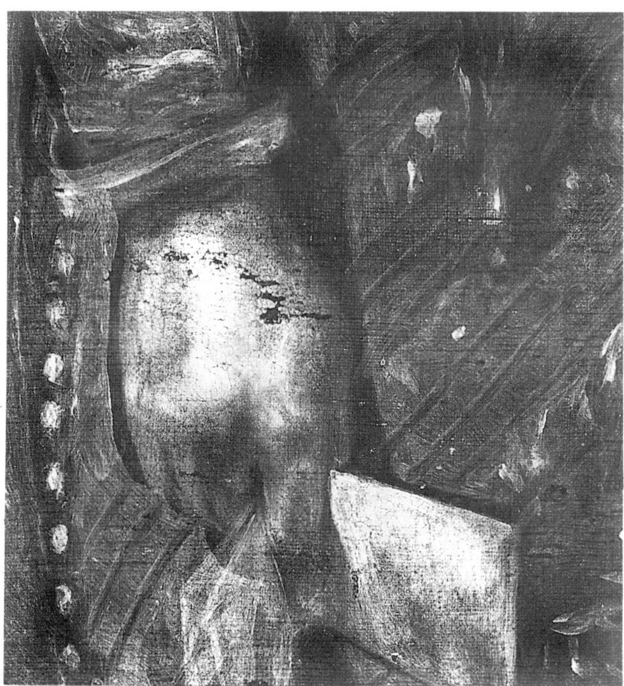

Fig. 16. X radiograph of right hand, *Portrait of Philip IV,* Isabella Stewart Gardner Museum, Boston.

References

1. B.I. Gilman, "Report of Facts and Opinions Regarding the New Velazquez," *Bulletin of the Museum of Fine Arts, Boston,* 1905. The Museum of Fine Arts files contain extensive records detailing the controvesy.

2. Jose Lopez-Rey, "The Reattributed Velazquez: Faulty Connoisseurship," *Art News,* pp. 50-2, 1973. The Altman painting has since been judged by Jose Lopez-Rey as a replica by Velazquez of an earlier painting, and the rediscovery of a receipt for the sale of the painting to Dona Antonia de Impenarrita signed by the artist firmly establishes the connection of the Metropolitan Museum portrait with Velazquez.

3. Jose Lopez-Rey, *Velazquez, A Catalog Raisonné of his Oeuvre,* (New York: Faber and Faber, 1963). The five paintings are illustrated in this book along with radiographs of the Prado portrait of Philip IV and of other paintings by Velazquez.

4. Jose Guidol, *Velazques,* (New York: The Viking Press, 1974). Gudiol is among the few contemporary scholars to accept the Museum of Fine Arts painting as being by the hand of Velasquez.

5. Radiographs of the Museum of Fine Arts and Gardner paintings were taken by Pamela England of the Museum of Fine Arts Research Laboratory; radiographs of the Metropolitan Museum painting were taken by Mark Leonard, formerly of the Metropolitan Museum Paintings Conservation staff. Radiographs of the Meadows Museum portrait were taken by James Roth formerly of the Nelson Gallery of Art, Atkins Museum of Fine Arts in Kansas City and radiographs of the Prado paintings, taken by the National Gallery of Sweden and previously published by Lopez-Rey, were provided by the Prado Museum. The cooperation and assistance of these people and of these institutions is greatly appreciated.

6. Francisco Pacheco, *Arte de la Pintura,* ed. F.J. Sanchez Canton (Madrid, 1956).

7. Jose Lopez-Rey "Velazquez' Philip IV", *Art News,* Summer 1968.

8. In this instance the combined factors of method and construction, choice of materials, and artist's handling produce a rather unusual and characteristic appearance.

9. Two pigment samples from the painting were taken by Pamela England of the Museum of Fine Arts' Research Laboratory in order to identify the ground and the yellow used in the gold chain. The examination report states as follows: "Using laser excited emission spectroscopy, the sample of red ground was shown to contain the elements typical of an earth pigment, namely silicon, aluminum, calcium, magnesium, iron, and titanium. Lead-tin yellow was confirmed as the pale yellow."

10. Richard Newman and Gridley Mckin Smith, "Observations on the Materials and Painting Technique of Diego Velazquez," *American Institute for Conservation Conference Proceedings,* 1982. Findings of the examination of the Prado full-length portrait of Philip IV which indicate that a mica-rich red ochre ground was found and the yellow pigment lead-tin yellow was also encountered in the painting.

11. Jose Lopez-Rey, "Velazquez' Philip IV", *Art News,* Summer 1968. In the x radiographs of the Prado picture, Philip IV's face appears fuller than it does in other representations. Lopez-Ray believes that this is a comparatively literal record of the King's features whereas in subsequent portraits his appearance is idealized. However, the mutual interference which results from the overlapping of two similar and radiographically dense images inevitably produces misleading distortions. For this reason caution must be exercised in drawing conclusions from this x radiographic detail and in comparing this head to those of the related versions.